Engineering Mechanics: STATICS

Engineering Mechanics:

R.C. Hibbeler

STATICS

SECOND EDITION

Macmillan Publishing Co., Inc.
New York
Collier Macmillan Publishers
London

Macmillan Publishing Co., Inc.
866 Third Avenue, New York, New York 10022

Collier Macmillan Canada, Ltd.

Library of Congress Cataloging in Publication Data

Hibbeler, R   C
    Engineering mechanics, statics.

    Includes index.
    1. Statics. I. Title
TA351.H5 1978   620.1'03   77-23191
ISBN 0-02-354020-6 (Hardbound)
ISBN 0-02-978980-X (International Edition)

Printing: 5 6 7 8   Year: 1 2 3 4

Preface

The purpose of this book is to provide the student with a clear and thorough presentation of the theory and application of the principles of engineering mechanics. Emphasis is placed on developing the student's ability to analyze problems—a most important skill for any engineer. Furthermore, the Système International or SI system of units is used for numerical work since this system is intended in time to become the worldwide standard for measurement.

The contents of each chapter are organized into well-defined sections. Selected groups of sections contain the development and explanation of specific topics, illustrative example problems, and a set of problems designed to test the student's ability to apply the theory. Many of the problems depict realistic situations encountered in engineering practice. It is hoped that this realism will both stimulate the student's interest in engineering mechanics and provide a means for developing the skill to reduce any such problem from its physical description to a model or symbolic representation to which the principles of mechanics may be applied. In any set, the problems are arranged in order of increasing difficulty. Furthermore, the answers to all but every fourth problem, which is indicated by an asterisk, are listed in the back of the book. SI units are used in all the numerical examples and problems; however, for the convenience of some instructors, every fifth problem is stated *twice,* once in SI units and again in FPS units.

Besides a change from FPS to SI units and the addition of many new problems, this book differs from the author's first edition, *Engineering Mechanics: Statics,* in many respects. Most of the text material has been completely rewritten so that topics within each section are categorized into subgroups, defined by boldface titles. The purpose of this is to present a structured method for introducing each new definition or concept, and to provide a convenient means for later reference or review of the material.

Another unique feature used throughout this book is the "Procedure for Analysis." This guide to problem solving, which was initially presented in Sec. 9-3 of the first edition, is essentially a step-by-step set of instructions which provide the student with a logical and orderly method to follow when applying the theory. As in the first edition, the example problems are solved using this outlined method in order to clarify application of the steps.

Since mathematics provides a systematic means of applying the principles of mechanics, the student is expected to have prior knowledge of algebra, geometry, trigonometry, and, for complete coverage, some calculus. Vector analysis is introduced at points where it is most applicable. Its use often provides a convenient means for presenting concise derivations of the theory, and it makes possible a simple and systematic solution of many complicated three-dimensional problems. Occasionally, the example problems are solved using several different methods of analysis so that the student develops the ability to use mathematics as a tool whereby the solution of any problem may be carried out in the most direct and effective manner.

The book is divided into 11 chapters, in which the principles introduced are first applied to simple situations. Specifically, each principle is applied first to a particle, then to a rigid body subjected to a coplanar system of forces, and finally to the most general case of spatial force systems acting on a rigid body.

In particular, an introduction to mechanics and a discussion of units is outlined in Chapter 1. The notion of a vector and the properties of a concurrent force system are introduced in Chapter 2. This theory is then applied to the equilibrium of particles in Chapter 3. Chapter 4 contains a general discussion of both concentrated and distributed force systems and the methods used to simplify them. The principles of rigid-body equilibrium are developed in Chapter 5. These principles are applied to specific problems involving the equilibrium of trusses, frames, and machines in Chapter 6, and to the analysis of internal forces in beams and cables in Chapter 7. Applications to problems involving frictional forces are discussed in Chapter 8; and topics related to the centroid and the center of gravity are given in Chapter 9. If time permits, sections concerning more advanced topics, indicated by stars, may be covered. Some topics in Chapter 10 ("Moments of Inertia for an Area") and Chapter 11 ("Virtual Work") may be omitted from the basic course. Note, however, that this more advanced material provides a suitable reference for basic principles when it is discussed in more advanced courses.

At the discretion of the instructor, some of the material may be presented in a different sequence with no loss in continuity. For example, it is possible to introduce the concept of a force and all the necessary methods of vector analysis by first covering Chapter 2 and Secs. 4-2 and 4-11. Then, after covering the rest of Chapter 4 (force and moment systems), the equilibrium methods in Chapters 3 and 5 can be discussed. Furthermore, Chapter 9 may be covered after Sec. 4-10 (distributed force systems), since understanding of this material does not depend upon the methods of equilibrium.

The author has endeavored to write this book so that it will appeal to both the student and the instructor. In doing so, it must be admitted that many people helped in its development. In this regard, I wish to acknowledge the valuable suggestions and comments made by M. H. Clayton, North Carolina State University; D. Krajcinovic, University of Illinois at Chicago Circle; W. Lee, United States Naval Academy; G. Mavrigian, Youngstown State University; W. C. Van Buskirk, Tulane University; and P. K. Mallick, Illinois Institute of Technology. Many thanks are also extended to all of the author's students and to the professionals who have provided suggestions and comments. Although the list is too long to mention, it is hoped that those who have given help will accept this anonymous recognition. Lastly, I should like to acknowledge the able assistance of my wife, Cornelie, who has furnished a great deal of her time and energy in helping to prepare the manuscript for publication.

Russell C. Hibbeler

Contents

1

General Principles

1–1. Mechanics

In general, *mechanics* is that branch of the physical sciences concerned with the state of rest or motion of bodies that are subjected to the action of forces. A thorough knowledge of this subject is required for further study of the physical sciences involving structural engineering, machine design, fluid flow, electrical instrumentation, and even the molecular and atomic behavior of elements.

Divisions of Mechanics.   Depending upon the nature of the problem being studied, mechanics is generally subdivided into three branches: *rigid-body or classical mechanics, deformable-body mechanics, and fluid mechanics.* In this book only rigid-body mechanics will be studied, since this subject forms a suitable basis for the design and analysis of many engineering problems, and provides the necessary background for a study of the mechanics of deformable bodies and the mechanics of fluids.

Rigid-body mechanics is generally divided into two areas: statics and dynamics. *Statics* deals with the equilibrium of bodies, that is, those which are either at rest or move with a constant velocity; whereas *dynamics* is concerned with the accelerated motion of bodies. Although statics can be considered as a special case of dynamics in which the acceleration is zero, statics deserves separate treatment in engineering education, since most structures are designed with the intention that they remain in equilibrium.

Historical Development.   The subject of statics developed very early in history, because the principles involved could be formulated simply from measurements of geometry and force. For example, the writings of Archimedes (287–212 B.C.) provide an explanation of the equilibrium of the lever. Studies of the pulley, inclined plane, and wrench are also recorded in ancient writings—at times when the requirements of engineering were limited primarily to building construction.

Since the principles of dynamics depend upon an accurate measurement of time, this subject developed much later. Galileo Galilei (1564–1642) was one of the first major contributors to this field. His work consisted of experiments using pendulums and falling bodies. The most significant contributions in dynamics, however, were made by Isaac Newton (1642–1727), who is noted for his formulation of the three fundamental laws of motion and the law of universal gravitational attraction. Shortly after these laws were postulated, important techniques for their application were developed by Euler, D'Alembert, Lagrange, and others.

Newton's Three Laws of Motion.   The entire structure of rigid-body or classical mechanics is formulated on the basis of Newton's three laws of motion. These laws, which apply to the motion of a particle, may be briefly stated as follows.

First Law.   A particle originally at rest, or moving in a straight line at a constant velocity, will continue to remain in this state provided the particle is not subjected to an unbalanced force.

Second Law.   A particle acted upon by an unbalanced force $\mathbf{F}$ receives an acceleration $\mathbf{a}$ in the direction of the force. The acceleration is directly proportional to the force and inversely proportional to the particle mass m. This law is commonly expressed in mathematical terms as

$$\mathbf{F} = m\mathbf{a} \tag{1-1}$$

Third Law.   For every force acting on a particle, the particle exerts an equal, opposite, and collinear reactive force.

Newton's Law of Gravitational Attraction.   Shortly after formulating his three laws of motion for a particle, Newton postulated a law governing the mutual attraction between any two particles. This law can be expressed mathematically as

$$F = G\frac{m_1 m_2}{r^2} \tag{1-2}$$

where
- F = force of attraction between the two particles
- G = universal constant of gravitation; according to experimental evidence, $G = 6.673(10^{-11})$ m^3/(kg · s^2)
- m_1, m_2 = masses of each of the two particles
- r = distance between the centers of both particles

Any two particles or bodies have a mutual attractive (gravitational) force acting between them. In the case of a particle located at or near the surface of the earth, however, the only attractive force having any sizable magnitude is that of the earth's gravitation. Consequently, this force, termed the *weight,* will be the only gravitational force considered.

Four fundamental quantities commonly used in mechanics are length, time, force, and mass. In general, the magnitude of each of these quantities is defined by an arbitrarily chosen *unit* or "standard."

Length. The concept of *length* is needed to locate the position of a point in space and thereby to describe the size of a physical system. The standard unit of length measurement is the *metre* (m), which is represented by 1 650 763.73 wavelengths of light produced from the orange-red line of the spectrum of krypton 86. All other units of length are defined in terms of this standard. For example, 1 foot (ft) is equal to 0.3048 m.

Time. The concept of *time* is conceived by a succession of events. The standard unit used for its measurement is the second (s), which is based on the duration of 9 192 631 770 cycles of vibration of an isotope cesium 133.

Force. In general, *force* is considered as a "push" or "pull" exerted by one body on another. This interaction can occur when there is either direct contact between bodies, such as a person pushing on a wall, or it can occur through a distance by which the bodies are physically separated. Examples of the latter type include gravitational, electrical, and magnetic forces. In any case, a force is completely characterized by its magnitude, direction, and point of application. Most often, engineers define the standard unit of force using either the newton (N) or the pound (lb). Each of these units can be *measured* with a spring balance to determine the amount of gravitational pull exerted by the earth upon an object. Since this force, defined as the *weight* of a body, *changes* with respect to the distance r from the center of the earth, Eq. 1–2, it is important to make measurements of weight at a specified latitude and height above sea level.

Mass. The *mass* of a body is regarded as a quantitative property of matter used to measure the resistance of matter to a change in velocity. Unlike weight, the mass of a body is *constant* regardless of its location. For this reason, comparison of masses is usually made by means of a lever-arm balance. The standard unit of mass is the *kilogram* (kg), defined by a bar of platinum–iridium alloy kept at the International Bureau of Weights and Measures in Sèvres, France.

Systems of Units. The four fundamental quantities—length, time, force, and mass—are not all independent from one another; instead, they are related by Newton's second law of motion—force is proportional to the product of mass and acceleration; i.e., $\mathbf{F} = m\mathbf{a}$. Hence, the units used to define the magnitude of force, mass, length, and time cannot *all* be selected arbitrarily. The equality $\mathbf{F} = m\mathbf{a}$ is maintained only if three of the four units, called *base units,* are *arbitrarily defined* and the fourth unit is *derived* from the equation.

Absolute System.   A system of units defined on the basis of length, time, and mass is referred to as an *absolute system,* since the measurements of all these quantities can be made at *any location.* As shown in Table 1–1, the International System of Units (SI) is absolute, since it specifies length in metres (m), time in seconds (s), and mass in kilograms (kg). The unit of force, called a newton (N), is derived from $\mathbf{F} = m\mathbf{a}.$ Thus, 1 newton is equal to a force required to give 1 kilogram of mass an acceleration of 1 m/s^2 ($\text{N} = \text{kg} \cdot \text{m/s}^2$).

Gravitational System.   A system of units defined on the basis of length, time, and force is referred to as a *gravitational system.* This is because force is measured in a gravitational field, and hence its magnitude depends upon where the measurement is made. In the FPS system of units, Table 1–1, length is in feet (ft), time is in seconds (s), and force is in pounds (lb). The unit of mass, called a *slug,* is derived from $\mathbf{F} = m\mathbf{a}.$ Hence, 1 slug is equal to the amount of matter accelerated at 1 ft/s^2 when acted upon by a force of 1 lb ($\text{slug} = \text{lb} \cdot \text{s}^2/\text{ft}$).

Table 1–1   System of Units

Type of System	*Name of System*	*Length*	*Time*	*Mass*	*Force*
Absolute	International System of Units (SI)	metre (m)	second (s)	kilogram (kg)	newton* (N) $\left(\dfrac{\text{kg} \cdot \text{m}}{\text{s}^2} \right)$
Gravitational	British Gravitational (FPS)	foot (ft)	second (s)	slug* $\left(\dfrac{\text{lb} \cdot \text{s}^2}{\text{ft}} \right)$	pound (lb)

*Derived unit.

1–3. The International System of Units

The International System of units, abbreviated SI after the French "Système International d'Unités," is a modern version of the metric system which has received worldwide recognition at the 11th International Conference of Weights and Measures in 1960. This absolute system of units is used throughout this book since it is intended to become the worldwide standard for measurement.

Base and Derived Units.   Only the seven arbitrarily defined *base units* listed in Table 1–2 exist in the SI system. All other units are *derived* from these. For example, as previously stated, the unit of force, the newton (N), is derived from Newton's law of motion ($N = kg \cdot m/s^2$). Another derived unit used in statics is the pascal (Pa), defined as the pressure caused by a force of 1 newton acting over an area of 1 square metre ($Pa = N/m^2$).

Table 1-2   Primary SI Units

Quantity	Base Unit	SI Symbol
Length	metre	m
Mass	kilogram	kg
Time	second	s
Electrical current	ampere	A
Amount of substance	mole	mol
Temperature	kelvin	K
Luminous intensity	candela	cd

Prefixes.   Since units often measure quantities that may vary considerably in magnitude, prefixes representing multiples and submultiples often are used to modify units.* In the SI system, prefixes are increments of three digits, such as those shown in Table 1-3. Attaching a prefix to a unit in effect creates a new unit; thus if a multiple or submultiple unit is raised to a power, the power applies to this new unit, not just to the original unit *without* the multiple or submultiple. For example, $(2\,kN)^2 = (2000\,N)^2 = 4(10^6)\,N^2$. Also, $1\,mm^2 = 1\,(mm)^2$ *not* $1\,m(m^2)$. Note that the SI system does not include the multiple deca (10) or the submultiple centi (0.01), which form part of the old metric system. Except for some volume or area measurements, the use of these prefixes is to be avoided in science and engineering.

*The kilogram is the only *base unit* that is defined with a prefix.

Table 1-3   Prefixes

Multiple	Exponential Form	Prefix	SI Symbol
1 000 000 000	10^9	giga	G
1 000 000	10^6	mega	M
1 000	10^3	kilo	k
Submultiple			
0.001	10^{-3}	milli	m
0.000 001	10^{-6}	micro	μ
0.000 000 001	10^{-9}	nano	n

Rules for Use. The following rules are given for the proper use of the various SI symbols:

1. A symbol is *never* written with a plural "s," since it may be confused with the unit for second (s).
2. Symbols are always written in lowercase letters, with two exceptions: symbols for the two largest prefixes shown in Table 1–3, giga and mega, are capitalized as G and M, respectively; and symbols named after an individual are capitalized, e.g., N and Pa.
3. Quantities defined by several units which are multiples of one another are separated by a *dot* to avoid confusion with prefix notation, as illustrated by $N = kg \cdot m/s^2 = kg \cdot m \cdot s^{-2}$. Also, m $\cdot$ s (metre-second); whereas ms (milli-second).
4. Physical constants or numbers having several digits on either side of the decimal point should be reported with a *space* between every three digits rather than with a comma, e.g., 73 569.213 427. In the case of four digits on either side of the decimal, the spacing is optional, e.g., 8537 or 8 537. Furthermore, always try to use decimals and avoid fractions; that is, write 15.25, *not* $15\frac{1}{4}$.
5. Compound prefixes should not be used, e.g., kμs (kilo-micro-second) should be expressed as ms (milli-second). It is also best to keep numerical values between 0.1 and 1000; otherwise, a suitable prefix should be chosen. For example, a force of 50 000 N is written as 50 kN.
6. With the exception of the base unit, the kilogram, in general avoid the use of a prefix in the denominator of a composite unit. For example, do not write N/mm, but rather kN/m.
7. Although not expressed in multiples of 10, the minute, hour, etc., are retained for practical purposes as multiples of the second. Furthermore, plane angular measurement is made using radians (rad). In this book, degrees will sometimes be used, where $360° = 2\pi$ rad. Fractions of a degree, however, should be expressed in decimal form rather than in minutes, as in 10.4°, not 10°24′.
8. When performing calculations, represent the numbers in terms of their *base or derived units* by converting any prefixes to powers of 10. The final result should then be expressed using a *single prefix*. For example,

$$(50 \text{ kN})(60 \text{ nm}) = [50(10^3) \text{ N}][60(10^{-9}) \text{ m}]$$
$$= 3000(10^{-6}) \text{ N} \cdot m = 3 \text{ mN} \cdot m.$$

Weight. Two terms often confused in the SI system are weight and mass. Specifically, *weight* is the force due to gravity acting on the mass of a body. When the body is allowed to fall freely, the weight acts as an unbalanced force which gives the body an acceleration. Hence, if a body has a mass of m kg and is located at a point where the acceleration of the body due to gravity is $a = g$ m/s², then, since $F = ma$, the weight W measured in newtons $(N = kg \cdot m/s^2)$ is

$$W = mg \qquad \text{(1-3)}$$

In particular, when a freely falling body is located at sea level and at a latitude of 45° (considered the standard location), the acceleration due to gravity is $g = 9.806\ 65$ m/s². For calculations, the value $g = 9.81$ m/s² will be used; so that from Eq. 1-3, 1 kg mass exerts a force or has a weight of 9.81 N, 2 kg weighs 19.62 N, and so on.

When learning to use SI units, it is generally agreed that one should *not* think in terms of conversion factors between systems. Instead, it is better to think *only* in terms of SI units. A "feeling" for these units can only be gained through experience. Study, for example, the geometry and loads acting on the structures and machines illustrated in the problems throughout this book. As a memory aid, it might be helpful to recall that a standard flashlight battery or a small apple weighs about 1 newton. Your body is a suitable reference for small distances. For example, the millimetre scale in Fig. 1-1 can be used to measure, say, the width of three or four fingers pressed together, about 50 mm, or the width of the small fingernail, about 10 mm. For most people, a stretched walking pace is about 1 metre long.

Millimetre scale
Fig. 1-1

1-4. Dimensional Quantities

Dimensional Homogeneity. The terms of any equation used to describe a physical process must be *dimensionally homogeneous;* that is, each term must be expressed in the same units. Provided that this is the case, all the terms of an equation can then be combined if numerical values are substituted for the variables. Consider, for example, the equation $s = vt + \frac{1}{2}at^2$, where in SI units, s is the position in metres (m), t is time in seconds (s), v is velocity in m/s, and a is acceleration in m/s². Regardless of how this equation is evaluated, it maintains its dimensional homogeneity. In the form stated, each term can be expressed in metres [m, (m/s̸)s̸, (m/s̸²)s̸²], or solving for a, $a = 2s/t^2 - 2v/t$, the terms are each expressed in units of m/s² [m/s², m/s², (m/s)1/s].

Since mechanics problems involve the solution of dimensionally homogeneous equations, the fact that all terms of an equation are represented by a consistent set of units can be used as a partial check for algebraic manipulations of an equation.

Table 1–4 Conversion Factors

Quantity	Unit of Measurement (FPS)	To Convert FPS to SI, Multiply FPS Units by:	Unit of Measurement (SI)
Force	lb	4.4482	N
Mass	lb_{mass}	0.4536	kg
Length	ft	0.3048	m
Area	ft^2	0.09290	m^2
Volume	ft^3	0.02832	m^3
Area moment of inertia	ft^4	0.008631	m^4
Moment of a force ⎫ Couple ⎬ Torque ⎭	lb · ft	1.3558	N · m
Linear load intensity	lb/ft	14.5938	N/m
Surface load intensity ⎫ Pressure ⎬	lb/ft^2	47.8800	Pa
Density	lb_{mass}/ft^3	16.0187	kg/m^3

Conversion of Units. In some cases it may be necessary to convert from one system of units to another. In this regard, Table 1–4 provides a set of direct conversion factors between FPS and SI units for the important physical quantities encountered in statics.

When derived units are present, a general procedure using a simple cancellation technique should be applied when conversion factors are not available. Suppose, for example, that it becomes necessary to convert a speed of 2 km/h to units of m/s. Since 1 km = 1000 m and 1 h = 3600 s, these factors of conversion can be arranged in a form for cancellation as follows:

$$2 \text{ km/h} = \frac{2 \cancel{\text{km}}}{\cancel{\text{h}}} \left(\frac{1 \cancel{\text{h}}}{3600 \text{ s}} \right) \left(\frac{1000 \text{ m}}{1 \cancel{\text{km}}} \right) = \frac{2000 \text{ m}}{3600 \text{ s}} = 0.556 \text{ m/s}$$

To minimize the chance of error, this technique should be used in all problem work requiring an adjustment of units.

Accuracy. In practice, forces are estimated to be about 90 per cent accurate, and often the mass of a body and its measurements are approximated. Hence, accuracy obtained from the solution of a problem can generally never be better than the accuracy of the problem data. This, of course, is what is to be expected, but often computers or hand calculators seem to involve more figures in the answer than the numbers used for the data. Since accuracy is often lost when subtracting numbers that are approximately equal, numerical work for problem solving should be carried out as accurately as possible, and the final answers should be rounded off to a value that reflects the accuracy of the original data.

One of the most effective ways of studying engineering mechanics is to *solve problems*. Merely studying concepts or general principles about some theory will be of little use if this knowledge is not applied in a meaningful way.

In solving problems it is very important to present the work in a *logical* and *orderly* manner. For this reason, the following steps should be performed:

1. Read the problem carefully. List the data given and the results required.
2. Draw and label any necessary diagrams needed for the solution.
3. List all the relevant principles, generally in mathematical form.
4. Think about the problem in terms of the actual *physical situation*. Try to correlate this knowledge with each mathematical expression that is written out.
5. Solve the necessary equations algebraically as far as is practical, then complete the solution numerically.
6. Check through the problem, making sure that the equations used are dimensionally homogeneous and that the units of the numerical data used in the solution are consistent.
7. Study the answer with technical judgment and common sense to determine whether or not it seems reasonable.
8. Once the solution has been completed, review the problem. Try to think of other ways of obtaining the same solution.

In applying this general procedure, do the work as neatly as possible. Being neat generally stimulates clear and orderly thinking, and vice versa.

Problems

1–1. Determine the area of a rectangle having a length of 0.12 km and a width of 200 mm.
$0.12 \times 10^3 \, m \times 200 \times 10^{-3} = 24 \, m^2$

1–2. Represent each of the following with units having an appropriate prefix: (a) 6540 m; (b) 5200 kN; (c) 0.0621 ms.

1–3. Evaluate each of the following and express with units having an appropriate prefix: (a) $(4 \text{ kN})^2$; (b) $(0.03 \text{ mm})^2$; (c) $(200 \text{ s})^3$.

* 1–4. What is the weight in newtons of an object that has a mass of: (a) 8 kg; (b) 200 g; (c) 2560 kg?

1–5. Is there a difference between $m \cdot kg$ and mkg? Explain.

1–6. Using the base units of the SI system show that Eq. 1–2 is a dimensionally homogeneous equation which gives F in newtons. Compute the gravitational force acting between two spheres that are touching each other. The mass of each sphere is 100 kg and the radius is 150 mm.

1–7. Represent each of the following combinations of units in the correct SI form: (a) μMN; (b) $k\mu m/Ms$; (c) MN/ks^2.

*1-8. Using Table 1-4, determine your mass in kilograms, weight in newtons, and height in metres.

1-9. The pascal (Pa) is actually a very small unit of pressure. To show this, convert 1 Pa to lb/ft^2. Atmospheric pressure at sea level is $14.7 \, lb/in.^2$. How many pascals is this?

1-10. If a body weighs 100 lb on earth, specify: (a) its mass in slugs; (b) its mass in kilograms; (c) its weight in newtons. If the 100-lb body is placed on the moon, where the acceleration due to gravity is $g_m = 5.30 \, ft/s^2$, determine: (d) its mass in kilograms; (e) its weight in newtons.

1-11. Represent each of the following combinations of units in the correct SI form: (a) g/ms; (b) N/nm; (c) $mm/(kg \cdot \mu s)$.

*1-12. Convert: (a) $200 \, lb \cdot ft$ to $N \cdot m$; (b) $300 \, lb/ft^3$ to Mg/m^3; (c) 6 ft/h to $\mu m/s$.

2

Force and Position Vectors

2–1. Scalars and Vectors

Many of the physical quantities in mechanics can be expressed mathematically by means of scalars and vectors.

Scalar. A quantity possessing only a magnitude is called a *scalar*. Mass, volume, and length are scalar quantities commonly used in statics. In this book, scalars will be indicated by letters in italic type, such as the scalar A. The mathematical operations involving scalars follow the same rules as those of elementary algebra.

Vector. A *vector* is a quantity that has both a magnitude and direction and "adds" according to the parallelogram law. This law utilizes a form of construction that accounts for the combined magnitude and direction of the vector. Vector quantities commonly used in the study of statics are position, force, and moment vectors.

For analytical work, a vector is represented by a letter with an arrow written over it, such as $\vec{A}$. The magnitude is designated by $|\vec{A}|$, or simply A. In this book vectors will be symbolized in boldface type; for example, **A** is used to designate the vector A, and its magnitude will be symbolized in italic type, A.

Graphically, a vector is represented by an arrow, used to define its magnitude, line of action, and direction. The *magnitude* of the vector is indicated by the *length* of the arrow, and its *direction* is indicated by an *arrowhead*. For example, the vector **A** shown in Fig. 2–1 has a magnitude of 4 units and is directed upward, along its line of action, which is 20° above the horizontal. The point O is called the *origin* or *tail* of the vector; the point P is the *terminus* or *tip*.

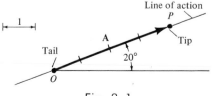

Fig. 2–1

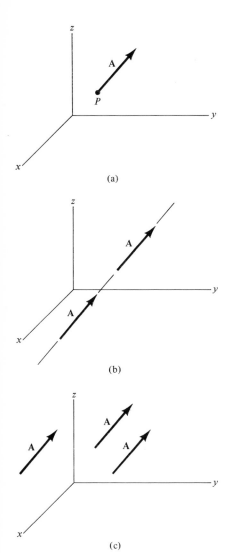

Fig. 2–2

Types of Vectors. Various types of vectors are used in statics:

1. A *fixed vector* is one that acts at a particular point P in space, Fig. 2–2a.
2. A *sliding vector* may be applied at any point along its line of action, Fig. 2–2b.
3. A *free vector* can act anywhere in space; it is only necessary that it preserve its magnitude and direction, Fig. 2–2c.
4. *Coplanar vectors* lie in the same plane, Fig. 2–6.
5. *Collinear vectors* have the same direction and the same line of action, Fig. 2–7.
6. *Concurrent vectors* have lines of action that pass through the same point, Fig. 2–6.
7. *Equal vectors* have the same magnitude and direction. Although it is not necessary that they originate from the same point, equal vectors must have magnitudes measured with the same units. Borrowing the equal sign from scalar algebra, we may write the equality of two vectors, **A** and **B**, Fig. 2–3, as **A** = **B**.

A = B

Fig. 2–3

8. A *negative vector* has a direction opposite to its positive counterpart but has the same magnitude, Fig. 2–4.

Fig. 2–4

9. A *null vector* has zero magnitude and therefore no specified direction. This is expressed as **A** = **0.**

2–2. Addition and Subtraction of Vectors

Vector Addition. Two vectors **A** and **B**, Fig. 2–5a, may be added to form a third vector **R** = **A** + **B** by using either the "parallelogram" or "triangle" law. Vector **R** is called the *resultant* and vectors **A** and **B** are called *components.*

Parallelogram Law. To form the sum **R** = **A** + **B** using the *parallelogram law,* the components **A** and **B** are joined at their tails, forming adjacent sides of a parallelogram constructed as shown in Fig. 2–5b. The resultant **R** is then determined by extending an arrow from the tails of **A** and **B** along the diagonal of the parallelogram to the opposite corner.

Triangle Law. The *triangle law* of addition is a special case of the parallelogram law, whereby vector **B** is added to vector **A** in a "tip-to-tail" fashion, i.e., by placing the tail of **B** at the tip of **A,** Fig. 2–5*c.* For the addition, **B** is treated as a free vector. The resultant **R** extends from the tail of **A** to the tip of **B.** In a similar manner, vector **R** can also be obtained by adding **A** to **B,** Fig. 2–5*d.* Here **A** is treated as a free vector. Figures 2–5*c* and 2–5*d* illustrate that vector addition is commutative; in other words, the vectors can be added in *any* order, i.e., $\mathbf{R} = \mathbf{A} + \mathbf{B} = \mathbf{B} + \mathbf{A}.$

Note that the addition of vectors must be carried out using either the parallelogram or triangle laws, as shown. Furthermore, the vectors added must have magnitudes measured with a scale having a consistent set of units.

Using the proper method of vector addition, it should be realized that the *same* resultant vector **R** can be formed from an infinite number of two-component combinations. For example, as shown in Fig. 2–6, $\mathbf{R} = \mathbf{A} + \mathbf{B} = \mathbf{C} + \mathbf{D} = \mathbf{E} + \mathbf{F}.$ If the two component vectors are *collinear,* the parallelogram or triangle law reduces to a single scalar or algebraic addition of the components, as shown in Fig. 2–7.

Vector Subtraction. Defining the (resultant) *difference* of two vectors **A** and **B,** one may write

$$\mathbf{R} = \mathbf{A} - \mathbf{B} = \mathbf{A} + (-\mathbf{B})$$

and then form the vector sum, Fig. 2–8. Subtraction is therefore defined as a special case of addition, so that the rules for vector addition will also apply for vector subtraction.

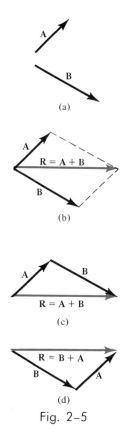

Fig. 2–5

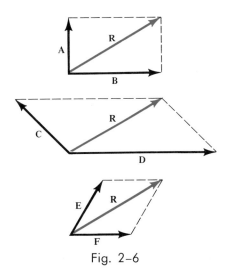

Fig. 2–6

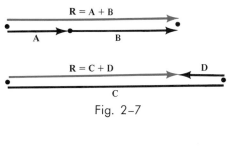

Fig. 2–7

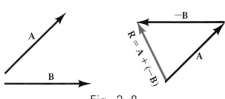

Fig. 2–8

13

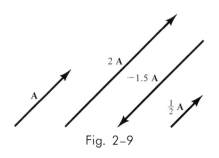

Fig. 2–9

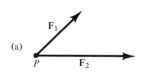

(a)

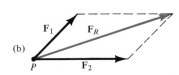

(b)

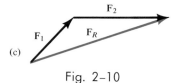

(c)

Fig. 2–10

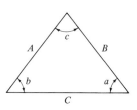

Sine law:
$$\frac{A}{\sin a} = \frac{B}{\sin b} = \frac{C}{\sin c}$$

Cosine law:
$$C = \sqrt{A^2 + B^2 - 2AB \cos c}$$

Fig. 2–11

2–3. Multiplication and Division of a Vector by a Scalar

The product of vector **A** and scalar m, yielding $m\mathbf{A}$, is defined as a vector having a magnitude mA. The direction of $m\mathbf{A}$ is the same as **A** provided m is positive; it is opposite to **A** if m is negative. The negative of a vector can thus be defined on the basis of multiplication of the vector by the scalar (-1). Since the vector *magnitude* is always a *positive quantity, a minus sign in front of a vector simply means that its direction is reversed.*

Division of a vector by a scalar can be defined using the laws of multiplication, since

$$\frac{\mathbf{A}}{m} = \frac{1}{m}\mathbf{A}, \qquad m \neq 0$$

Graphic examples of multiplication and division of vectors by scalars are shown in Fig. 2–9.

2–4. Force as a Vector Quantity

When any external load is applied to the surface of a body, it is actually distributed over a finite area. In cases where this area is *small* compared to the size of the body, it may become difficult to measure the exact area of contact or to determine the loading distribution. As a result, computations regarding the effect of the load can be greatly simplified if the load is represented as a single *concentrated force* which acts only at a *point* on the body. Experimentally it has been shown that a concentrated force is a vector quantity, since it adds according to the parallelogram law; and when it is applied to a body, it has a specified magnitude, direction, and point of application. The units of force magnitude are commonly expressed in newtons (N) or pounds (lb).

A common problem in statics involves the determination of the *resultant force*, $\mathbf{F}_R$, given two *component forces* $\mathbf{F}_1$ and $\mathbf{F}_2$ which act at some point P, Fig. 2–10a. The resultant $\mathbf{F}_R = \mathbf{F}_1 + \mathbf{F}_2$ can be determined by using either the parallelogram law, Fig. 2–10b, or the triangle law, Fig. 2–10c. If more than two forces act at P, successive applications of either of these two laws are required. For example, if three forces $\mathbf{F}_1$, $\mathbf{F}_2$, $\mathbf{F}_3$, act at P, the resultant of any two of the forces is found, say $\mathbf{F}_1 + \mathbf{F}_2$, and then this resultant is added to the third force, yielding the resultant of all three forces, i.e., $\mathbf{F}_R = (\mathbf{F}_1 + \mathbf{F}_2) + \mathbf{F}_3$.

PROCEDURE FOR ANALYSIS

Problems that involve the addition of two force components contain *two unknowns,* represented as the magnitude and direction of the resultant force $\mathbf{F}_R$. In some problems, however, $\mathbf{F}_R$ may be known and the magnitudes or directions of the components must then be determined. In any case, the following three-step procedure should be used for the solution.

Step 1: Make a sketch showing the vector addition, using the parallelogram law. If possible, determine the interior angles of the parallelogram from the geometry of the problem. Unknown angles, along with known and unknown force magnitudes, should clearly be labeled on this sketch.

Step 2: Redraw a portion of the constructed parallelogram to illustrate the tip-to-tail addition of the components by the triangle law.

Step 3: By using trigonometry, the two unknowns can be determined from the data listed on the triangle in *Step 2.* If the triangle does *not* contain a 90° angle, the law of sines and the law of cosines may be used for the solution. These formulas are given in Fig. 2–11 for the triangle *ABC*.

The following examples numerically illustrate this three-step method for solution.

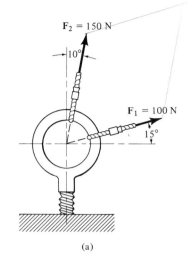

(a)

Example 2–1

The screw eye in Fig. 2–12a is subjected to two forces, $\mathbf{F}_1$ and $\mathbf{F}_2$. Determine the magnitude and direction of the resultant force.

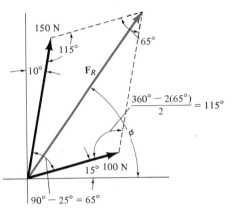

(b)

Solution

Step 1: The parallelogram law of addition is shown in Fig. 2–12b. The two unknowns are represented as the magnitude F_R and direction ϕ (phi) of the resultant force $\mathbf{F}_R$.

Step 2: From Fig. 2–12b, the force triangle, Fig. 2–12c, is constructed.

Step 3: F_R is determined by means of the law of cosines:

$$F_R = \sqrt{(100)^2 + (150)^2 - 2(100)(150)\cos 115°}$$
$$= \sqrt{10\,000 + 22\,500 - 30\,000(-0.423)}$$
$$= 212.6 \text{ N} \qquad\qquad Ans.$$

The angle θ (theta) is determined by applying the law of sines, using the computed value of F_R.

$$\frac{150}{\sin \theta} = \frac{212.6}{\sin 115°}$$

$$\sin \theta = \frac{150}{212.6}(0.906) = 0.639$$

$$\theta = 39.7°$$

Thus, ϕ is

$$\phi = 39.7° + 15.0° = 54.7° \qquad\qquad Ans.$$

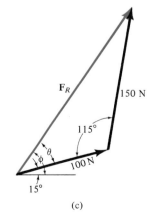

(c)

Fig. 2–12

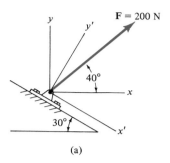

(a)

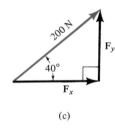

(b)

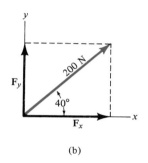

(c)

F = 200 N

Example 2–2

Resolve the 200-N force shown acting on the inclined pin, Fig. 2–13a, into components in the (a) x and y directions; (b) x' and y' directions; (c) x' and y directions.

Solution

In each case the parallelogram law is used to resolve **F** into its two components, and then the triangle law is applied to determine the numerical results by trigonometry.

Part (a). The vector addition $\mathbf{F} = \mathbf{F}_x + \mathbf{F}_y$ is shown in Fig. 2–13b. From the vector triangle, Fig. 2–13c,

$$F_x = 200 \cos 40° = 153.2 \text{ N} \qquad \textit{Ans.}$$
$$F_y = 200 \sin 40° = 128.6 \text{ N} \qquad \textit{Ans.}$$

Part (b). The vector addition $\mathbf{F} = \mathbf{F}_{x'} + \mathbf{F}_{y'}$ is shown in Fig. 2–13d. From the vector triangle, Fig. 2–13e,

$$F_{x'} = 200 \cos 70° = 68.4 \text{ N} \qquad \textit{Ans.}$$
$$F_{y'} = 200 \sin 70° = 187.9 \text{ N} \qquad \textit{Ans.}$$

Part (c). The vector addition $\mathbf{F} = \mathbf{F}_{x'} + \mathbf{F}_y$ is shown in Fig. 2–13f. Applying the law of sines and using the data listed on the vector triangle, Fig. 2–13g, yields

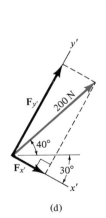

(d)

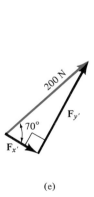

(e)

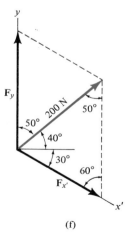

(f)

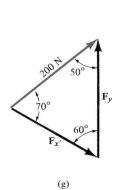

(g)

Fig. 2–13

$$\frac{F_{x'}}{\sin 50°} = \frac{200}{\sin 60°}$$

$$F_{x'} = 200\left(\frac{\sin 50°}{\sin 60°}\right) = 176.9 \text{ N} \qquad Ans.$$

$$\frac{F_y}{\sin 70°} = \frac{200}{\sin 60°}$$

$$F_y = 200\left(\frac{\sin 70°}{\sin 60°}\right) = 217.0 \text{ N} \qquad Ans.$$

Example 2–3

The force **F** acting on the frame shown in Fig. 2–14a has a magnitude of 500 N and is to be resolved into two components acting along struts AB and AC. Determine the angle θ so that the component $\mathbf{F}_{AC}$ is directed from A towards C and has a magnitude of 400 N.

Solution

By the parallelogram law, the vector addition of the two specified components yielding the resultant is shown in Fig. 2–14b. The corresponding triangle law of addition is shown in Fig. 2–14c. The angle ϕ can be determined by using the law of sines:

$$\frac{400}{\sin \phi} = \frac{500}{\sin 60°}$$

$$\sin \phi = \left(\frac{400}{500}\right)\sin 60° = 0.693$$

$$\phi = 43.9°$$

Hence,

$$\theta = 180° - 60° - 43.9° = 76.1° \qquad Ans.$$

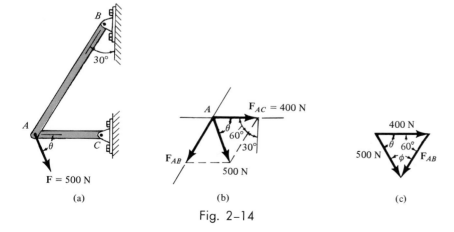

Fig. 2–14

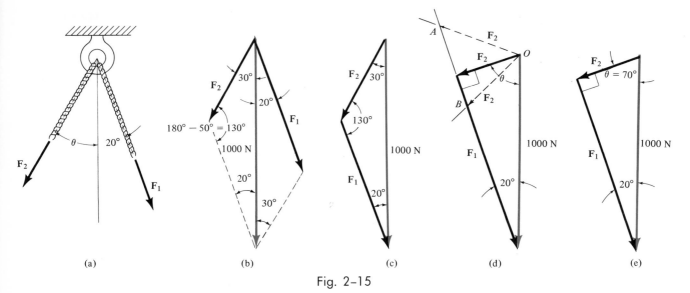

(a) (b) (c) (d) (e)

Fig. 2–15

Example 2–4

The ring shown in Fig. 2–15a is subjected to two forces, F_1 and F_2. If it is required that the resultant force have a magnitude of 1 kN and be directed vertically downward, determine: (a) the magnitudes of F_1 and F_2 provided $\theta = 30°$; (b) the magnitudes of F_1 and F_2 when F_2 is a minimum.

Solution

Part (a). A sketch of the vector addition, according to the parallelogram law, is shown in Fig. 2–15b. From the vector triangle constructed in Fig. 2–15c the unknown magnitudes F_1 and F_2 can be determined using the law of sines.

$$\frac{F_1}{\sin 30°} = \frac{1000}{\sin 130°}$$

$$F_1 = 652.7 \text{ N} \hspace{3cm} Ans.$$

$$\frac{F_2}{\sin 20°} = \frac{1000}{\sin 130°}$$

$$F_2 = 446.5 \text{ N} \hspace{3cm} Ans.$$

Part (b). By the triangle law, vector F_2 may be added to F_1 in various ways to yield the resultant F_R, Fig. 2–15d. In particular, the *minimum* length or magnitude of F_2 will occur when the line of action of F_2 is *perpendicular* to F_1. Any other direction, such as OA or OB, yields a larger value for F_2. Hence, when $\theta = 90° - 20° = 70°$, the value of F_2 is minimum. From the triangle shown in Fig. 2–15e, it is seen that

$$F_1 = 1000 \sin 70° = 939.7 \text{ N} \hspace{2cm} Ans.$$

$$F_2 = 1000 \sin 20° = 342.0 \text{ N} \hspace{2cm} Ans.$$

18

Problems

2-1. The boom collar resists the pull of the 12-kN force. Resolve this force into components acting along the x and y axes.

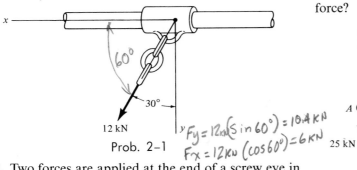

$Fy = 12_{kN}(Sin 60°) = 10.4 KN$
$Fx = 12_{KN}(cos 60°) = 6 KN$

Prob. 2-1

2-2. Two forces are applied at the end of a screw eye in order to remove the post. Determine the angle θ and the magnitude of force **F** so that the resultant force acting on the post is directed vertically upward and has a magnitude of 750 N.

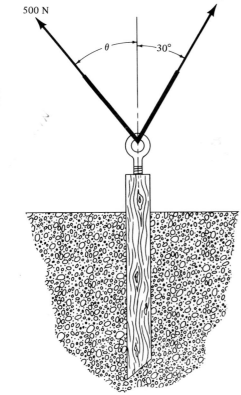

$\frac{750}{Sin\theta} = \frac{500}{Sin30°}$

$\theta = \frac{750}{500}Sin 30°$

$\theta = 48.6°$

$= 180 - 30 - \theta$

$= 101.4°$

$\frac{R}{(01.4°)} = \frac{500}{sin 30°}$

$F = 980.2 N$

Prob. 2-2

2-3. Determine the magnitude of **F** such that the resultant force acts along the axis *AB* of the boom, 12° from the horizontal. What is the magnitude of this resultant force?

$tan 12° = \frac{25}{F}$

$0.213 = \frac{25}{F} \Rightarrow F = 117.62 KN$

$R = \sqrt{117.62^2 + 25^2}$

$R = 120.24$

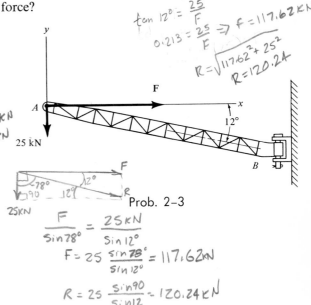

Prob. 2-3

$\frac{F}{Sin 78°} = \frac{25 KN}{Sin 12°}$

$F = 25 \frac{Sin 78°}{Sin 12°} = 117.62 KN$

$R = 25 \frac{Sin 90}{Sin 12} = 120.24 KN$

*2-4. Determine the magnitude of the resultant force acting on the end of the bracket. Specify its direction, measured from the x axis.

$\frac{R}{Sin 60°} = \frac{400}{Sin 50°}$

$R = 452.2 KN$

$\phi = 10°$

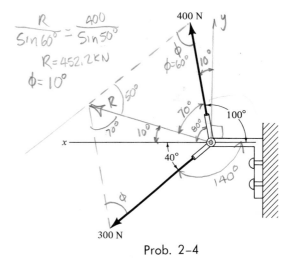

Prob. 2-4

19

2-5. Determine the magnitude of the resultant force acting on the pin and specify its direction, measured from the x axis. Set $F = 5$ kN and $P = 7$ kN.

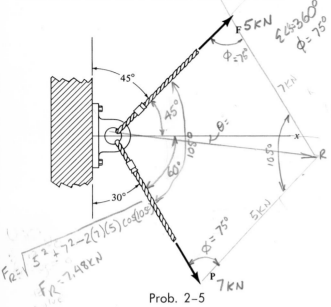

$F \; 5KN$

$2 \beta = 360°$
$\phi = 75°$
$\phi = 75°$

$7KN$

$45°$

$45°$

$\theta =$

$60°$

$30°$

$105°$

$5KN$

$\phi = 75°$

$F_R = \sqrt{5^2 + 7^2 - 2(7)(5)\cos 105°}$

$F_R = 7.48 KN$

$7KN$
P

Prob. 2-5

2-5a. Solve Prob. 2-5 with $F = 5$ lb and $P = 10$ lb.

2-6. The bracket supports two forces. Determine the angle θ so that the line of action of the resultant force is along the x axis. What is the magnitude of the resultant?

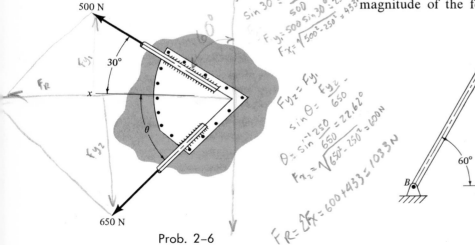

500 N

30°

F_R

F_{y_1}

x

θ

F_{y_2}

650 N

Prob. 2-6

$\sin 30° = \frac{F_{y_1}}{500}$

$F_{y_1} = 500 \sin 30° = 250N$

$F_{x_1} = \sqrt{500^2 - 250^2} = 433N$

$F_{y_2} = F_{y_1}$

$\sin \theta = \frac{650}{650}$

$\theta = \sin^{-1} \frac{250}{650} = 22.62°$

$F_{x_2} = \sqrt{650^2 - 250^2} = 600N$

$F_R = \Sigma F_x = 600 + 433 = 1033N$

2-7. Resolve the force **F** into two components, one acting parallel and the other acting perpendicular to the *aa* axis. **F** is coplanar with the *aa* axis and the vertical.

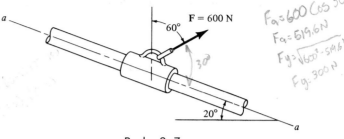

a

$F = 600$ N

60°

30

20°

a

Prob. 2-7

$F_a = 600 \cos 30°$
$F_a = 519.6N$
$F_y = \sqrt{600^2 - 519.6^2}$
$F_y = 300 N$

*****2-8.** Determine the magnitude of the resultant force if: (a) $\mathbf{F}_R = \mathbf{F}_1 + \mathbf{F}_2$; (b) $\mathbf{F}'_R = \mathbf{F}_1 - \mathbf{F}_2$.

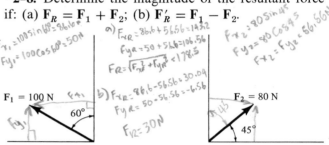

$F_1 = 100$ N

$F_{x_1} = 100 \sin 60° = 86.16$
$F_{y_1} = 100 \cos 60° = 50N$

60°

F_{x_1}

F_{y_1}

$F_2 = 80$ N

45°

F_{x_2}

$F_{x_2} = 80 \sin 45$
$F_{y_2} = 80 \cos 45$
$F_{x_2} = F_{y_2} = 56.56N$

a) $F_{xR} = 86.16 + 56.56 = 142.2$
$F_{yR} = 50 + 56.56 = 106.56$
$F_R = \sqrt{F_{xR}^2 + F_{yR}^2} = 178.5$

b) $F_{xR} = 86.16 - 56.56 = 30.04$
$F_{yR} = 50 - 56.56 = -6.56$
$F_{xR} = 30N$

Prob. 2-8

2-9. Determine the angle θ of the 500-N force such that when the force is resolved into two components acting along struts AB and AC, the component of force along AC is 300 N, directed from A towards C. What is the magnitude of the force component acting along AB?

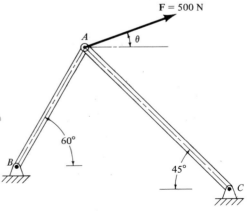

$F = 500$ N

A

θ

B

60°

45°

C

Prob. 2-9

2-10. The vertical force **F** acts downward at A on the two-member frame. Determine the magnitudes of the two components of **F** directed along the axes of AB and AC. Set $F = 500$ N.

*2-12. If $\theta = 30°$ and $\phi = 45°$, determine the magnitudes of $\mathbf{F}_1$ and $\mathbf{F}_2$ such that the resultant force $\mathbf{F}_R$ has a magnitude of 900 N and is directed along the positive x' axis.

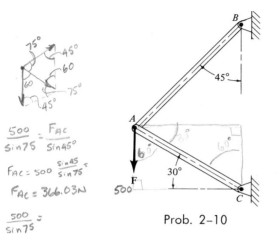

$\dfrac{500}{\sin 75} = \dfrac{F_{AC}}{\sin 45°}$

$F_{AC} = 500 \dfrac{\sin 45}{\sin 75} =$

$F_{AC} = 366.03 N$

$\dfrac{500}{\sin 75} =$

Prob. 2-10

2-10a. Solve Prob. 2-10 with $\mathbf{F} = 350$ lb.

2-11. Resolve the 200-N force acting on the bracket into two components acting: (a) along the x' and y' axes; (b) along the x and y' axes.

a) $F_{y'} = 200 \cos 10°$
$= 196.96 N$
$F_{x'} = \sqrt{200^2 - 196.96^2} = 34.7 N$

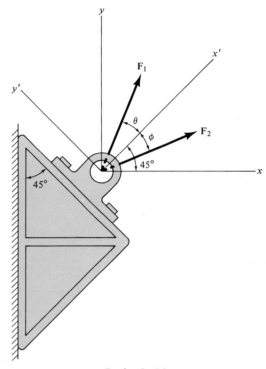

Prob. 2-12

2-13. If $F_1 = F_2 = 900$ N in Prob. 2-12, determine the angles θ and ϕ such that the resultant force $F_R = 900$ N and is directed along the positive x' axis.

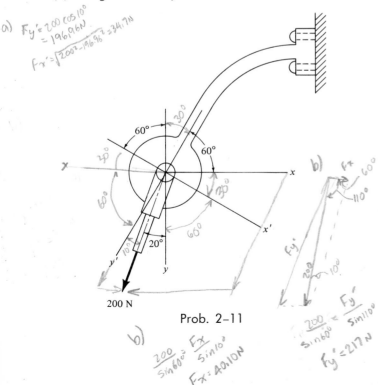

200 N

Prob. 2-11

b) $\dfrac{200}{\sin 60°} = \dfrac{F_x}{\sin 10°}$
$F_x = 40.10 N$

b) $\dfrac{200}{\sin 60°} = \dfrac{F_y}{\sin 110°}$
$F_y = 217 N$

2-14. Two forces act at the end of the boom. Determine the magnitude of **F** and its direction θ such that: (a) the resultant of the two forces, $\mathbf{F}_R$, has a magnitude of 1.5 kN and is directed along the x axis; (b) the resultant $\mathbf{F}_R = \mathbf{0}$.

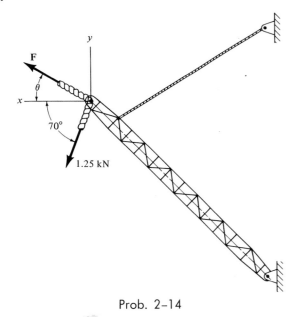

70°

1.25 kN

Prob. 2–14

2-15a. Solve Prob. 2–15 with $\theta = 12°$ and $F_R = 1200$ lb.

*2-16. If the resultant $\mathbf{F}_R$ of the two forces acting on the aircraft in Prob. 2–15 is to be directed along the x axis and have a magnitude of 4 kN, determine the angle θ for cable AB such that the force in this cable is a minimum. What is the magnitude of the force in each cable for this situation?

2-17. Determine the magnitude of the resultant $\mathbf{F}_R = \mathbf{F}_1 + \mathbf{F}_2 + \mathbf{F}_3$ of the three forces by first finding the resultant $\mathbf{F}' = \mathbf{F}_1 + \mathbf{F}_2$ and then forming $\mathbf{F}_R = \mathbf{F}' + \mathbf{F}_3$.

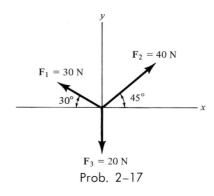

$F_1 = 30$ N

$F_2 = 40$ N

30°   45°

$F_3 = 20$ N

Prob. 2–17

2-15. The jet aircraft is subjected to the towing forces developed in cables AB and AC. If it is required that the resultant $\mathbf{F}_R$ of these two forces be directed along the x axis, determine the magnitudes of the two cable forces provided $\theta = 15°$ and $F_R = 6$ kN.

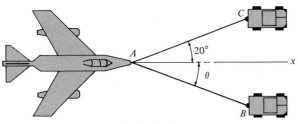

Prob. 2–15

Law of Sines

F_{AC}

20°

15°

F_{AB}

$F_R = 6$ kN

6 KN

20°

15°   145°

$\dfrac{6\,KN}{\sin 145°} = \dfrac{F_{AB}}{\sin 20°}$

$F_{AB} = 6\,\dfrac{\sin 20}{\sin 145} = 3.58\,N$ ✓

$F_{AC} = 6\,\dfrac{\sin 15}{\sin 145} = 2.71\,N$ ✓

2–5. Cartesian Vectors

Right-Handed Coordinate Systems. A right-handed Cartesian coordinate system will be used in developing the theory of vector algebra that follows. A coordinate system is said to be right-handed provided the thumb of the right hand points in the direction of the positive z axis when the right-hand fingers are curled from the positive x to the positive y axis, Fig. 2–16. Furthermore, according to this rule, the z axis in Fig. 2–17a is directed outward, perpendicular to the page.

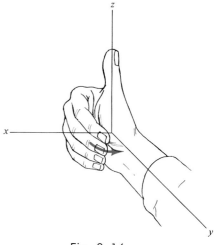

Fig. 2–16

Rectangular Components of a Vector. A vector may have one, two, or three rectangular components, depending upon how the vector is oriented relative to the rectangular x,y,z coordinate axes. For example:

1. If **A** is directed along the x axis, Fig. 2–17a, then

$$\mathbf{A} = \mathbf{A}_x$$

2. If **A** lies in the x-y plane, then the two components $\mathbf{A}_x$ and $\mathbf{A}_y$ are determined by using the parallelogram law, Fig. 2–17b, where

$$\mathbf{A} = \mathbf{A}_x + \mathbf{A}_y$$

3. If **A** is directed within an octant of the x,y,z frame, Fig. 2–17c, then by vector addition, $\mathbf{A} = \mathbf{A}' + \mathbf{A}_z$ and $\mathbf{A}' = \mathbf{A}_x + \mathbf{A}_y$. Combining these equations, **A** is represented by the vector sum of its *three* rectangular components,

$$\mathbf{A} = \mathbf{A}_x + \mathbf{A}_y + \mathbf{A}_z \qquad (2\text{–}1)$$

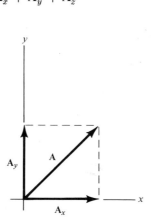

(a)

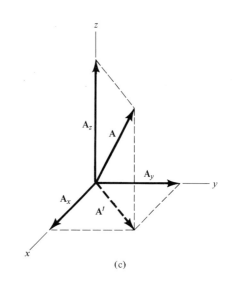

(c)

(b)

Fig. 2–17

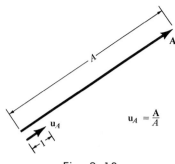

Fig. 2–18

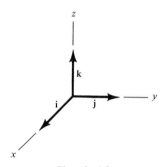

Fig. 2–19

Unit Vectors. A *unit vector* is a free vector having a magnitude of 1. If **A** is a vector having a magnitude $A \neq 0$, then a unit vector having the *same direction* as **A** is represented by

$$\mathbf{u}_A = \frac{\mathbf{A}}{A} \qquad (2\text{–}2)$$

Rewriting this equation gives

$$\mathbf{A} = A\mathbf{u}_A \qquad (2\text{–}3)$$

Since vector **A** is of a certain kind, e.g., a force vector, it is customary to use the proper set of units for its description. The magnitude A also has this same set of units; hence, from Eq. 2–2, the *unit vector will be dimensionless* since the units will cancel out. Equation 2–3 indicates that vector **A** may therefore be expressed in terms of both its magnitude and direction *separately;* i.e., A (a scalar) expresses the *magnitude* of **A,** and $\mathbf{u}_A$ (a dimensionless vector) expresses the *directional sense* of **A,** Fig. 2–18.

Cartesian Unit Vectors. In order to simplify vector-algebra operations, a set of *Cartesian unit vectors,* **i, j,** and **k,** will be used to define the *directions* of the *positive x, y,* and *z* axes, respectively, Fig. 2–19. Then using Eq. 2–3 and these unit vectors, the vector components of Fig. 2–17 may be written in Cartesian vector form. For example:

1. If **A** is directed along the positive *x* axis, Fig. 2–20*a*, then it has a magnitude of A_x and acts in the **i** direction. Therefore, Eq. 2–3 may be written as

$$\mathbf{A} = A_x\mathbf{i}$$

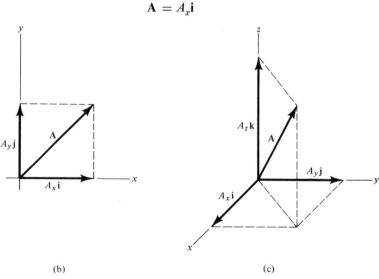

(a)

(b)

(c)

Fig. 2–20

24

2. If A lies in the x-y plane, Fig. 2–20b, then

$$\mathbf{A} = A_x\mathbf{i} + A_y\mathbf{j}$$

3. If $\mathbf{A}$ is directed within an octant of the x,y,z frame, Fig. 2–20c, then

$$\mathbf{A} = A_x\mathbf{i} + A_y\mathbf{j} + A_z\mathbf{k} \qquad (2\text{–}4)$$

There is a distinct advantage to writing vectors in terms of their Cartesian components. In doing so, the *magnitude* and *direction* of each *component vector* are *separated*, and this will simplify the operations of vector algebra, particularly in three dimensions.

Magnitude of a Cartesian Vector. It is always possible to obtain the magnitude of a vector when the vector is expressed in terms of its Cartesian components. For example:

1. If $\mathbf{A}$ is directed along the positive x axis, Fig. 2–21a, $\mathbf{A} = A_x\mathbf{i}$, so that the magnitude of $\mathbf{A}$ is

$$A = A_x$$

2. If $\mathbf{A}$ lies in the x-y plane, Fig. 2–21b, the magnitude of $\mathbf{A}$ is found by simply applying the Pythagorean theorem. From the right triangle OBC, we have

$$A = \sqrt{(A_x)^2 + (A_y)^2}$$

3. If $\mathbf{A}$ is directed within the octant of the x,y,z frame, where $\mathbf{A} = A_x\mathbf{i} + A_y\mathbf{j} + A_z\mathbf{k}$, then A can be found using the construction shown in Fig. 2–21c. From the right triangle ODB, the magnitude of line segment OB is $OB = \sqrt{(A_x)^2 + (A_y)^2}$. Similarly, from the right triangle OBC, $A = \sqrt{(OB)^2 + (A_z)^2}$. Combining both of these equations yields

$$A = \sqrt{(A_x)^2 + (A_y)^2 + (A_z)^2} \qquad (2\text{–}5)$$

Hence, in all cases, the *magnitude of* $\mathbf{A}$ *is determined by taking the square root of the sum of the squares of the magnitudes of its components.*

Direction of a Cartesian Vector. The *direction* of vector $\mathbf{A}$ is defined by the *coordinate direction angles* α (alpha), β (beta), and γ (gamma), measured between the *tail* of $\mathbf{A}$ and the *positive x, y*, and *z* axes, Fig. 2–22. Note that regardless of where $\mathbf{A}$ is directed, α, β, or γ will *never* be greater than 180°. For example:

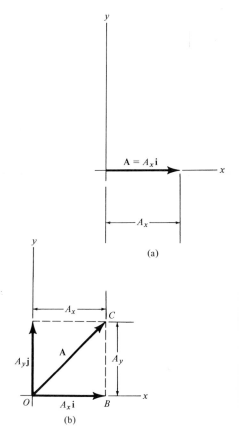

(a)

(b)

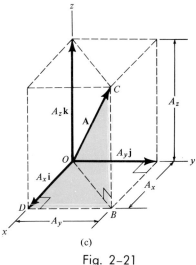

(c)

Fig. 2–21

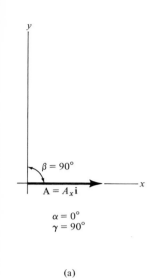

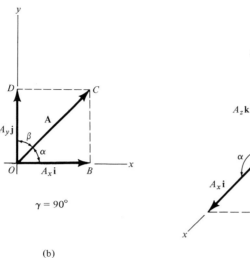

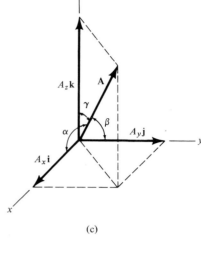

(a)

(b)

(c)

Fig. 2–22

1. If $\mathbf{A}$ is directed along the positive x axis, Fig. 2–22a, where $\mathbf{A} = A_x\mathbf{i}$, the direction of $\mathbf{A}$ is defined by

$$\alpha = 0°, \ \beta = 90°, \ \gamma = 90°$$

2. If $\mathbf{A}$ lies in the x-y plane, Fig. 2–22b, again $\gamma = 90°$ and the angles α and β in this case are determined by trigonometry. For example, from right triangles OBC and ODC, we have

$$\cos \alpha = \frac{A_x}{A}, \qquad \cos \beta = \frac{A_y}{A}, \qquad \gamma = 90°$$

3. If $\mathbf{A}$ is directed within the octant of the x,y,z frame, Fig. 2–22c, then it can be projected onto the x,y,z axes to yield each of its components as shown in Fig. 2–23. Referring to Fig. 2–23a, it is necessary that ODB be a right triangle having a right angle at D. Thus, provided the magnitude of $\mathbf{A}$ is calculated from Eq. 2–5, the angle α can be determined from $\cos \alpha = OD/OB = A_x/A$. In a similar manner, using right triangles OCB, Fig. 2–23b, and OEB, Fig 2–23c, we obtain $\cos \beta = A_y/A$ and $\cos \gamma = A_z/A$. The numbers

$$\cos \alpha = \frac{A_x}{A}, \qquad \cos \beta = \frac{A_y}{A}, \qquad \cos \gamma = \frac{A_z}{A} \qquad (2\text{-}6)$$

are known as the *direction cosines* of vector $\mathbf{A}$. The coordinate direction angles α, β, and γ are therefore defined from the relations

$$\alpha = \cos^{-1}\left(\frac{A_x}{A}\right), \qquad \beta = \cos^{-1}\left(\frac{A_y}{A}\right), \qquad \gamma = \cos^{-1}\left(\frac{A_z}{A}\right) \quad (2\text{-}7)$$

*An easy way of obtaining the direction cosines of **A** is to form a unit vector in the direction of **A**.* Provided **A** is expressed in Cartesian vector form as $\mathbf{A} = A_x\mathbf{i} + A_y\mathbf{j} + A_z\mathbf{k}$ (Eq. 2–4), we have

$$\mathbf{u}_A = \frac{\mathbf{A}}{A} = \frac{A_x}{A}\mathbf{i} + \frac{A_y}{A}\mathbf{j} + \frac{A_z}{A}\mathbf{k} \qquad (2\text{–}8)$$

where $A = \sqrt{(A_x)^2 + (A_y)^2 + (A_z)^2}$ (Eq. 2–5). By comparison with Eqs. 2–6, it is seen that *the* **i, j,** *and* **k** *components of* $\mathbf{u}_A$ *represent the direction cosines* of **A**, i.e.,

$$\mathbf{u}_A = \cos\alpha\,\mathbf{i} + \cos\beta\,\mathbf{j} + \cos\gamma\,\mathbf{k} \qquad (2\text{–}9)$$

Since the magnitude of a vector is equal to the square root of the sum of the squares of the magnitudes of its components, and $\mathbf{u}_A$ has a magnitude of 1, then from Eq. 2–9 an important relation between the direction cosines can be formulated as,

$$\cos^2\alpha + \cos^2\beta + \cos^2\gamma = 1 \qquad (2\text{–}10)$$

This equation is useful for determining one of the coordinate direction angles if the other *two* are *known*.

If the magnitude and direction angles of **A** are given, **A** may be expressed in Cartesian vector form as

$$\begin{aligned} \mathbf{A} &= A\mathbf{u}_A \\ &= A\cos\alpha\,\mathbf{i} + A\cos\beta\,\mathbf{j} + A\cos\gamma\,\mathbf{k} \qquad (2\text{–}11) \\ &= A_x\mathbf{i} + A_y\mathbf{j} + A_z\mathbf{k} \end{aligned}$$

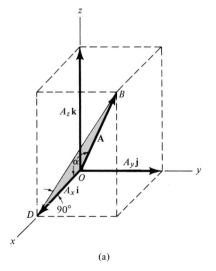

(a)

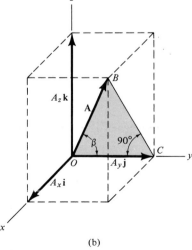

(b)

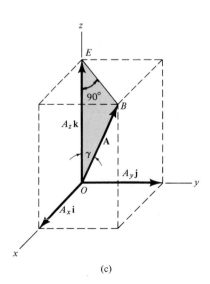

(c)

Fig. 2–23

2-6. Addition and Subtraction of Cartesian Vectors

The vector operations of addition and subtraction of two or more vectors are greatly simplified if the vectors are expressed in terms of their Cartesian components. For example, in the case of two vectors **A** and **B**:

1. If both vectors are directed along the positive x axis, Fig. 2–24a, then $\mathbf{A} = A_x\mathbf{i}$, $\mathbf{B} = B_x\mathbf{i}$, and $\mathbf{R} = \mathbf{A} + \mathbf{B} = A_x\mathbf{i} + B_x\mathbf{i}$. By factoring out the unit vector **i**, the resultant vector can be written as

$$\mathbf{R} = \mathbf{A} + \mathbf{B} = (A_x + B_x)\mathbf{i}$$

2. If both vectors lie in the x-y plane, Fig. 2–24b, then $\mathbf{A} = A_x\mathbf{i} + A_y\mathbf{j}$, $\mathbf{B} = B_x\mathbf{i} + B_y\mathbf{j}$, and $\mathbf{R} = \mathbf{A} + \mathbf{B} = A_x\mathbf{i} + A_y\mathbf{j} + B_x\mathbf{i} + B_y\mathbf{j}$ or

$$\mathbf{R} = \mathbf{A} + \mathbf{B} = (A_x + B_x)\mathbf{i} + (A_y + B_y)\mathbf{j}$$

3. If both vectors are directed within an octant of the x,y,z frame, Fig. 2–24c, then $\mathbf{A} = A_x\mathbf{i} + A_y\mathbf{j} + A_z\mathbf{k}$ and $\mathbf{B} = B_x\mathbf{i} + B_y\mathbf{j} + B_z\mathbf{k}$, and the resultant vector, **R,** has components which represent the scalar sums of the components of **A** and **B,** i.e.,

$$\mathbf{R} = \mathbf{A} + \mathbf{B} = (A_x + B_x)\mathbf{i} + (A_y + B_y)\mathbf{j} + (A_z + B_z)\mathbf{k}$$

Vector subtraction, being a special case of vector addition, simply requires a scalar subtraction of the respective **i, j,** and **k** components of **A** and **B.** For example

$$\mathbf{R} = \mathbf{A} - \mathbf{B} = (A_x - B_x)\mathbf{i} + (A_y - B_y)\mathbf{j} + (A_z - B_z)\mathbf{k}$$

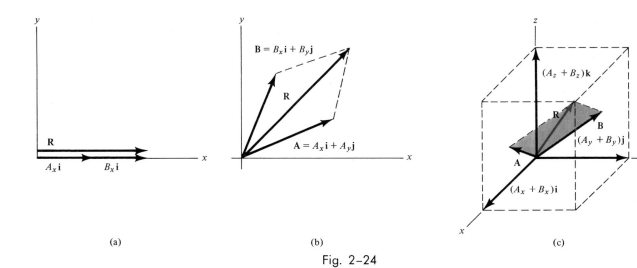

(a)   (b)   (c)

Fig. 2–24

Concurrent Force Systems. In particular, the concept of vector addition may be generalized and applied to a system of several concurrent forces for which all the forces have Cartesian components acting along the x, y, and z axes. In this case, the force resultant of the system is the vector sum of all the forces in the system and can be written as

$$\mathbf{F}_R = \Sigma \mathbf{F} = \Sigma F_x \mathbf{i} + \Sigma F_y \mathbf{j} + \Sigma F_z \mathbf{k} \qquad (2\text{--}12)$$

Here $\Sigma F_x \mathbf{i}$, $\Sigma F_y \mathbf{j}$, and $\Sigma F_z \mathbf{k}$ represent the sums of the Cartesian components of each force in the system.

The following examples numerically illustrate the methods used to apply the above theory to the solution of problems involving force as a vector quantity.

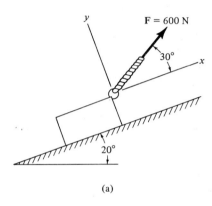

(a)

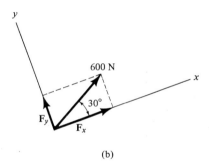

(b)

Fig. 2–25

Example 2-5 (Two Dimensions)

The rope in Fig. 2–25a is subjected to a tension of 600 N when used to pull the crate up the 20° incline. Express this force as a Cartesian vector having components parallel and perpendicular to the incline.

Solution

As shown in Fig. 2–25b, the force is resolved into its Cartesian components using the parallelogram law. From this figure,

$$F_x = 600 \cos 30° = 519.6 \text{ N}$$
$$F_y = 600 \sin 30° = 300.0 \text{ N}$$

Since the magnitudes of the $\mathbf{i}$ and $\mathbf{j}$ components of $\mathbf{F}$ are now known, the force may be expressed as

$$\mathbf{F} = \{519.6\mathbf{i} + 300.0\mathbf{j}\} \text{ N} \qquad\qquad Ans.$$

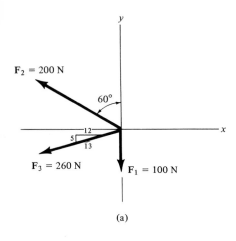

(a)

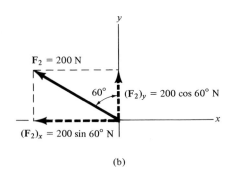

(b)

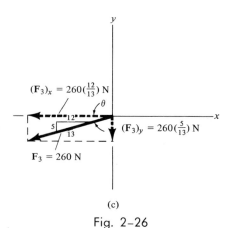

(c)

Fig. 2-26

Example 2-6 (Two Dimensions)

Express each of the concurrent forces F_1, F_2, and F_3, shown in Fig. 2-26a, in Cartesian vector form.

Solution

Since F_1 acts in the $-j$ direction, i.e., along the negative y axis, and the magnitude of F_1 is 100 N,

$$F_1 = F_1(-j) = \{-100j\} \text{ N} \qquad Ans.$$

By the parallelogram law, F_2 is resolved into x and y components, Fig. 2-26b. The magnitude of each component is determined by trigonometry. Since $(F_2)_x$ acts in the $-i$ direction, and $(F_2)_y$ acts in the $+j$ direction,

$$F_2 = 200 \sin 60°(-i) + 200 \cos 60°j$$
$$= \{-173.2i + 100.0j\} \text{ N} \qquad Ans.$$

The magnitudes of the components of F_3, Fig. 2-26c, can be obtained by solving for the angle θ using the "slope triangle," i.e., $\theta = \tan^{-1} \frac{5}{12}$, and then proceeding in the same manner as for F_2. An easier method, however, consists of using proportional parts of similar triangles, i.e.,

$$\frac{(F_3)_x}{260} = \frac{12}{13} \qquad (F_3)_x = 260\left(\frac{12}{13}\right) = 240 \text{ N}$$

Similarly,

$$(F_3)_y = 260\left(\frac{5}{13}\right) = 100 \text{ N}$$

Notice that the magnitude of the *horizontal component*, $(F_3)_x$, was obtained by multiplying the force magnitude by the ratio of the *horizontal leg* of the slope triangle divided by the hypotenuse; whereas the magnitude of the *vertical component*, $(F_3)_y$, was obtained by multiplying the force magnitude by the ratio of the *vertical leg* divided by the hypotenuse. Hence,

$$F_3 = 240(-i) + 100(-j)$$
$$= \{-240i - 100j\} \text{ N} \qquad Ans.$$

Example 2-7 (Two Dimensions)

The gusset plate shown in Fig. 2-27a is subjected to four forces, which are concurrent at point O. Using Cartesian vectors, determine the magnitude and direction of the resultant of these forces.

Solution

The resultant force is the vector sum of all the forces, $F_R = \Sigma F$

(Eq. 2–12). For the solution each force will first be expressed in Cartesian vector form and then the respective **i** and **j** components will be added. The forces $\mathbf{F}_1$ and $\mathbf{F}_3$ are determined by inspection, with the $+x$ axis directed to the right and the $+y$ axis directed upward,

$$\mathbf{F}_1 = (400 \text{ N})(-\mathbf{i}) = \{-400\mathbf{i}\} \text{ N}$$
$$\mathbf{F}_3 = (500 \text{ N})(-\mathbf{j}) = \{-500\mathbf{j}\} \text{ N}$$

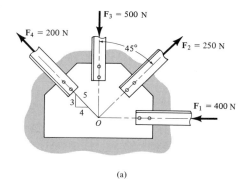

(a)

Forces $\mathbf{F}_2$ and $\mathbf{F}_4$ are resolved into components as shown in Figs. 2–27b and 2–27c. Hence,

$$\mathbf{F}_2 = 250 \sin 45°\mathbf{i} + 250 \cos 45°\mathbf{j} = \{176.8\mathbf{i} + 176.8\mathbf{j}\} \text{ N}$$
$$\mathbf{F}_4 = 200(\tfrac{4}{5})(-\mathbf{i}) + 200(\tfrac{3}{5})\mathbf{j} = \{-160\mathbf{i} + 120\mathbf{j}\} \text{ N}$$

The resultant force is therefore,

$$\mathbf{F}_R = \Sigma\mathbf{F} = \mathbf{F}_1 + \mathbf{F}_2 + \mathbf{F}_3 + \mathbf{F}_4$$
$$= -400\mathbf{i} + (176.8\mathbf{i} + 176.8\mathbf{j}) + (-500\mathbf{j}) + (-160\mathbf{i} + 120\mathbf{j})$$
$$= (-400 + 176.8 - 160)\mathbf{i} + (176.8 - 500 + 120)\mathbf{j}$$
$$= \{-383.2\mathbf{i} - 203.2\mathbf{j}\} \text{ N} \qquad\qquad Ans.$$

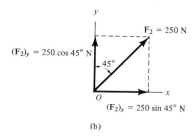

(b)

This result, which indicates the effect of all four forces on the gusset plate, is shown in Fig. 2–27d. From Eq. 2–5 the magnitude of $\mathbf{F}_R$ is

$$F_R = \sqrt{F_x^2 + F_y^2}$$
$$= \sqrt{(-383.2)^2 + (-203.2)^2} = 433.7 \text{ N} \qquad Ans.$$

The direction cosines of $\mathbf{F}_R$ can be determined from the *components* of the *unit vector* acting in the direction of $\mathbf{F}_R$. Thus, using Eq. 2–8 gives

$$\mathbf{u}_R = \frac{\mathbf{F}_R}{F_R} = -\frac{383.2}{433.7}\mathbf{i} - \frac{203.2}{433.7}\mathbf{j} = -0.883\mathbf{i} - 0.468\mathbf{j}$$

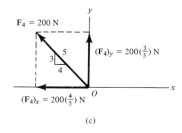

(c)

Hence, from Eq. 2–9,

$$\cos \alpha = -0.883$$
$$\cos \beta = -0.468$$
$$\cos \gamma = 0$$

or

$$\alpha = 152.1° \qquad\qquad Ans.$$
$$\beta = 117.9° \qquad\qquad Ans.$$
$$\gamma = 90° \qquad\qquad Ans.$$

By definition, the coordinate direction angles α and β as indicated in Fig. 2–27d are measured between the *positive x* and *y* axes and the *tail* of the force. These two angles can also be determined by first finding θ in Fig. 2–27d, i.e., $\theta = \tan^{-1}(203.2/383.2) = 27.9°$, so that $\alpha = 180° - \theta = 152.1°$ and $\beta = 90° + \theta = 117.9°$.

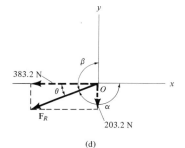

(d)

Fig. 2–27

Example 2-8 (Three Dimensions)

Express (a) force $\mathbf{F}_1$ in Fig. 2–28a and (b) $\mathbf{F}_2$ in Fig. 2–28b in Cartesian vector form.

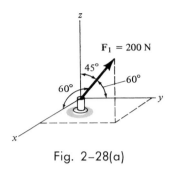

Fig. 2–28(a)

Solution

Part (a).   Since the coordinate direction angles α, β, and γ are given and the magnitude of $\mathbf{F}_1$ is 200 N, Eq. 2–11 can be used. Hence,

$$\mathbf{F}_1 = F_1 \cos \alpha \mathbf{i} + F_1 \cos \beta \mathbf{j} + F_1 \cos \gamma \mathbf{k}$$
$$= 200 \cos 60°\mathbf{i} + 200 \cos 60°\mathbf{j} + 200 \cos 45°\mathbf{k}$$
$$= \{100\mathbf{i} + 100\mathbf{j} + 141.4\mathbf{k}\} \text{ N} \hspace{2cm} Ans.$$

Note that the coordinate direction angles for $\mathbf{F}_1$ satisfy Eq. 2–10, i.e.,

$$\cos^2 \alpha + \cos^2 \beta + \cos^2 \gamma = 1$$
$$\cos^2 60° + \cos^2 60° + \cos^2 45° = 1$$

Furthermore, the magnitude of $\mathbf{F}_1$ is 200 N, which can be shown by applying Eq. 2–5.

$$F_1 = \sqrt{(F_1)_x^2 + (F_1)_y^2 + (F_1)_z^2}$$
$$= \sqrt{(100)^2 + (100)^2 + (141.4)^2} = 200 \text{ N}$$

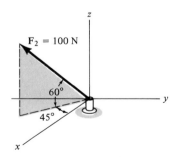

Fig. 2–28(b)

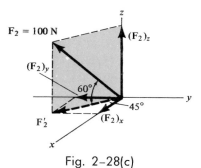

$F_2 = 100$ N

$(F_2)_z$

$(F_2)_y$

60°

45°

F_2'

$(F_2)_x$

z

y

x

Fig. 2–28(c)

Part (b). The angles of 60° and 45° defining the direction of F_2 are not coordinate direction angles. Why? By two successive applications of the parallelogram law, however, F_2 can be resolved into its x, y, and z components as shown in Fig. 2–28c. By trigonometry, the magnitudes of the components are,

$$(F_2)_z = 100 \sin 60° = 86.6 \text{ N}$$
$$F_2' = 100 \cos 60° = 50 \text{ N}$$
$$(F_2)_x = 50 \cos 45° = 35.4 \text{ N}$$
$$(F_2)_y = 50 \sin 45° = 35.4 \text{ N}$$

Realizing that $(F_2)_y$ has a direction defined by $-j$, we have

$$F_2 = (F_2)_x + (F_2)_y + (F_2)_z$$
$$F_2 = \{35.4i - 35.4j + 86.6k\} \text{ N} \hspace{2cm} Ans.$$

To show that the magnitude of this vector is indeed 100 N, apply Eq. 2–5,

$$F_2 = \sqrt{(F_2)_x^2 + (F_2)_y^2 + (F_2)_z^2}$$
$$= \sqrt{(35.4)^2 + (-35.4)^2 + (86.6)^2} = 100 \text{ N}$$

If needed, the coordinate direction angles of F_2 can be determined from the components of the unit vector acting in the direction of F_2. Hence,

$$\mathbf{u}_{F_2} = \frac{F_2}{F_2} = \frac{(F_2)_x}{F_2}i + \frac{(F_2)_y}{F_2}j + \frac{(F_2)_z}{F_2}k$$
$$= \frac{35.4}{100}i - \frac{35.4}{100}j + \frac{86.6}{100}k$$
$$= 0.354i - 0.354j + 0.866k$$

so that

$$\cos \alpha = 0.354, \hspace{0.5cm} \text{or} \hspace{0.5cm} \alpha = \cos^{-1}(0.354) = 69.3°$$
$$\cos \beta = -0.354, \hspace{0.5cm} \text{or} \hspace{0.5cm} \beta = \cos^{-1}(-0.354) = 110.7°$$
$$\cos \gamma = 0.866, \hspace{0.5cm} \text{or} \hspace{0.5cm} \gamma = \cos^{-1}(0.866) = 30.0°$$

Indicate these angles on Fig. 2–28b.

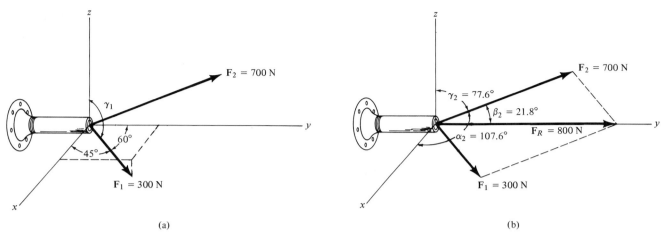

Fig. 2-29

Example 2-9 (Three Dimensions)

Two forces act on the pipe shown in Fig. 2-29a. Specify the direction of $\mathbf{F}_2$ so that the resultant force $\mathbf{F}_R$ acts along the positive y axis and has a magnitude of 800 N.

Solution

To solve this problem, the resultant force and its two components, $\mathbf{F}_1$ and $\mathbf{F}_2$, will each be expressed in Cartesian vector form. Then, as shown in Fig. 2-29b, it is necessary that $\mathbf{F}_R = \mathbf{F}_1 + \mathbf{F}_2$.

Since only two direction angles for $\mathbf{F}_1$ have been specified, Fig. 2-29a, the third angle, γ_1, can be determined from Eq. 2-10, i.e.,

$$\cos^2 \alpha_1 + \cos^2 \beta_1 + \cos^2 \gamma_1 = 1$$
$$\cos^2 45° + \cos^2 60° + \cos^2 \gamma_1 = 1$$
$$\cos \gamma_1 = \pm \sqrt{1 - (0.707)^2 - (0.5)^2} = \pm 0.50$$

Hence,

$$\gamma_1 = \cos^{-1}(0.5) = 60° \quad \text{or} \quad \gamma_1 = \cos^{-1}(-0.5) = 120°$$

From Fig. 2-29a, however, it is necessary that $\gamma_1 = 120° > 90°$. Thus, from Eq. 2-11,

$$\mathbf{F}_1 = F_1 \mathbf{u}_{F_1} = F_1 \cos \alpha_1 \mathbf{i} + F_1 \cos \beta_1 \mathbf{j} + F_1 \cos \gamma_1 \mathbf{k}$$
$$= 300 \cos 45° \mathbf{i} + 300 \cos 60° \mathbf{j} + 300 \cos 120° \mathbf{k}$$
$$= \{212.1\mathbf{i} + 150\mathbf{j} - 150\mathbf{k}\} \text{ N}$$

Since $F_2 = 700$ N, then by Eq. 2-11,

$$\mathbf{F}_2 = F_2 \mathbf{u}_{F_2} = 700 \cos \alpha_2 \mathbf{i} + 700 \cos \beta_2 \mathbf{j} + 700 \cos \gamma_2 \mathbf{k}$$

According to the problem statement, the resultant force $\mathbf{F}_R$ has a magnitude of 800 N and acts in the $+\mathbf{j}$ direction. Hence,

$$F_R = (800 \text{ N})(+\mathbf{j}) = \{800\mathbf{j}\} \text{ N}$$

It is required that

$$F_R = F_1 + F_2$$
$$800\mathbf{j} = 212.1\mathbf{i} + 150\mathbf{j} - 150\mathbf{k} + 700 \cos \alpha_2\mathbf{i} + 700 \cos \beta_2\mathbf{j} + 700 \cos \gamma_2\mathbf{k}$$
$$= (212.1 + 700 \cos \alpha_2)\mathbf{i} + (150 + 700 \cos \beta_2)\mathbf{j} + (-150 + 700 \cos \gamma_2)\mathbf{k}$$

To satisfy this equation, the corresponding $\mathbf{i}$, $\mathbf{j}$, and $\mathbf{k}$ components on the left and right sides must be equal. This is equivalent to stating that the x, y, and z components of F_R be equal to the corresponding x, y, and z components of $(F_1 + F_2)$. Hence,

$$0 = 212.1 + 700 \cos \alpha_2 \qquad \alpha_2 = \cos^{-1}\left(\frac{-212.1}{700}\right) = 107.6° \qquad Ans.$$

$$800 = 150 + 700 \cos \beta_2 \qquad \beta_2 = \cos^{-1}\left(\frac{650}{700}\right) = 21.8° \qquad Ans.$$

$$0 = -150 + 700 \cos \gamma_2 \qquad \gamma_2 = \cos^{-1}\left(\frac{150}{700}\right) = 77.6° \qquad Ans.$$

The results are shown in Fig. 2–29b.

Problems

2-18. Express the force $\mathbf{F}$ acting on the bracket as a Cartesian vector.

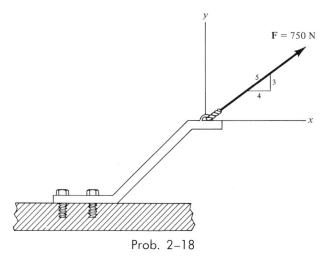

Prob. 2–18

2-19. Determine the magnitude and direction of: (a) $F_1 = \{60\mathbf{i} - 50\mathbf{j}\}$ N; (b) $F_2 = \{-400\mathbf{i} - 850\mathbf{j}\}$ N. Sketch each force using an x-y coordinate reference.

*2-20. Express each force acting on the pin as a Cartesian vector with reference to the x and y axes shown. Set $F_1 = 100$ N, $F_2 = 200$ N, and $F_3 = 250$ N.

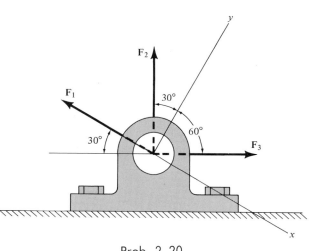

Prob. 2–20

*2-20a. Solve Prob. 2-20 with $F_1 = 60$ lb, $F_2 = 130$ lb, and $F_3 = 170$ lb.

2-21. Express the forces $\mathbf{F}_1$ and $\mathbf{F}_2$ in Cartesian vector notation with respect to the x and y axes. Determine the magnitude of the resultant force and its direction measured from the positive x axis.

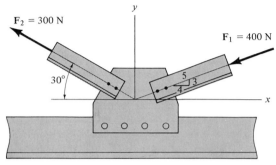

F$_2$ = 300 N

F$_1$ = 400 N

30°

5
3
4

y

x

Prob. 2-21

2-22. Express each of the three forces acting on the bracket in Cartesian vector form, and compute the magnitude and direction of the resultant force.

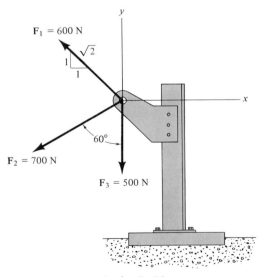

$F_1 = 600$ N

$\sqrt{2}$

1

1

y

x

60°

$F_2 = 700$ N

$F_3 = 500$ N

Prob. 2-22

2-23. Each of the three springs is subjected to a tension force having a magnitude of 40 N. Express these three spring forces $\mathbf{F}_A$, $\mathbf{F}_B$, and $\mathbf{F}_C$ as Cartesian vectors, and find the resultant force acting on O.

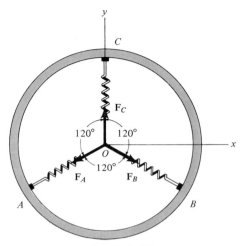

y

C

F_C

120°   120°

O

120°

F_A F_B

x

A   B

Prob. 2-23

*2-24. Determine the magnitude and direction of:
(a) $\mathbf{F}_1 = \{200\mathbf{i} + 250\mathbf{j} - 150\mathbf{k}\}$ N; (b) $\mathbf{F}_2 = \{2.5\mathbf{i} - 1.5\mathbf{j} + 4\mathbf{k}\}$ kN.

2-25. Express each of the three forces acting on the bracket in Cartesian vector form. Determine the magnitude and direction θ of $\mathbf{F}_1$, so that the resultant force $\mathbf{F}_R$ is directed along the positive x' axis of the bracket and $F_R = 500$ N. Set $F_2 = 200$ N and $F_3 = 100$ N.

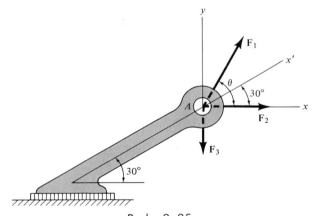

y

F_1

x'

θ

30°

A

F_2

x

F_3

30°

Prob. 2-25

2-25a. Solve Prob. 2-25 so that $F_R = 300$ lb. Set $F_2 = 70$ lb and $F_3 = 45$ lb.

2-26. The ball joint is subjected to the three forces shown. Express each force in Cartesian vector form and determine the magnitude and direction of the resultant force.

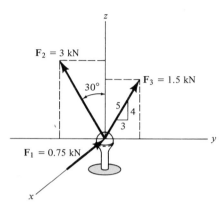

Prob. 2-26

2-27. The magnitudes of the y and z components of the force $\mathbf{F}$ are known to be $F_y = 60$ N and $F_z = 15$ N, respectively. If the angle α that $\mathbf{F}$ makes with the x axis is $45°$, determine the magnitude and the direction of $\mathbf{F}$. What is the magnitude of $\mathbf{F}_x$?

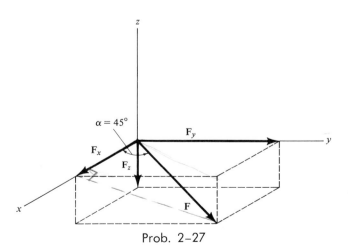

Prob. 2-27

*2-28. The shaft S exerts three force components on the die D. Find the magnitude and direction of the resultant force. Force $\mathbf{F}_2$ acts in the octant shown.

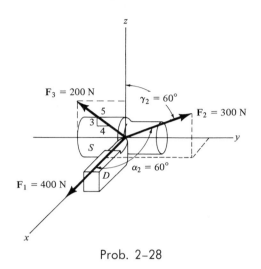

Prob. 2-28

2-29. Force $\mathbf{F}$ acts on the peg A such that one of its components, lying in the x-y plane, has a magnitude of 300 N. Express $\mathbf{F}$ as a Cartesian vector and determine its magnitude.

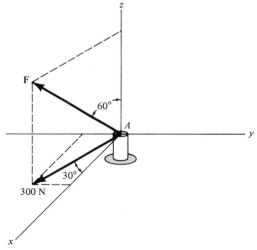

Prob. 2-29

37

2-30. Specify the magnitude F_3 and directions α_3, β_3, and γ_3 of $\mathbf{F}_3$ so that the resultant force of the three forces is $\mathbf{F}_R = \{9\mathbf{j}\}$ kN. Set $F_1 = 12$ kN and $F_2 = 10$ kN.

2-30a. Solve Prob. 2–30 so that $\mathbf{F}_R = \{600\mathbf{j}\}$ lb. Set $F_1 = 300$ lb and $F_2 = 260$ lb.

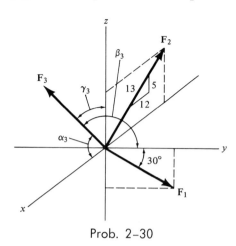

Prob. 2–30

2-7. Cartesian Position Vectors

Most often the x, y, and z coordinate axes are used as a frame of reference in space, in which case the three axes are measured in units of length. Using these axes the *position vector* $\mathbf{r}$ will be defined as a fixed vector which locates a point in space relative to another point. For example, if $\mathbf{r}$ extends from the origin of coordinates, O, to point $P(x, y, z)$, Fig. 2–30a, then $\mathbf{r}$ can be expressed in Cartesian vector form as

$$\mathbf{r} = x\mathbf{i} + y\mathbf{j} + z\mathbf{k} \qquad (2\text{–}13)$$

The tip-to-tail vector addition of the three components, which yields vector $\mathbf{r}$, is shown in Fig. 2–30b.

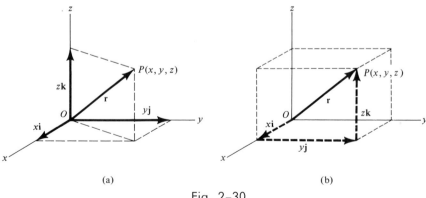

(a)   (b)

Fig. 2–30

In the more general case, the *position vector* **r** is directed from point A to point B in space, Fig. 2–31a. This vector can be expressed in Cartesian vector form using the coordinates of the tail $A(x_A, y_A, z_A)$ and tip $B(x_B, y_B, z_B)$ of the vector. From the figure, by the triangle law of vector addition, we require that

$$\mathbf{r}_A + \mathbf{r} = \mathbf{r}_B$$

Solving for **r** and expressing $\mathbf{r}_A$ and $\mathbf{r}_B$ in Cartesian vector form, yields

$$\mathbf{r} = \mathbf{r}_B - \mathbf{r}_A = (x_B\mathbf{i} + y_B\mathbf{j} + z_B\mathbf{k}) - (x_A\mathbf{i} + y_A\mathbf{j} + z_A\mathbf{k})$$

or

$$\mathbf{r} = (x_B - x_A)\mathbf{i} + (y_B - y_A)\mathbf{j} + (z_B - z_A)\mathbf{k} \qquad (2\text{–}14)$$

Thus, the **i**, **j**, *and* **k** *components of the position vector* **r** *may be formed by taking the coordinates of the tip of the vector,* $B(x_B, y_B, z_B)$, *and subtracting from them the corresponding coordinates of the tail,* $A(x_A, y_A, z_A)$. *The tip-to-tail addition of these three components, which yields* **r**, *is shown in Fig. 2–31b.*

Magnitude of r. Following the method used in Sec. 2–5, the *magnitude* of **r** is simply

$$r = \sqrt{(x_B - x_A)^2 + (y_B - y_A)^2 + (z_B - z_A)^2} \qquad (2\text{–}15)$$

Direction of r. The direction of **r** can be determined by first constructing a unit vector **u** that is directed from point A towards point B, Fig. 2–32.

$$\mathbf{u} = \frac{\mathbf{r}}{r} = \frac{x_B - x_A}{r}\mathbf{i} + \frac{y_B - y_A}{r}\mathbf{j} + \frac{z_B - z_A}{r}\mathbf{k} \qquad (2\text{–}16)$$

The coordinate direction angles α, β, and γ which **r** makes with the positive x, y, and z axes are found from the **i**, **j**, and **k** *components* of this unit vector, i.e.,

$$\alpha = \cos^{-1}\left(\frac{x_B - x_A}{r}\right)$$

$$\beta = \cos^{-1}\left(\frac{y_B - y_A}{r}\right) \qquad (2\text{–}17)$$

$$\gamma = \cos^{-1}\left(\frac{z_B - z_A}{r}\right)$$

As shown in Fig. 2–32, these angles are measured from the *tail* of **u** to the *positive, x, y,* and z axes.

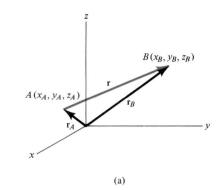

(a)

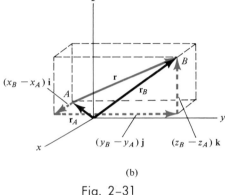

(b)

Fig. 2–31

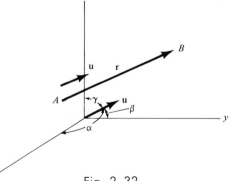

Fig. 2–32

2–8. Cartesian Force Vectors

The unit vector **u,** which is defined by Eq. 2–16, can be used to construct a Cartesian force vector that has the *same direction* as **r.** For example, suppose that points A and B represent the end points of a cord that is attached at A to a fixed support, Fig. 2–33. If the cord is subjected to a tension force **F** acting at B, this force can be expressed as a Cartesian vector by noting that the *magnitude* of **F** is F and the *direction* of **F** is specified by the unit vector **u,** Eq. 2–16. Hence,

$$\mathbf{F} = F\mathbf{u} = \left\{ F\frac{x_B - x_A}{r}\mathbf{i} + F\frac{y_B - y_A}{r}\mathbf{j} + F\frac{z_B - z_A}{r}\mathbf{k} \right\} \quad (2\text{–}18)$$

In particular, note that the force **F** *does not extend from A to B; only the position vector* **r** *does.* This is because both the coordinate axes and **r** have units of length, e.g., metres; whereas **F** has units of force, e.g., newtons.

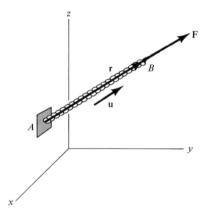

Fig. 2–33

PROCEDURE FOR ANALYSIS

Provided **F** is directed along a line which extends from point A to point B, **F** can be expressed as a Cartesian vector as follows:

Step 1:   Find the coordinates of points A and B.

Step 2:   Determine the position vector **r** (Eq. 2–14) and compute its magnitude r (Eq. 2–15).

Step 3:   Determine a unit vector $\mathbf{u} = \mathbf{r}/r$ (Eq. 2–16) which defines the *directions* of *both* **r** and **F.**

Step 4:   Determine **F** by combining its magnitude F and direction **u,** i.e., $\mathbf{F} = F\mathbf{u}$ (Eq. 2–18).

This procedure is illustrated numerically in the following example problems.

Example 2–10

The man shown in Fig. 2–34a pulls on the cord with a force of 175 N. Represent this force, acting on the support A, as a Cartesian vector and determine its direction.

Solution

Force $\mathbf{F}$ is shown in Fig. 2–34b. The direction of this vector, $\mathbf{u}$, is determined from the position vector $\mathbf{r}$, which extends from A to B, Fig. 2–34b. To formulate $\mathbf{F}$ as a Cartesian vector use the four-step procedure outlined above.

Step 1: The coordinates of the end points of the cord are $A(0, 0, 7.5)$ and $B(3, -2, 1.5)$.

Step 2: Forming the position vector $\mathbf{r}$ by subtracting the corresponding x, y, and z coordinates of B from those of A, we have

$$\mathbf{r} = (3 - 0)\mathbf{i} + (-2 - 0)\mathbf{j} + (1.5 - 7.5)\mathbf{k}$$
$$= \{3\mathbf{i} - 2\mathbf{j} - 6\mathbf{k}\} \text{ m}$$

Indicate the tip-to-tail addition of these $\mathbf{i}$, $\mathbf{j}$, and $\mathbf{k}$ components on Fig. 2–34a and show that they yield $\mathbf{r}$. The magnitude of $\mathbf{r}$, which represents the *length* of cord AB, is

$$r = \sqrt{(3)^2 + (-2)^2 + (-6)^2} = 7 \text{ m}$$

Step 3: Forming the unit vector that defines the direction of both $\mathbf{r}$ and $\mathbf{F}$, yields

$$\mathbf{u} = \frac{\mathbf{r}}{r} = \frac{3}{7}\mathbf{i} - \frac{2}{7}\mathbf{j} - \frac{6}{7}\mathbf{k}$$

As shown in Fig. 2–34b, the coordinate direction angles, which should *always* be measured between the *positive axes* and the *tail* of $\mathbf{u}$, are

$$\alpha = \cos^{-1}\left(\frac{3}{7}\right) = 64.6° \qquad Ans.$$

$$\beta = \cos^{-1}\left(\frac{-2}{7}\right) = 106.6° \qquad Ans.$$

$$\gamma = \cos^{-1}\left(\frac{-6}{7}\right) = 149.0° \qquad Ans.$$

Step 4: Since $\mathbf{F}$ has a *magnitude* of 175 N and a *direction* specified by $\mathbf{u}$, then

$$\mathbf{F} = F\mathbf{u} = 175 \text{ N}\left(\frac{3}{7}\mathbf{i} - \frac{2}{7}\mathbf{j} - \frac{6}{7}\mathbf{k}\right)$$

$$= \{75\mathbf{i} - 50\mathbf{j} - 150\mathbf{k}\} \text{ N} \qquad Ans.$$

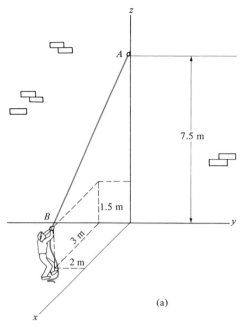

(a)

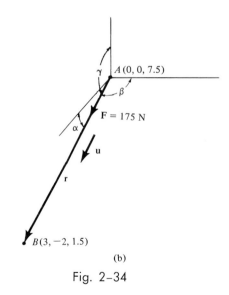

(b)

Fig. 2–34

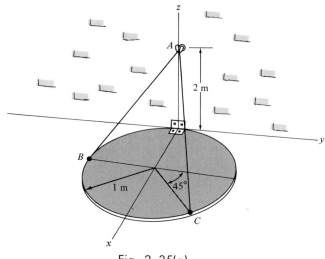

Fig. 2–35(a)

Example 2–11

The circular plate shown in Fig. 2–35a is partially supported by two cables AB and AC. If the magnitude of force in each cable is $F_B = 300$ N and $F_C = 500$ N, express each force as a Cartesian vector acting on the hook at A and determine the magnitude and direction of the resultant force.

Solution

The directions of the cable forces $\mathbf{F}_B$ and $\mathbf{F}_C$ shown in Fig. 2–35b are specified by forming unit vectors $\mathbf{u}_B$ and $\mathbf{u}_C$ along the cables. These unit vectors are obtained from the associated position vectors $\mathbf{r}_B$ and $\mathbf{r}_C$. With reference to Fig. 2–35b, for $\mathbf{F}_B$, we have

$$B(1, -1, 0), \quad A(0, 0, 2)$$
$$\mathbf{r}_B = (1 - 0)\mathbf{i} + (-1 - 0)\mathbf{j} + (0 - 2)\mathbf{k}$$
$$= \{1\mathbf{i} - 1\mathbf{j} - 2\mathbf{k}\} \text{ m};$$
$$r_B = \sqrt{(1)^2 + (-1)^2 + (-2)^2} = 2.45 \text{ m}$$
$$\mathbf{u}_B = \frac{\mathbf{r}_B}{r_B} = \frac{1}{2.45}\mathbf{i} - \frac{1}{2.45}\mathbf{j} - \frac{2}{2.45}\mathbf{k}$$
$$= 0.408\mathbf{i} - 0.408\mathbf{j} - 0.816\mathbf{k}$$
$$\mathbf{F}_B = F_B\mathbf{u}_B = 300 \text{ N}(0.408\mathbf{i} - 0.408\mathbf{j} - 0.816\mathbf{k})$$
$$= \{122.4\mathbf{i} - 122.4\mathbf{j} - 244.8\mathbf{k}\} \text{ N} \qquad \textit{Ans.}$$

In a similar manner, for $\mathbf{F}_C$,

$$C(1.707, 0.707, 0), \quad A(0, 0, 2)$$
$$\mathbf{r}_C = (1.707 - 0)\mathbf{i} + (0.707 - 0)\mathbf{j} + (0 - 2)\mathbf{k}$$
$$\mathbf{r}_C = \{1.707\mathbf{i} + 0.707\mathbf{j} - 2\mathbf{k}\} \text{ m},$$

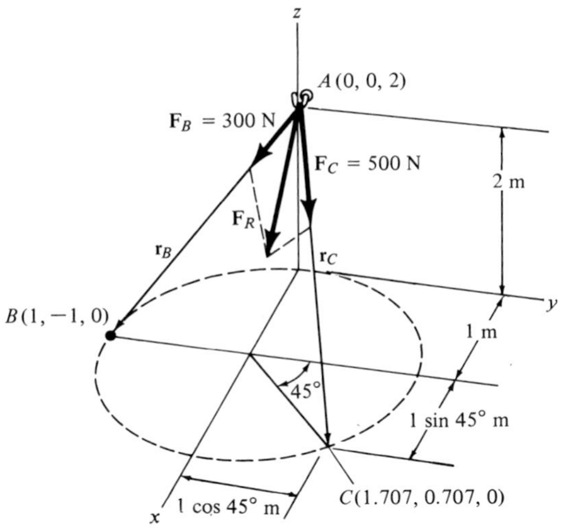

Fig. 2-35(b)

$$r_C = \sqrt{(1.707)^2 + (0.707)^2 + (-2)^2} = 2.72 \text{ m}$$

$$\mathbf{u}_C = \frac{\mathbf{r}_C}{r_C} = \frac{1.707}{2.72}\mathbf{i} + \frac{0.707}{2.72}\mathbf{j} - \frac{2}{2.72}\mathbf{k}$$

$$= 0.627\mathbf{i} + 0.260\mathbf{j} - 0.735\mathbf{k}$$

$$\mathbf{F}_C = F_C\mathbf{u}_C = 500 \text{ N}(0.627\mathbf{i} + 0.260\mathbf{j} - 0.735\mathbf{k})$$

$$= \{313.5\mathbf{i} + 130.0\mathbf{j} - 367.5\mathbf{k}\} \text{ N} \qquad \textit{Ans.}$$

The resultant force $\mathbf{F}_R$ acting at A is determined by the vector sum, shown graphically in Fig. 2-35*b*.

$$\mathbf{F}_R = \Sigma\mathbf{F};$$

$$\mathbf{F}_R = \mathbf{F}_B + \mathbf{F}_C$$

$$= (122.4\mathbf{i} - 122.4\mathbf{j} - 244.8\mathbf{k}) + (313.5\mathbf{i} + 130.0\mathbf{j} - 367.5\mathbf{k})$$

$$= \{435.9\mathbf{i} + 7.6\mathbf{j} - 612.3\mathbf{k}\} \text{ N}$$

The *magnitude* of $\mathbf{F}_R$ is

$$F_R = \sqrt{(435.9)^2 + (7.6)^2 + (-612.3)^2} = 751.7 \text{ N} \qquad \textit{Ans.}$$

The *direction* is determined from the components of the unit vector of $\mathbf{F}_R$.

$$\mathbf{u}_R = \frac{\mathbf{F}_R}{F_R} = \frac{435.9}{751.7}\mathbf{i} + \frac{7.6}{751.7}\mathbf{j} - \frac{612.3}{751.7}\mathbf{k}$$

$$= 0.580\mathbf{i} + 0.0101\mathbf{j} - 0.815\mathbf{k}$$

so

$$\alpha = \cos^{-1}(0.580) = 54.6° \qquad \textit{Ans.}$$

$$\beta = \cos^{-1}(0.0101) = 89.4° \qquad \textit{Ans.}$$

$$\gamma = \cos^{-1}(-0.815) = 144.5° \qquad \textit{Ans.}$$

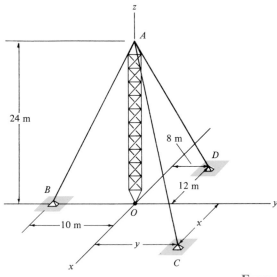

(a)

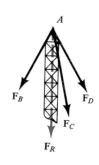

Fig. 2-36   (b)

Example 2-12

Three guy cables are attached to the top of a mast as shown in Fig. 2-36a. The tension in each of these cables is as follows: $F_B = 260$ N, $F_C = 600$ N, and $F_D = 560$ N. Determine the $(x, y, 0)$ coordinate position of point C so that the resultant of these three forces is directed downward along the axis of the mast, i.e., acting in a direction from point A towards point O. What is the magnitude of the resultant force?

Solution

The tension force in each cable is directed away from point A as shown in Fig. 2-36b. The resultant $\mathbf{F}_R$ is equal to the sum of these three forces, i.e.,

$$\mathbf{F}_R = \Sigma \mathbf{F}; \qquad \mathbf{F}_R = \mathbf{F}_B + \mathbf{F}_C + \mathbf{F}_D \qquad (1)$$

For the solution, each force will be expressed in Cartesian vector form. Since $\mathbf{F}_R$ has an unknown magnitude F_R and acts in the $-\mathbf{k}$ direction, then

$$\mathbf{F}_R = -F_R \mathbf{k}$$

The direction of each of the cable forces is obtained from the unit vectors directed along the cables. With reference to Fig. 2-36a,

$$B(0, -10, 0), \quad A(0, 0, 24)$$

$$\mathbf{r}_B = \{-10\mathbf{j} - 24\mathbf{k}\} \text{ m}; \qquad r_B = 26 \text{ m}$$

$$\mathbf{u}_B = -\frac{10}{26}\mathbf{j} - \frac{24}{26}\mathbf{k}$$

$$\mathbf{F}_B = (260 \text{ N})\mathbf{u}_B = \{-100\mathbf{j} - 240\mathbf{k}\} \text{ N}$$

$$C(x, y, 0), \quad A(0, 0, 24)$$

$$\mathbf{r}_C = \{x\mathbf{i} + y\mathbf{j} - 24\mathbf{k}\} \text{ m}, \qquad r_C = \sqrt{x^2 + y^2 + (-24)^2}$$

44

$$\mathbf{u}_C = \frac{x\mathbf{i} + y\mathbf{j} - 24\mathbf{k}}{\sqrt{x^2 + y^2 + (-24)^2}}$$

$$\mathbf{F}_C = 600 \text{ N } \mathbf{u}_C = \left\{ \frac{600x\mathbf{i} + 600y\mathbf{j} - 14\,400\mathbf{k}}{\sqrt{x^2 + y^2 + (-24)^2}} \right\} \text{N}$$

$$D(-12, 8, 0), \quad A(0, 0, 24)$$

$$\mathbf{r}_D = \{-12\mathbf{i} + 8\mathbf{j} - 24\mathbf{k}\} \text{ m}; \qquad r_D = 28 \text{ m}$$

$$\mathbf{u}_D = -\frac{12}{28}\mathbf{i} + \frac{8}{28}\mathbf{j} - \frac{24}{28}\mathbf{k}$$

$$\mathbf{F}_D = 560 \text{ N } \mathbf{u}_D = \{-240\mathbf{i} + 160\mathbf{j} - 480\mathbf{k}\} \text{ N}$$

Substituting these results into Eq. (1) yields

$$-F_R\mathbf{k} = (-100\mathbf{j} - 240\mathbf{k}) + \left[\frac{600x\mathbf{i} + 600y\mathbf{j} - 14\,400\mathbf{k}}{\sqrt{x^2 + y^2 + (-24)^2}} \right]$$
$$+ (-240\mathbf{i} + 160\mathbf{j} - 480\mathbf{k})$$

To satisfy this vector equation, the magnitude of the $\mathbf{i}$, $\mathbf{j}$, and $\mathbf{k}$ components of the vector on the left side of the equation must be *equal* to the corresponding magnitudes of the $\mathbf{i}$, $\mathbf{j}$, and $\mathbf{k}$ components of the vectors on the right. Hence,

$$0 = \frac{600x}{\sqrt{x^2 + y^2 + (-24)^2}} - 240 \qquad (2)$$

$$0 = -100 + \frac{600y}{\sqrt{x^2 + y^2 + (-24)^2}} + 160 \qquad (3)$$

$$-F_R = -240 - \frac{14\,400}{\sqrt{x^2 + y^2 + (-24)^2}} - 480 \qquad (4)$$

After placing the constants in Eqs. (2) and (3) on the left side of the equal sign, squaring and rearranging terms, we have

$$y^2 = 5.25x^2 - 576$$

and

$$x^2 = 99y^2 - 576$$

Substituting one of these equations into the other and solving yields $x = \pm 10.54$ m and $y = \pm 2.63$ m. The squaring of Eqs. (2) and (3) has introduced extraneous roots in the solution. By substitution, however, the only two of the four roots that satisfy Eqs. (2) and (3) are

$$x = 10.54 \text{ m} \qquad\qquad Ans.$$
$$y = -2.63 \text{ m} \qquad\qquad Ans.$$

Hence, point C has coordinates of $C(10.54, -2.63, 0)$. Substituting these values into Eq. (4) yields

$$F_R = 1267 \text{ N} = 1.27 \text{ kN} \qquad\qquad Ans.$$

Problems

2-31. A position vector extends from the origin to the point (2 m, 3 m, 6 m). Determine the angles α, β, and γ which the tail of the vector makes with the x, y, and z axes, respectively.

*2-32. Represent the position vector $\mathbf{r}$ acting from point $A(3\text{ m},\ 5\text{ m},\ 6\text{ m})$ to point $B(5\text{ m},\ -2\text{ m},\ 1\text{ m})$ in Cartesian vector form. Give its direction cosines, and find the distance between points A and B.

2-33. If $\mathbf{r}_1 = \{3\mathbf{i} - 4\mathbf{j} + 3\mathbf{k}\}$ m, $\mathbf{r}_2 = \{4\mathbf{i} - 5\mathbf{k}\}$ m, $\mathbf{r}_3 = \{3\mathbf{i} - 2\mathbf{j} + 5\mathbf{k}\}$ m, determine the magnitude and direction of $\mathbf{r} = 2\mathbf{r}_1 - \mathbf{r}_2 + 3\mathbf{r}_3$.

2-34. Given the three position vectors

$$\mathbf{r}_1 = \{2\mathbf{i} + 5\mathbf{j} + 4\mathbf{k}\}\text{ m}$$
$$\mathbf{r}_2 = \{3\mathbf{i} + 2\mathbf{k}\}\text{ m}$$
$$\mathbf{r}_3 = \{-2\mathbf{i} + 4\mathbf{j}\}\text{ m}$$

find the magnitude and direction of $\mathbf{r} = \mathbf{r}_1 - \mathbf{r}_2 + \frac{1}{2}\mathbf{r}_3$.

2-35. The hinged plate is supported by the cord AB. If the force in the cord is $F = 450$ N, express this force, directed from A towards B, as a Cartesian vector. What is the length of the cord? Set $a = 2$ m, $b = 1$ m, and $c = 2$ m.

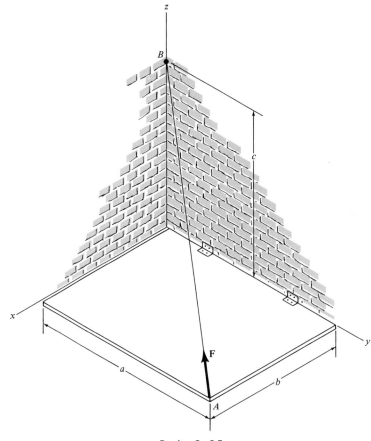

Prob. 2-35

2-35a. Solve Prob. 2-35 with $F = 125$ lb, $a = 6$ ft, $b = 4$ ft, and $c = 5$ ft.

*2-36. The crate, supported by the shear-leg derrick, creates a force of $F_B = 600$ N along strut AB and $F_C = 900$ N along the cable AC. If the forces are directed as shown, represent each force in Cartesian vector form.

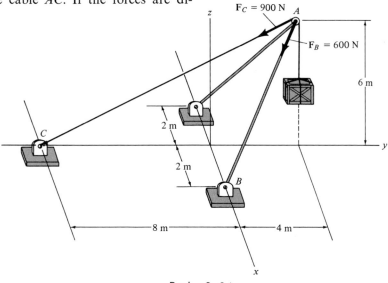

Prob. 2-36

2-37. The load at A creates a force of 200 N in wire AB. Express this force in Cartesian vector form, acting on A and directed towards B.

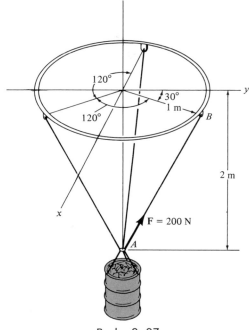

Prob. 2-37

2-38. The cable OA exerts a force on point O of $F = \{40i + 60j + 70k\}$ N. If the length of the cable is 3 m, what are the coordinates (x, y, z) of point A?

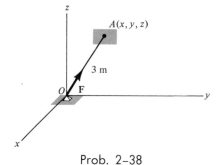

Prob. 2-38

2-39. A position vector r has a magnitude of 10 m and a direction defined by the angles $\alpha = 30°$, $\gamma = 75°$. If it is known that the y component of r is negative, determine the angle β and the components of r.

*2-40. A wire is bent into the shape shown. If it is known that a position vector, acting from point A to point B, is represented as $r = \{15i + 25j - 30k\}$ mm, determine the length of OA and its coordinate direction angles. Set $a = 40$ mm and $b = 30$ mm.

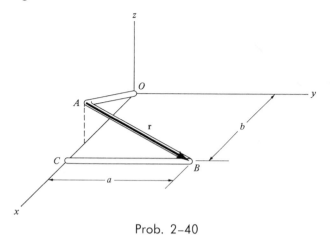

Prob. 2–40

*2–40a. Solve Prob. 2–40 with $\mathbf{r} = \{1\mathbf{i} + 3\mathbf{j} - 5\mathbf{k}\}$ in., $a = 4$ in., and $b = 3$ in.

2–41. The antenna cables CA and CB exert forces of $F_A = 500$ N and $F_B = 700$ N, respectively, on point C. Express the resultant of these two forces, acting on C, as a vector. What is the magnitude and direction of this resultant?

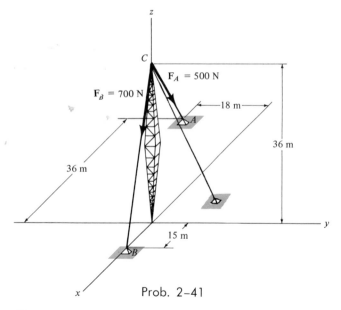

Prob. 2–41

2–42. The door is held open by means of two chains. If the tension in AB and CD is $F_A = 300$ N and $F_C = 250$ N, respectively, express each of these forces in Cartesian vector form.

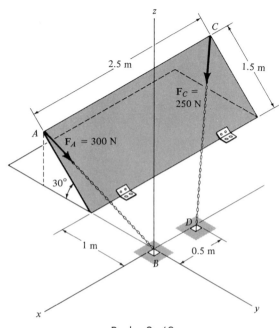

Prob. 2–42

Equilibrium of a Particle

3–1. Conditions for the Equilibrium of a Particle

In this section both the necessary and sufficient conditions required for the equilibrium of a particle will be developed. A *particle* may be defined as being a relatively small portion of matter such that its dimensions or size will be of no consequence in the analysis of the physical problem. Thus, objects considered as particles can only be subjected to a system of *concurrent forces*. Since a body can be thought of as a combination of many particles, the study of the equilibrium of particles is a necessary prerequisite to the study of body equilibrium.

The necessary and sufficient conditions for particle equilibrium are based upon a balance of forces. Formally stated as Newton's first law of motion: If the *resultant force* acting on a particle is *zero,* the particle is in equilibrium. Hence, equilibrium requires that a particle either be at rest, if originally at rest, or move with constant velocity, if it is originally moving with a velocity. Most often, however, the term "equilibrium" or more specifically "static equilibrium" is used to describe an object at rest.

Using Eq. 2–12, $\mathbf{F}_R = \Sigma\mathbf{F},$ the condition for particle equilibrium may be stated mathematically as

$$\Sigma\mathbf{F} = \mathbf{0} \qquad (3\text{–}1)$$

where $\Sigma\mathbf{F}$ is determined by the *vector sum* of *all the forces* acting on the particle. If each of these forces is resolved into its respective $\mathbf{i}$, $\mathbf{j}$, and $\mathbf{k}$ components, Eq. 3–1 may be written as

$$\Sigma F_x\mathbf{i} + \Sigma F_y\mathbf{j} + \Sigma F_z\mathbf{k} = \mathbf{0} \qquad (3\text{–}2)$$

Here ΣF_x, ΣF_y, and ΣF_z represent the *algebraic sums* of the x, y, and z

force components acting on the particle. For this vector equation to be equal to zero, each of its respective **i, j,** and **k** components must be equal to zero, i.e.,

$$\Sigma F_x = 0 \qquad \Sigma F_y = 0 \qquad \Sigma F_z = 0 \qquad (3\text{-}3)$$

When satisfied, vector equation 3–1, or the three scalar equations 3–3, provide both necessary and sufficient conditions for particle equilibrium.

3–2. Free-Body Diagrams

All equilibrium problems should be solved by *first* drawing a *free-body diagram* of the particle. Provided the free-body diagram is correctly drawn, the effect of all the forces acting on the particle will be taken into account when the equations of equilibrium are written.

PROCEDURE FOR DRAWING A FREE-BODY DIAGRAM

To construct a free-body diagram, the following three steps must be performed:

Step 1: Imagine the particle to be *isolated* from its surroundings by drawing (sketching) an outlined shape of the particle.

Step 2: Indicate on this sketch *all* the forces that act *on the particle.* These forces will either be *active forces,* which tend to set the particle in motion, e.g., weight, or magnetic and electrostatic interaction; or *reactive forces,* such as those caused by the constraints or supports that tend to prevent motion.

Step 3: The forces that are *known* should be labeled with their proper magnitudes and directions. Letters are used to represent the magnitudes and direction angles of forces that are unknown. In particular, if a force has a known line of action but unknown magnitude, the "arrowhead," which defines the directional sense of the force, can be assumed. The correctness of the directional sense will become apparent after solving the equilibrium equations for the unknown magnitude. By definition, the *magnitude* of a force is *always positive,* so that if the solution yields a "negative" magnitude, the *minus sign* indicates that the arrowhead or directional sense of the force is opposite to that which was originally assumed.

When drawing a free-body diagram it is most important to *show only the forces acting on the object being considered.* For example, following the above procedure, the free-body diagram of the block in Fig. 3–1*a* is shown in Fig. 3–1d. There are only two forces acting on the block, its weight **W** and the tension **T**$_D$ of cord *BD*. Notice that **T**$_D$ acts at point *D*, where the cord is attached. Since the cords *BA* and *BC* are not attached to the block, they do *not* exert forces *on the block* and are therefore not represented on this free-body diagram. Instead, these forces are shown on the free-body diagram of the ring at *B*, Fig. 3–1b. When drawing the free-body diagrams of *contacting bodies,* it is necessary to carefully observe Newton's third law of motion—that for every action there is an equal but opposite force reaction. For example, the free-body diagram for a segment of cord *BD* is shown in Fig. 3–1c, where in this case **T**$_D$ at *D* acts downward and represents the effect of the block *on cord BD*. Likewise, a free-body diagram of the ring at *B*, Fig. 3–1b, shows the force **T**$_D$, the effect of cord *BD on the ring,* acting downward.

Types of Supports and Connections. The following types of supports and connections are often encountered in particle equilibrium problems.

Cables and Pulleys. Throughout this book, except in Sec. 7–4, all cables are assumed to have negligible weight and they cannot be stretched. A cable carries only a tension force, and this force is always directed along the cable. In Chapter 5 it will be shown that the tension force developed in a *continuous cable,* which passes over a frictionless pulley, must have a *constant* magnitude to keep the cable in equilibrium. Hence, for any angle θ, shown in Fig. 3–2, the cable is subjected to a constant tension **T** throughout its length.

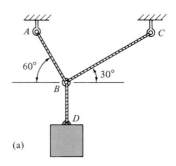

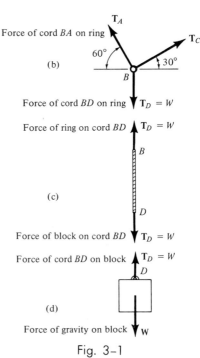

Fig. 3–1

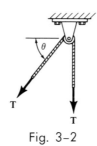

Fig. 3–2

Springs. If a *linear elastic spring* is used for a connection, the length of the spring will change in direct proportion to the force acting on it. A characteristic that defines the "elasticity" of a spring is the *stiffness k*. Specifically, the magnitude of force developed by a linear elastic spring which has a stiffness k, and is deformed (compressed or elongated) a distance x measured from its unloaded position, is

$$F = kx \qquad (3\text{–}4)$$

51

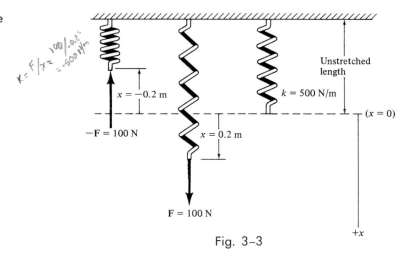

$k = F/x = \dfrac{100/.02}{\sim} = -500 \, N/m$

Unstretched length

$x = -0.2$ m

$k = 500$ N/m

$(x = 0)$

$-F = 100$ N

$x = 0.2$ m

$F = 100$ N

$+x$

Fig. 3–3

For example, the spring shown in Fig. 3–3 has a stiffness of $k = 500$ N/m, so that to stretch or compress the spring a distance of $x = 0.2$ m, a force of $F = kx = 500$ N/m$(0.2$ m$) = 100$ N is needed.

Concurrent Force Systems Acting on a Rigid Body.   A special case of equilibrium occurs when a *rigid body is subjected to a system of concurrent forces*. As an example, consider the sphere shown in Fig. 3–4*a*, which has a mass *m* and is supported by the rope *AB* and the smooth plane. The free-body diagram is shown in Fig. 3–4*b*. There are three forces acting *on* the sphere:

1. The force $\mathbf{T}_B$ has an unknown magnitude T_B and is directed away from the sphere, along the axis of the rope.
2. The reactive force $\mathbf{N}_P$ of the plane has an unknown magnitude N_P. Since the plane is *smooth,** the direction of $\mathbf{N}_P$ is upward, *normal or perpendicular to the tangent line at the point of contact.*†
3. The weight $\mathbf{W}$ of the sphere, which represents the effect of gravity, acts downward towards the center of the earth. The weight of each particle of the sphere contributes to the total weight and provided the material is the same throughout, i.e., homogeneous, the total weight may be represented as a concentrated force acting through the sphere's geometric center *O*.‡ Since the mass of the sphere is *m*, using Eq. 1–3, the weight has a magnitude of $W = mg$, where $g = 9.81$ m/s^2.

---

*Rough surfaces of contact contribute frictional forces which act tangent to the surface at points of contact. This type of force will be discussed in Chapter 8.

†For problem solving, the magnitude of the normal force will be designated by the symbol *N with a subscript,* to distinguish it from the symbol N used to specify its magnitude in newtons.

‡The methods used to determine the geometric center of an object will be discussed in Chapter 9.

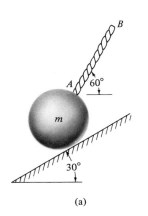

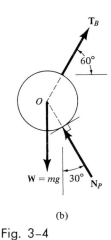

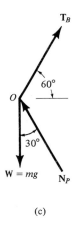

(a)   (b)   (c)

Fig. 3-4

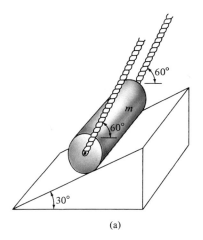

As shown in Fig. 3-4b, the lines of action of each of the three forces intersect at the common point O. Hence, for equilibrium of the sphere, the force system may be considered *concurrent* and coplanar at point O, as shown in Fig. 3-4c, and must therefore satisfy the same equilibrium condition as for a particle, i.e., $\Sigma\mathbf{F} = \mathbf{0}$.

In cases where a three-dimensional concurrent force system acts on a rigid body, the forces can be represented as a system of coplanar forces provided the force system is *symmetric* with respect to a plane passing midway through the body. For example, consider the cylinder of mass m, resting on an inclined plane and supported by two ropes, Fig. 3-5a. The forces acting on the cylinder are shown on the free-body diagram, Fig. 3-5b. The plane, in this case, exerts a *distributed load* along the line of contact of the cylinder with the plane. This loading can, by virtue of symmetry, be represented by a general *resultant force* $\mathbf{N}_P$ acting through the center of the cylinder and having an unknown magnitude.* Symmetry also requires the cable tension forces $\mathbf{T}$ to be the same, as indicated. Representing these forces by the single unknown $T_B = 2T$, the free-body diagram of the cylinder can then be viewed from the side, which yields the *same* free-body diagram as that of the sphere in Fig. 3-4b, or point O, Fig. 3-4c.

(a)

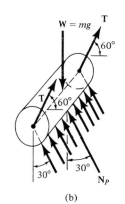

(b)

Fig. 3-5

3-3. Coplanar Force Systems Acting on a Particle

In some cases a coplanar force system may act on a particle. If the force system lies in the x-y plane, then from Eqs. 3-3, the following *two scalar equations must be satisfied for equilibrium:*

*When the magnitude of $\mathbf{N}_P$ is determined, the intensity of the distributed loading can be obtained by using the methods of Sec. 4-9.

$$\Sigma F_x = 0$$
$$\Sigma F_y = 0$$

(3-5)

In this case, the equilibrium equations require that the *sum* of the x and y components of all the forces acting on the particle be equal to zero. As a result, these two equations can be solved for, at most, two unknowns, generally represented as angles or magnitudes of forces shown on the free-body diagram of the particle.

PROCEDURE FOR ANALYSIS

The following two-step procedure should be used for solving coplanar-force equilibrium problems:

Step 1:   Draw a free-body diagram of the particle. As outlined in Sec. 3-2, this requires that all the known and unknown force magnitudes and angles be labeled on the diagram. The directional sense of a force having an unknown magnitude and known line of action can be assumed.

Step 2:   Apply the two equations of equilibrium, $\Sigma F_x = 0$ and $\Sigma F_y = 0$, to the force system shown on the free-body diagram. If more than two unknowns exist, and the problem involves a spring, apply $F = kx$ (Eq. 3-4) to relate the spring force on the free-body diagram to the deformation x of the spring.

If the solution of the equations yields a *negative* force magnitude, it indicates that the direction of the force shown on the free-body diagram is *opposite* to that which was *assumed*.

The following example problems numerically illustrate this solution procedure.

Example 3-1

A block having a mass of 10 kg is held in equilibrium by two cords, as shown in Fig. 3-6a. The cord *ABC* is continuous and passes over a

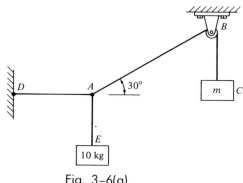

Fig. 3-6(a)

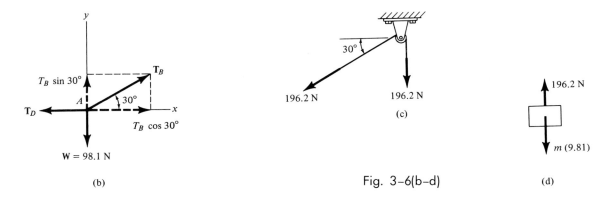

(b)

Fig. 3–6(b–d)

(c)

(d)

frictionless pulley at B. Determine the tension in cord AD and the mass m of the block at C for equilibrium.

Solution

Step 1: The tension in cord AD and the mass m of C can be obtained by investigating the equilibrium of point A. As shown on the free-body diagram, Fig. 3–6b, there are three concurrent forces acting *on* this point. Cord AE exerts a downward force on A equal to the weight of the block, $W = (10 \text{ kg})(9.81 \text{ m/s}^2) = 98.1 \text{ N}$. The tension forces in AD and ABC have unknown magnitudes but known directions.

Step 2: Since the equilibrium equations require a summation of the x and y components of each force, $\mathbf{T}_B$ must be resolved into x and y components. These components, shown dashed on the free-body diagram, have magnitudes of $T_B \cos 30°$ and $T_B \sin 30°$, respectively. Equations 3–5 will be applied by assuming that "positive" force components act along the positive x and y axes. Indicating these directions alongside the equations, we have

$\xrightarrow{+} \Sigma F_x = 0;$ $\qquad\qquad T_B \cos 30° - T_D = 0$ $\qquad\qquad$ (1)

$+\uparrow \Sigma F_y = 0;$ $\qquad\qquad T_B \sin 30° - 98.1 = 0$ $\qquad\qquad$ (2)

Solving Eq. (2) for T_B and substituting into Eq. (1) to obtain T_D yields

$$T_B = 196.2 \text{ N}$$
$$T_D = 169.9 \text{ N}$$
Ans.

Since cord ABC passes over a frictionless pulley, the force $T_B = 196.2 \text{ N}$ is constant throughout the cord, Fig. 3–6c. By Newton's third law of motion—for every action there is an equal and opposite reaction—the 196.2-N force acts on block C as shown in Fig. 3–6d. For equilibrium of block C we require that

$+\uparrow \Sigma F_y = 0;$ $\qquad\qquad 196.2 - m(9.81) = 0$

$$m = \frac{196.2}{9.81} = 20 \text{ kg} \qquad\qquad Ans.$$

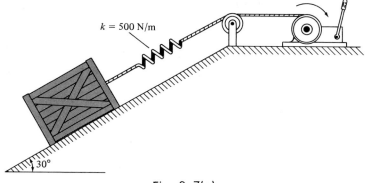

$k = 500$ N/m

30°

Fig. 3–7(a)

Example 3–2

A crate has a mass of 50 kg and is being pulled with *constant velocity* up a smooth 30° inclined plane as shown in Fig. 3–7a. Determine the stretch developed in the spring (shock absorber) of the towing cable required for equilibrium.

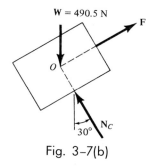

$W = 490.5$ N

F

O

N_C

30°

Fig. 3–7(b)

Solution

Step 1: The stretch x of the spring is related to the force $\mathbf{F}$ in the cable by $F = kx$. The cable force $\mathbf{F}$ can be determined by investigating the equilibrium of the crate. As shown on the free-body diagram, Fig. 3–7b, the weight $W = 50(9.81) = 490.5$ N, normal force $\mathbf{N}_C$, and cable force $\mathbf{F}$ are all concurrent at point O; therefore, a free-body diagram of this point can be considered, Fig. 3–7c. The two unknowns are represented as the magnitudes of $\mathbf{N}_C$ and $\mathbf{F}$.

Step 2: If the x and y axes are oriented in the horizontal and vertical directions, Fig. 3–7c, forces $\mathbf{N}_C$ and $\mathbf{F}$ must be resolved into x and y components by using trigonometry. Applying Eqs. 3–5, we have

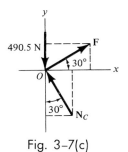

Fig. 3-7(c)

$\xrightarrow{+}\Sigma F_x = 0;$ $F\cos 30° - N_C \sin 30° = 0$

$+\uparrow\Sigma F_y = 0;$ $F\sin 30° + N_C \cos 30° - 490.5 = 0$

Solving these equations simultaneously yields

$$F = 245.3 \text{ N} \qquad N_C = 424.8 \text{ N}$$

Note that it is also possible to obtain these results in a more direct fashion by orienting the x and y axes along the incline, Fig. 3-7d. In this case, only the weight must be resolved into x and y components, so that,

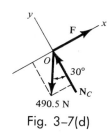

Fig. 3-7(d)

$+\nearrow\Sigma F_x = 0;$ $F - 490.5\sin 30° = 0$ $F = 245.3 \text{ N}$

$+\nwarrow\Sigma F_y = 0;$ $N_C - 490.5\cos 30° = 0$ $N_C = 424.8 \text{ N}$

The solution here is easily obtained since **F** and **N**$_C$ *do not* have components along *both* the x and y axes. It should also be noted that the magnitude of the towing force **F** is *less* than the weight of the crate. This indicates that an inclined plane acts as a "simple machine," since it allows a larger load to be balanced by a smaller one.

Knowing the force in the cable and the spring stiffness k, the amount of stretch x is

$F = kx;$ $245.3 \text{ N} = (500 \text{ N/m})x$

$$x = \frac{245.3 \text{ N}}{500 \text{ N/m}} = 0.4905 \text{ m} = 490.5 \text{ mm} \qquad \qquad Ans.$$

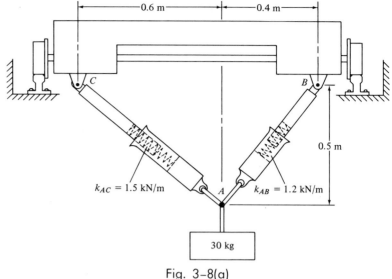

Fig. 3–8(a)

Example 3–3

The spring-loaded supporting rods AB and AC in Fig. 3–8a are used as shock absorbers for transporting fragile loads. Knowing that the elastic-spring constants for the rods AB and AC are $k_{AB} = 1.2$ kN/m and $k_{AC} = 1.5$ kN/m, respectively, determine the unstretched length of each rod after the 30-kg block is removed. Neglect the weight of the rods.

Solution

Step 1: If the forces $\mathbf{F}_B$ and $\mathbf{F}_C$ developed in rods AB and AC are known, the *extension* x of each rod, when the rods support the 30-kg block, $(F = kx)$ can be determined. Using geometry, it is then possible to calculate the unstretched length of each rod when the block is removed.

The magnitudes of $\mathbf{F}_B$ and $\mathbf{F}_C$ can be obtained by investigating a free-body diagram of point A, Fig. 3–8b. As shown, the block has a weight of $W = 30(9.81) = 294.3$ N, and the forces $\mathbf{F}_B$ and $\mathbf{F}_C$ have slopes that are defined from the geometry of rods AB and AC.

Step 2: Applying the scalar equations of equilibrium, realizing that the components of the forces are proportional to the slope triangles, we have

$$\xrightarrow{+}\Sigma F_x = 0; \qquad \frac{4}{\sqrt{(4)^2 + (5)^2}}F_B - \frac{6}{\sqrt{(6)^2 + (5)^2}}F_C = 0$$

$$0.625F_B - 0.768F_C = 0 \qquad (1)$$

$$+\uparrow\Sigma F_y = 0;$$

$$\frac{5}{\sqrt{(4)^2 + (5)^2}}F_B + \frac{5}{\sqrt{(6)^2 + (5)^2}}F_C - 294.3 = 0$$

$$0.781F_B + 0.640F_C - 294.3 = 0 \qquad (2)$$

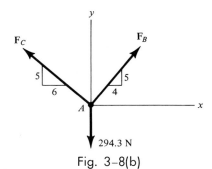

Fig. 3-8(b)

The tension forces in the rods can be determined by solving Eqs. (1) and (2) simultaneously; i.e., solving for F_B in Eq. (1),

$$F_B = 1.229F_C \qquad (3)$$

and substituting this into Eq. (2) and solving for F_C yields

$$0.781(1.229F_C) + 0.640F_C - 294.3 = 0$$
$$F_C = 184.0 \text{ N}$$

From Eq. (3),

$$F_B = 1.229(184.0 \text{ N})$$
$$= 226.1 \text{ N}$$

The forces shown on the free-body diagram of point A represent the effect of the rods and the block *on* point A. According to Newton's third law of motion, point A must exert *equal, but opposite forces* on the rods and block, as shown in Fig. 3-8c. The *extension* of each rod created by forces $\mathbf{F}_B$ and $\mathbf{F}_C$ is, therefore,

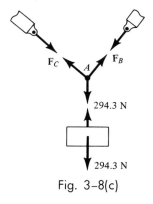

$F_B = k_{AB}x_{AB};$ $\qquad x_{AB} = \dfrac{226.1 \text{ N}}{1200 \text{ N/m}} = 0.188 \text{ m}$

$F_C = k_{AC}x_{AC};$ $\qquad x_{AC} = \dfrac{184.0 \text{ N}}{1500 \text{ N/m}} = 0.123 \text{ m}$

Using the Pythagorean theorem, the *extended length* of each rod when the rods are supporting the block can be determined. Referring to Fig. 3-8a, we have

$$l_{AB} = \sqrt{(0.4)^2 + (0.5)^2} = 0.640 \text{ m}$$
$$l_{AC} = \sqrt{(0.6)^2 + (0.5)^2} = 0.781 \text{ m}$$

Thus, the *unloaded length* of each rod becomes

$$l'_{AB} = l_{AB} - x_{AB} = 0.640 - 0.188$$
$$= 0.452 \text{ m} \qquad\qquad Ans.$$
$$l'_{AC} = l_{AC} - x_{AC} = 0.781 - 0.123$$
$$= 0.658 \text{ m} \qquad\qquad Ans.$$

Fig. 3-8(c)

Problems

3-1. The joint of a light metal truss is formed by riveting four angles to the *gusset plate*. Knowing the force in members A and C, determine the forces $\mathbf{F}_B$ and $\mathbf{F}_D$ acting on members B and D that are required for equilibrium. The force system is concurrent at point O.

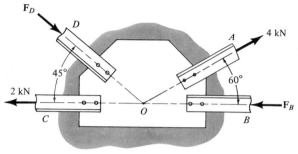

Prob. 3–1

3-2. Determine the magnitude and direction θ of $\mathbf{R}$, such that the particle P is in equilibrium when subjected to the three forces.

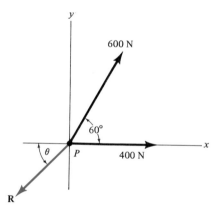

Prob. 3–2

3-3. The patella P located in the human knee joint is subjected to tendon forces $\mathbf{T}_1$ and $\mathbf{T}_2$, and a resultant force $\mathbf{F}$ exerted on the patella by the femoral articular A. If the directions of these forces are estimated from an X-ray as shown, determine the magnitudes of $\mathbf{T}_1$ and $\mathbf{F}$ when the tendon force $T_2 = 80$ N.

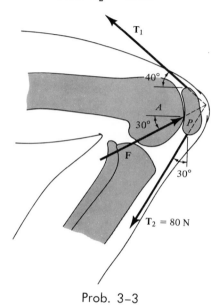

Prob. 3–3

*3-4. The members of a truss are connected to the gusset plate. If the forces are concurrent at point O, determine the magnitudes of **F** and **T** for equilibrium.

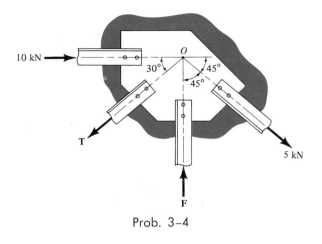

Prob. 3-4

3-6. Three blocks have masses as indicated and are connected by the cords shown. If the cords pass over small frictionless pulleys, determine the sag s for equilibrium.

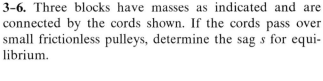

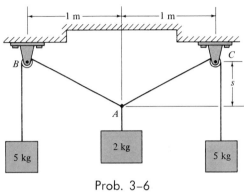

Prob. 3-6

3-5. If $\theta = 30°$, determine the magnitudes of $\mathbf{F}_A$ and $\mathbf{F}_B$ required to hold the sphere having a mass of $m = 10$ kg in equilibrium.

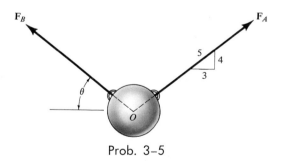

Prob. 3-5

3-5a. Solve Prob. 3-5 if $\theta = 45°$ and the sphere weighs $W = 50$ lb.

3-7. Determine the height h for the location of the framework AB so that a tension of 300 N is developed in each of the links OA and OB. It is assumed that the force in each link acts along the axis of the link.

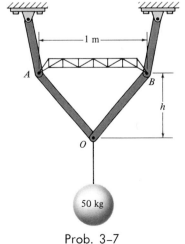

Prob. 3-7

*3-8. The 20-kg pipe rests between the two smooth inclined planes. If $P = 50$ N, determine the normal reactions that planes A and B exert on the pipe.

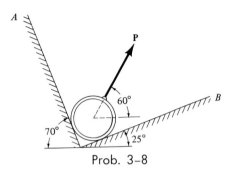

Prob. 3-8

3-10. The sphere has a mass of $m = 25$ kg, rests between two smooth inclined planes A and B, and is attached to a spring. If the unstretched length of the spring is $l_o = 200$ mm, determine the normal reactions of the planes acting on the sphere. Set $l = 500$ mm and $k = 600$ N/m.

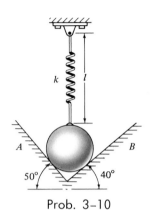

Prob. 3-10

3-9. The uniform 50-kg crate is suspended by using a 2-m-long cord that is attached to the sides of the crate and passes over the small pulley located at O. If the cord can be attached at either points A and B, or C and D, determine which attachment produces the least amount of tension in the cord and specify the cord tension in this case.

3-10a. Solve Prob. 3-10 if the sphere weighs $W = 50$ lb, and $l_o = 4$ in., $l = 10$ in., $k = 2.5$ lb/in.

3-11. The particle P is subjected to the action of four *coplanar* forces. Determine the magnitude of F and the angle θ for the 750-N force for equilibrium.

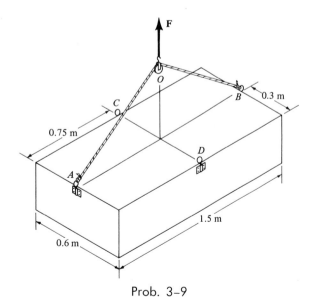

Prob. 3-9

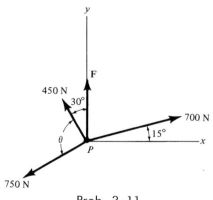

Prob. 3-11

*3-12. The screw of the adjustable C clamp exerts a vertical force of $F = 800$ N at A on the holding block B. Determine the forces that the block exerts on the smooth pipe at C and D and the force that the pipe exerts on the pad P. Neglect the weights of both the block and the pipe.

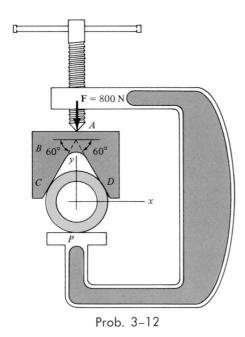

Prob. 3-12

the link. *Hint:* Analyze the forces at C to obtain F_{BC}, then analyze the forces at B.

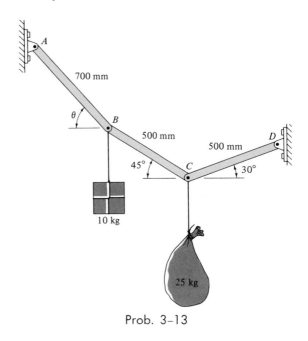

Prob. 3-13

3-14. Two spheres A and B have an equal mass and are electrostatically charged such that the repulsive force acting between them has a magnitude of 20 mN, and is directed along the dashed line. Determine the angle θ, the tension in the cords, and the mass m of each sphere.

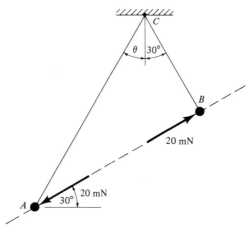

Prob. 3-14

3-13. The links AB, BC, and CD support the loading shown. Determine the tension in AB and the angle θ for equilibrium. The force in each link acts along the axis of

3-15. Determine the length of cord AC so that $\theta = 30°$ and the spring AB remains horizontal. The unstretched length of AB is $l_{AB} = 0.4$ m, $l = 2$ m, $k_{AB} = 300$ N/m, and the lamp has a mass of $m = 8$ kg.

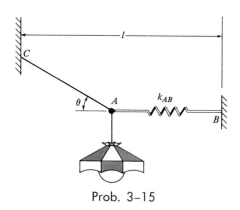

Prob. 3-15

3-15a. Solve Prob. 3-15 if $\theta = 20°$, $l_{AB} = 1.25$ ft, $l = 4$ ft, $k_{AB} = 30$ lb/ft, and the lamp weighs $W = 15$ lb.

*3-16. A 2-m-long cord is fixed to a pin at A and passes over *small* pulleys at B and C. When a 4-kg block D is suspended from the cord, the distance s is measured as 0.5 m. Compute the mass of block E.

3-17. A continuous cable of total length 4 m is wrapped around the *small* frictionless pulleys at A, B, C, and D. If the stiffness of each spring is $k = 500$ N/m and each spring is stretched 300 mm, determine the mass m of each block. Neglect the weight of the pulleys and cords. The springs are unstretched when $d = 2$ m.

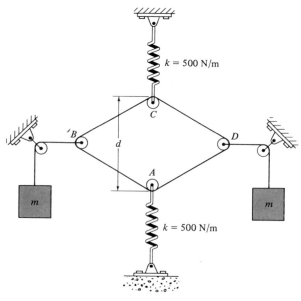

Prob. 3-17

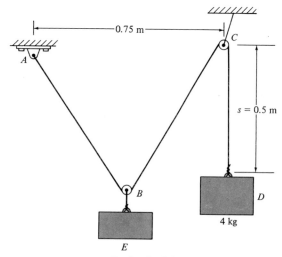

Prob. 3-16

3-18. The roller mechanism AB is at rest between the smooth inclines of the 30° wedge. If each of the rollers has a mass of 5 kg, and the undeformed length of each spring is 100 mm, determine the distance h for equilibrium of the mechanism. Assume that both rollers remain in the same horizontal plane aa. The stiffness of each spring is $k = 300$ N/m. Neglect the size of the rollers.

3-19. The roller has a mass of 1 kg and is supported by two springs. If the springs are unstretched when the roller is at the equilibrium position A, determine the stiffness k of each spring so that the roller will be in equilibrium when the height $h = 100$ mm. Assume that the surfaces of contact are smooth and neglect the size of the roller.

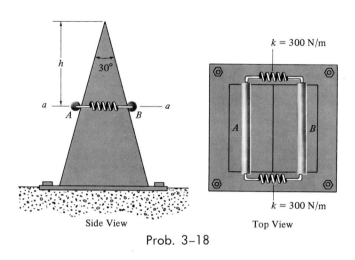

Prob. 3-18

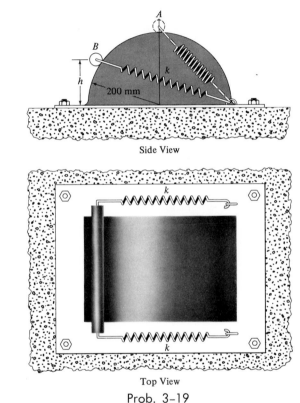

Prob. 3-19

*3–20. A sphere has a mass of $m = 4$ kg and is resting on the smooth parabolic surface. Determine the force it exerts on the surface and the mass m_B of block B needed to hold it in the equilibrium position shown; $a = 400$ mm.

*3–20a. Determine the weight W_B of block B in Prob. 3–20 if the sphere has a weight of $W = 10$ lb and $a = 2$ ft.

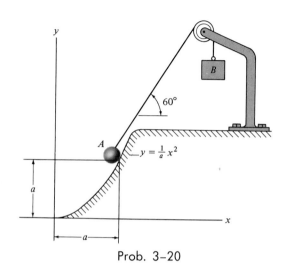

Prob. 3–20

3–21. The ends of the 6-m-long cable AB are attached to the fixed walls. If a bucket and its contents have a mass of 10 kg and are suspended from the cable by means of a *small* pulley as shown, determine the location x of the pulley for equilibrium.

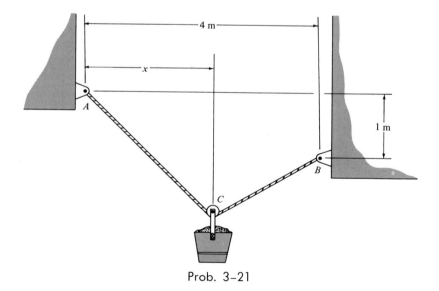

Prob. 3–21

3-22. Determine the angle θ at which $\mathbf{F}_B$ must act in Prob. 2–5 in order that the magnitude F_B is a minimum. The sphere has a mass of 10 kg and the forces are concurrent at the center O. Compute the magnitudes of $\mathbf{F}_A$ and $\mathbf{F}_B$ for this case.

3–4. Spatial Force Systems Acting on a Particle

Since spatial geometry is generally complicated, a Cartesian vector analysis is particularly helpful for solving three-dimensional particle equilibrium problems. In this case the equilibrium equation (Eq. 3–1) is first applied in vector form,

$$\Sigma \mathbf{F} = \mathbf{0} \qquad (3\text{--}6)$$

and then reduced to its three scalar component equations (Eq. 3–3), i.e.,

$$\begin{aligned} \Sigma F_x &= 0 \\ \Sigma F_y &= 0 \\ \Sigma F_z &= 0 \end{aligned} \qquad (3\text{--}7)$$

These equations can be solved for, at most, three unknowns, generally represented as angles or magnitudes of forces shown on the free-body diagram of the particle.

PROCEDURE FOR ANALYSIS

The following two-step procedure should be used for solving spatial force equilibrium problems:

Step 1: Draw a free-body diagram of the particle and label all the known and unknown forces on this diagram.

Step 2: Establish the x, y, and z coordinate axes with origin located at the particle and apply the equations of equilibrium. Use the three scalar equations 3–7 in cases where it is easy to resolve each force acting on the particle into its x, y, and z components. If the geometry of the problem is complicated, however, first express each force acting on the particle in Cartesian vector form, and then substitute these vectors into Eq. 3–6. By setting the respective $\mathbf{i}$, $\mathbf{j}$, and $\mathbf{k}$ components equal to zero, the three scalar equations 3–7 can be generated.

The following example problems numerically illustrate this solution procedure.

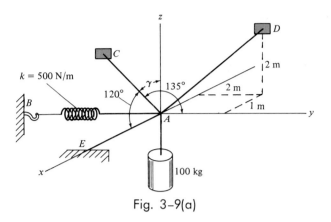

Fig. 3–9(a)

Example 3–4

The 100-kg cylinder shown in Fig. 3–9a is supported by four cords. If the spring in cord AB has a stiffness of $k = 500$ N/m and is stretched 0.5 m, determine the tension in each cord for equilibrium of the cylinder.

Solution

Step 1: The force in each cord can be determined by investigating the free-body diagram of point A, Fig. 3–9b. The weight of the cylinder is $W = 100(9.81) = 981$ N. Since the spring is stretched 0.5 m, the force in cord AB is

$$F_B = 500 \text{ N/m}(0.5 \text{ m}) = 250 \text{ N} \qquad \text{Ans.}$$

The other *three* unknown force magnitudes, F_C, F_D, and F_E, are determined from the equations of equilibrium.

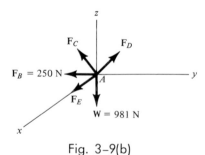

Fig. 3–9(b)

Step 2: Applying Eq. 3–1 to all the forces on the free-body diagram yields

$$\mathbf{\Sigma F} = \mathbf{0}; \qquad \mathbf{F}_B + \mathbf{F}_C + \mathbf{F}_D + \mathbf{F}_E + \mathbf{W} = \mathbf{0} \qquad (1)$$

Using the methods of Section 2–8, each of these forces can be expressed in

Cartesian vector form. Since $\mathbf{F}_B$, $\mathbf{F}_E$, and $\mathbf{W}$ all act along the coordinate axes, we have

$$\mathbf{F}_B = F_B(-\mathbf{j}) = -250\mathbf{j}$$
$$\mathbf{F}_E = F_E(\mathbf{i}) = F_E\mathbf{i}$$
$$\mathbf{W} = W(-\mathbf{k}) = -981\mathbf{k}$$

The direction of $\mathbf{F}_D$ is defined by the unit vector $\mathbf{u}_D$, which in turn is determined from the position vector $\mathbf{r}_D$. Since the coordinates of point D are $D(-1, 2, 2)$, then

$$\mathbf{r}_D = \{-1\mathbf{i} + 2\mathbf{j} + 2\mathbf{k}\}\text{ m}, \qquad r_D = \sqrt{(-1)^2 + (2)^2 + (2)^2} = 3\text{ m}$$

$$\mathbf{u}_D = \frac{\mathbf{r}_D}{r_D} = -\frac{1}{3}\mathbf{i} + \frac{2}{3}\mathbf{j} + \frac{2}{3}\mathbf{k} = -0.333\mathbf{i} + 0.667\mathbf{j} + 0.667\mathbf{k}$$

$$\mathbf{F}_D = F_D\mathbf{u}_D = -0.333F_D\mathbf{i} + 0.667F_D\mathbf{j} + 0.667F_D\mathbf{k}$$

Two coordinate direction angles, $\alpha = 120°$ and $\beta = 135°$, are given for $\mathbf{F}_C$. The third angle, γ, is determined from the identity, Eq. 2–10.

$$\cos^2 120° + \cos^2 135° + \cos^2 \gamma = 1$$
$$\cos \gamma = \pm 0.5$$
$$\gamma = \cos^{-1}(0.5) = 60° \quad \text{or} \quad \gamma = \cos^{-1}(-0.5) = 120°$$

By inspection of cord AC, Fig. 3–9a, $\gamma = 60°$. Hence, applying Eq. 2–11 yields.

$$\mathbf{F}_C = F_C \cos 120°\mathbf{i} + F_C \cos 135°\mathbf{j} + F_C \cos 60°\mathbf{k}$$
$$= -0.5F_C\mathbf{i} - 0.707F_C\mathbf{j} + 0.5F_C\mathbf{k}$$

Substituting the forces into Eq. (1), we have

$$-250\mathbf{j} - 0.5F_C\mathbf{i} - 0.707F_C\mathbf{j} + 0.5F_C\mathbf{k}$$
$$- 0.333F_D\mathbf{i} + 0.667F_D\mathbf{j} + 0.667F_D\mathbf{k} + F_E\mathbf{i} - 981\mathbf{k} = 0$$

Equating the respective components in the $\mathbf{i}$, $\mathbf{j}$, and $\mathbf{k}$ directions equal to zero, yields the three scalar equations

$\Sigma F_x = 0;$	$-0.5F_C - 0.333F_D + F_E = 0$	(2)
$\Sigma F_y = 0;$	$-0.707F_C + 0.667F_D - 250 = 0$	(3)
$\Sigma F_z = 0;$	$0.5F_C + 0.667F_D - 981 = 0$	(4)

The three unknowns are determined by simultaneous solution of Eqs. (3) and (4) for F_C and F_D. Substituting these results into Eq. (2) yields F_E. Hence,

$$F_C = 605.6\text{ N} \qquad\qquad Ans.$$
$$F_D = 1016.8\text{ N} \qquad\qquad Ans.$$
$$F_E = 641.4\text{ N} \qquad\qquad Ans.$$

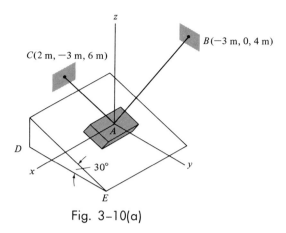

Fig. 3–10(a)

Example 3–5

The 5-kg block shown in Fig. 3–10a rests on a smooth inclined plane and is prevented from slipping down the plane by the two guy wires AC and AB. Determine the tension developed in these wires and the reactive force of the plane on the block. The coordinates of points B and C are given with respect to the origin of the x, y, and z axes at point A.

Solution

Step 1:   A free-body diagram of the block is shown in Fig. 3–10b. There are four forces acting *on* the block: the forces $\mathbf{F}_C$ and $\mathbf{F}_B$ developed by the wires, the weight $W = (5)(9.81) = 49.05$ N, and the normal force $\mathbf{N}_P$ which the smooth plane exerts *on* the block. Since all these forces are concurrent at point A, it is also possible to use the free-body diagram of this point for the analysis. As shown, the x and y axes lie in the horizontal plane such that the y axis is directed parallel to DE. Consequently, the normal force $\mathbf{N}_P$ has only two components, directed along the y and z axes. The three unknown magnitudes N_P, F_C, and F_B are determined from the equations of equilibrium.

Step 2:   Applying Eq. 3–1, we have

$$\Sigma \mathbf{F} = 0; \qquad\qquad \mathbf{F}_C + \mathbf{F}_B + \mathbf{N}_P + \mathbf{W} = 0 \qquad\qquad (1)$$

Using the methods of Sec. 2–8, each of these forces can be expressed as a Cartesian vector.

$$\mathbf{F}_C = F_C \mathbf{u}_C = F_C \frac{\mathbf{r}_C}{r_C} = F_C \left[\frac{2\mathbf{i} - 3\mathbf{j} + 6\mathbf{k}}{\sqrt{(2)^2 + (-3)^2 + (6)^2}} \right]$$

$$= 0.286 F_C \mathbf{i} - 0.429 F_C \mathbf{j} + 0.857 F_C \mathbf{k}$$

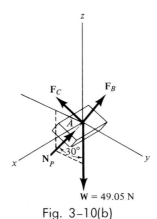

$\mathbf{W} = 49.05 \text{ N}$

Fig. 3–10(b)

$$\mathbf{F}_B = F_B \mathbf{u}_B = F_B \frac{\mathbf{r}_B}{r_B} = F_B \left[\frac{-3\mathbf{i} + 4\mathbf{k}}{\sqrt{(-3)^2 + (4)^2}} \right]$$
$$= -0.6F_B\mathbf{i} + 0.8F_B\mathbf{k}$$
$$\mathbf{N}_P = N_P \sin 30°\mathbf{j} + N_P \cos 30°\mathbf{k} = 0.5N_P\mathbf{j} + 0.866N_P\mathbf{k}$$
$$\mathbf{W} = \{-49.05\mathbf{k}\} \text{ N}$$

Substituting these forces into Eq. (1), yields

$$0.286F_C\mathbf{i} - 0.429F_C\mathbf{j} + 0.857F_C\mathbf{k} - 0.6F_B\mathbf{i} + 0.8F_B\mathbf{k}$$
$$+ 0.5N_P\mathbf{j} + 0.866N_P\mathbf{k} - 49.05\mathbf{k} = 0$$

Equating the respective components in the $\mathbf{i}$, $\mathbf{j}$, and $\mathbf{k}$ directions equal to zero we have

$$\Sigma F_x = 0; \qquad\qquad 0.286F_C - 0.6F_B = 0 \qquad\qquad (2)$$
$$\Sigma F_y = 0; \qquad\qquad -0.429F_C + 0.5N_P = 0 \qquad\qquad (3)$$
$$\Sigma F_z = 0; \qquad 0.857F_C + 0.8F_B + 0.866N_P - 49.05 = 0 \qquad (4)$$

The three unknowns in these equations are obtained by simultaneous solution. For example, solve for F_B in Eq. (2) and for N_P in Eq. (3), both in terms of F_C. Substitute these results in Eq. (4), thereby obtaining an equation in terms of the single unknown F_C. After solving for F_C, it is possible to obtain the values of F_B and N_P by resubstituting into Eqs. (2) and (3) the value of F_C. The results are:

$$F_C = 24.8 \text{ N} \qquad\qquad Ans.$$
$$F_B = 11.8 \text{ N} \qquad\qquad Ans.$$
$$N_P = 21.3 \text{ N} \qquad\qquad Ans.$$

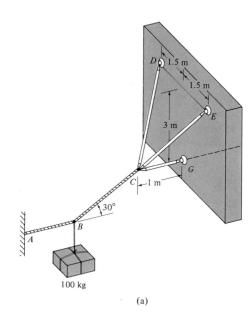

(a)

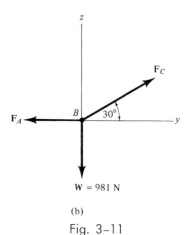

W = 981 N

(b)

Fig. 3–11

Example 3–6

A 100-kg crate is supported by three struts CG, CD, and CE and the cable arrangement shown in Fig. 3–11a. Determine the forces developed in the struts and the tension in the cables. Assume that the force developed in each strut passes along the axis of the strut. These forces may be either tensile or compressive.

Solution

Step 1: Since the weight of the crate, $W = 100(9.81) = 981$ N, acts downward at point B, the force in cables BA and BC can be obtained by investigating the equilibrium of this point. The free-body diagram is shown in Fig. 3–11b.

Step 2: The force system at B is coplanar, and therefore the unknowns can be determined *directly* from the two *scalar* equations of equilibrium,

$$\xrightarrow{+} \Sigma F_y = 0; \qquad\qquad F_C \cos 30° - F_A = 0$$
$$+\uparrow \Sigma F_z = 0; \qquad\qquad F_C \sin 30° - 981 = 0$$

Solving gives

$$F_C = 1962 \text{ N} \qquad\qquad Ans.$$
$$F_A = 1699.1 \text{ N} \qquad\qquad Ans.$$

Step 1: Since the force in cable BC is known, a free-body diagram of point C can now be drawn to "expose" the three unknown forces in the struts, Fig. 3–11c. As shown, $\mathbf{F}_D$ and $\mathbf{F}_E$ are assumed to be *tensile* forces, since they *pull* on C, and $\mathbf{F}_G$ is assumed to be *compressive*, since it *pushes* on C. The direction of $\mathbf{F}_B$ on point C is determined by the principle of action and equal but opposite reaction (Newton's third law), as shown on the free-body diagrams in Fig. 3–11d.

Step 2: Applying Eq. 3–1 to point C, Fig. 3–11c, we have

$$\Sigma \mathbf{F} = 0; \qquad\qquad \mathbf{F}_D + \mathbf{F}_E + \mathbf{F}_G + \mathbf{F}_B = 0 \qquad (1)$$

Since the coordinates of points D and E are $D(-1.5, 1, 3)$ and $E(1.5, 1, 3)$,

$$\mathbf{F}_D = F_D \mathbf{u}_D = F_D \frac{\mathbf{r}_D}{r_D} = F_D \left[\frac{-1.5\mathbf{i} + 1\mathbf{j} + 3\mathbf{k}}{\sqrt{(-1.5)^2 + (1)^2 + (3)^2}} \right]$$
$$= -0.429 F_D \mathbf{i} + 0.286 F_D \mathbf{j} + 0.857 F_D \mathbf{k}$$

$$\mathbf{F}_E = F_E \mathbf{u}_E = F_E \frac{\mathbf{r}_E}{r_E} = F_E \left[\frac{1.5\mathbf{i} + 1\mathbf{j} + 3\mathbf{k}}{\sqrt{(1.5)^2 + (1)^2 + (3)^2}} \right]$$
$$= 0.429 F_E \mathbf{i} + 0.286 F_E \mathbf{j} + 0.857 F_E \mathbf{k}$$

By inspection of Fig. 3–11c, note that $\mathbf{F}_G$ acts in the $-\mathbf{j}$ direction. Furthermore, using trigonometry $\mathbf{F}_B$ can be resolved into its two components.

Hence

$$\mathbf{F}_G = -F_G\mathbf{j}$$
$$\mathbf{F}_B = 1962\cos 30°(-\mathbf{j}) + 1962\sin 30°(-\mathbf{k})$$
$$= -1699.1\mathbf{j} - 981\mathbf{k}$$

Substituting these forces into Eq. (1), we have

$$-0.429F_D\mathbf{i} + 0.286F_D\mathbf{j} + 0.857F_D\mathbf{k} + 0.429F_E\mathbf{i} + 0.286F_E\mathbf{j}$$
$$+ 0.857F_E\mathbf{k} - F_G\mathbf{j} - 1699.1\mathbf{j} - 981\mathbf{k} = \mathbf{0}$$

Equating the respective **i**, **j**, and **k** components equal to zero yields

$$\Sigma F_x = 0; \qquad -0.429F_D + 0.429F_E = 0 \qquad\qquad (2)$$
$$\Sigma F_y = 0; \qquad 0.286F_D + 0.286F_E - F_G - 1699.1 = 0 \qquad (3)$$
$$\Sigma F_z = 0; \qquad 0.857F_D + 0.857F_E - 981 = 0 \qquad\qquad (4)$$

Solving Eqs. (2) and (4) for F_D and F_E and substituting the results into Eq. (3) to obtain F_G yields

$$F_D = F_E = 572.3 \text{ N} \qquad\qquad Ans.$$
$$F_G = -1371.7 \text{ N} \qquad\qquad Ans.$$

The negative sign for F_G indicates that force $\mathbf{F}_G$ was assumed to be acting in the wrong direction on the free-body diagram, Figs. 3–11c and 13–11d. Strut CG must therefore exert a force of 1371.7 N on point C, as shown in Fig. 3–11e. With reference to this figure, given the equal but opposite reactions, strut CG is subjected to a tensile (elongation) force of 1371.7 N caused by the weight of the crate.

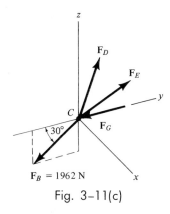

Fig. 3–11(c)

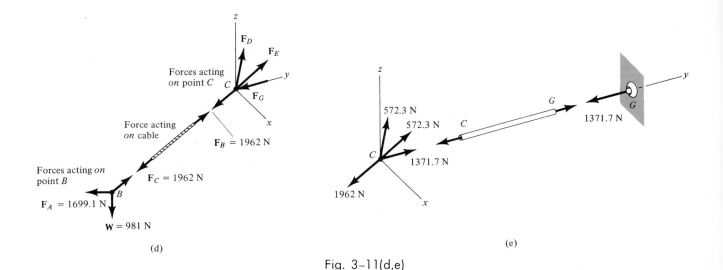

Fig. 3–11(d,e)

Problems

3-23. The sphere, having a mass of 10 kg, is forced up against the corner of three smooth mutually perpendicular planes A, B, and C. If the applied force **F** has a line of action that passes from point $P(1 \text{ m}, 2 \text{ m}, 2 \text{ m})$ through the center of the sphere, O, and a magnitude of $F = 60$ N, determine the normal reactions of the three planes acting on the sphere.

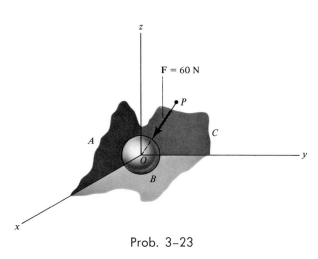

Prob. 3–23

3-25. The boom OA is used to support a crate having a mass of $m = 200$ kg. Determine the tension forces acting in the cables AB and AC and the compressive force in the boom which is directed along the boom axis OA. Set $a = 3$ m, $b = 4$ m, $c = 7$ m, and $d = 1.5$ m.

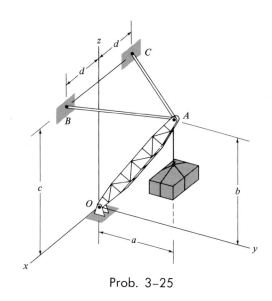

Prob. 3–25

*3-24. The joint O of a space frame is subjected to four forces. Strut OA lies in the x-y plane and strut OB lies in the y-z plane. Determine the forces acting in each of the three struts required for equilibrium of the joint.

3-25a. Solve Prob. 3–25 if the crate has a weight of $W = 350$ lb, and $a = 10$ ft, $b = 15$ ft, $c = 18$ ft, $d = 7$ ft.

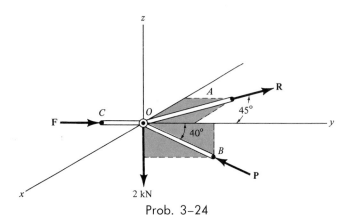

Prob. 3–24

3-26. A conical pivot bearing P is supported by three equally spaced smooth rollers. Determine the normal force that each of these rollers exerts on the bearing when the bearing is subjected to an axial force of 800 N.

*3-28. The ends of the three cables are attached to a ring at A and to the edge of a uniform 200-kg plate. Determine the tension in each of the cables for equilibrium.

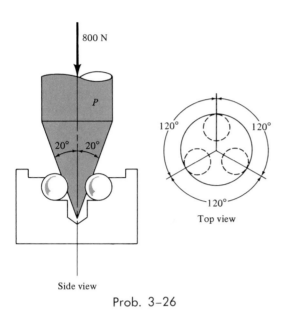

Side view

Prob. 3-26

3-27. Determine the magnitude of forces **P, R,** and **F** for equilibrium of the concurrent force system. The force **F** is located in the octant shown.

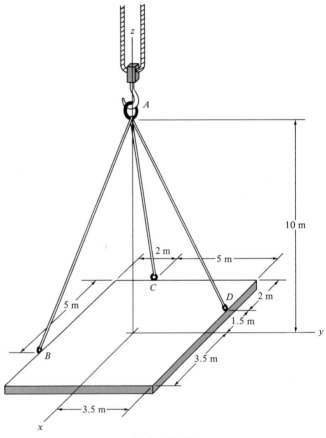

Prob. 3-28

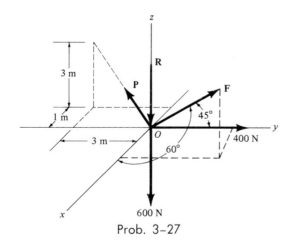

Prob. 3-27

3-29. The lamp has a mass of 10 kg and is supported by a pole AO and cables AB and AC. If the force in the pole acts along its axis, determine the forces in AO, AB, and AC for equilibrium.

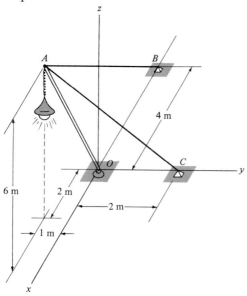

Prob. 3-29

3-30. Three outer blocks, each having a mass of $m = 2$ kg, and a central block E, having a mass of $m_E = 3$ kg, are suspended from the pulley and cable system shown. If the pulleys are frictionless and the weights of the cables are negligible, determine the sag s for equilibrium of the system. Set $a = AB = AC = BC = 1$ m.

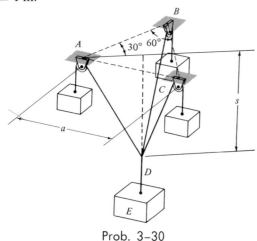

Prob. 3-30

3-30a. Solve Prob. 3-30 if each of the three outer blocks weighs $W = 10$ lb and the central block E weighs $W_E = 12$ lb. Set $a = AB = AC = BC = 1.5$ ft.

3-31. Determine the tension developed in the three cables required to support the traffic light, which has a mass of 10 kg.

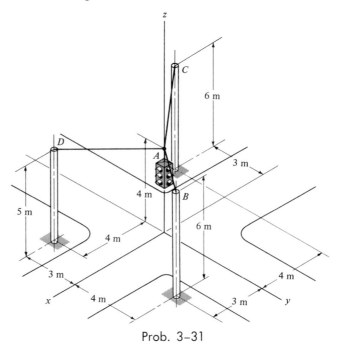

Prob. 3-31

* **3-32.** If the bucket and its contents have a mass of 25 kg, determine the force in cables DA and DC and the force acting along strut DB.

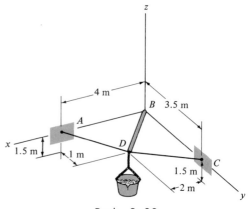

Prob. 3-32

3-33. Determine the magnitude and direction of the force **P** required to keep the concurrent force system in equilibrium.

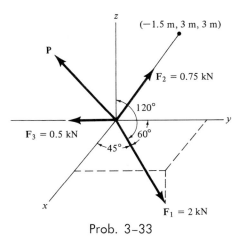

Prob. 3-33

3-34. The boom supports a bucket and its contents, which have a total mass of 200 kg. Determine the forces developed in struts AD and AE and the tension in cable AB for equilibrium. The force in each strut acts along its axis.

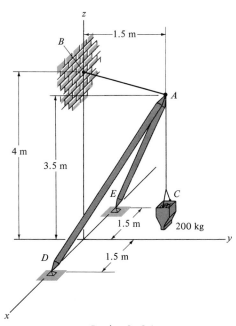

Prob. 3-34

3-35. Determine the tensile forces acting along the axes of struts DE and DF required to support the load, which has a mass of $m = 20$ kg. Set $a = 0.4$ m, $b = 0.6$ m, $c = 1.2$ m, $d = 1.5$ m, and $h = 2$ m. *Hint:* First determine the forces at C, then determine the forces at D.

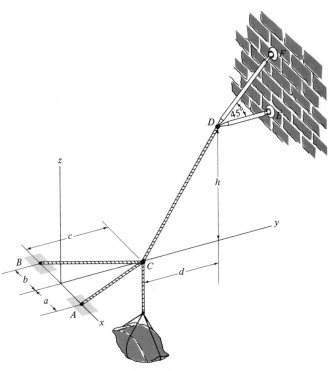

Prob. 3-35

3-35a. Solve Prob. 3-35 if the load weighs $W = 100$ lb, and $a = 2$ ft, $b = 3$ ft, $c = 5$ ft, $d = 4$ ft, $h = 6$ ft.

*3–36. Determine the tension developed in cables *OD* and *OB* and the strut *OC*, required to support the 50-kg crate. The spring *OA* has an unstretched length of 0.8 m and a stiffness $k_{OA} = 1.2$ kN/m. The force in the strut acts along the axis of the strut.

so that the tension in each of the cords, *OA*, *OB*, and *OC*, equals 20 N. The lamp has a mass of 5 kg. Set $AB = BC = AC = 0.5$ m.

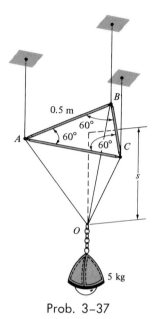

Prob. 3–37

3–38. Determine the force in each leg of the tripod. Neglect the weight of the legs and assume that the force acting in each leg is directed along the axis of the leg.

Prob. 3–36

3–37. The triangular frame *ABC* can be adjusted vertically between the three equal-length cords. If it remains in a horizontal plane, determine the required distance *s*

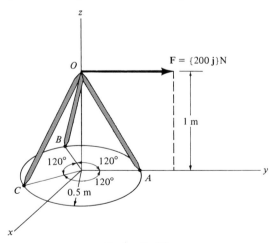

Prob. 3–38

3–39. The block having a mass of 50 kg is suspended from point A by the three struts AB, AC, and AD. Determine the force acting along the axis of each strut for equilibrium.

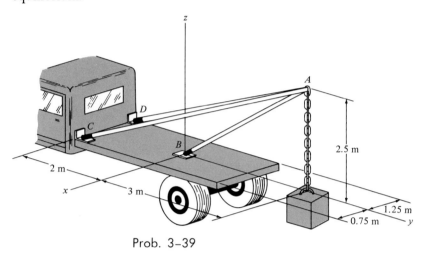

Prob. 3–39

*3–40. A sphere has a mass of $m = 2$ kg and rests between the 45° grooves A and B of a $\theta = 10°$ incline, and against a vertical wall at C. If all three surfaces of contact are smooth, determine the reactions of the surfaces on the sphere. *Suggestion:* Use the x, y, and z axes, with origin at the center of the sphere, and the z axis inclined as shown.

*3–40a. Solve Prob. 3–40 if the sphere has a weight of $W = 10$ lb and $\theta = 15°$.

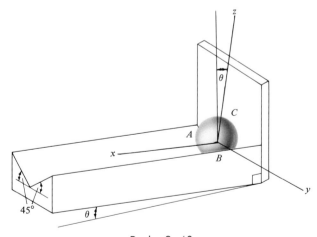

Prob. 3–40

3–41. Determine the force acting along the axes of struts *OA* and *OB* and the tension developed in cables *OC*, *OD*, and *DE* for equilibrium. *Hint:* First determine the forces at *D*, then determine the forces at *O*.

3–42. Knowing that cord *OB* has been stretched 20 mm and fixed in place as shown, determine the tension developed in each of the other three cables to hold the 20-kg sphere in equilibrium. Cable *OD* lies in the *x-y* plane.

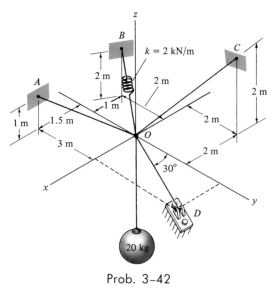

Prob. 3–42

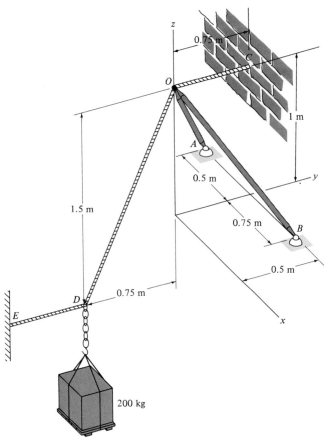

Prob. 3–41

3-43. A small peg P rests on a spring that is contained inside the smooth pipe. When the spring is compressed so that $s = 0.15$ m, the spring exerts an upward force of 60 N on the peg. Determine the point of attachment $A(x, y, 0)$ of cord PA so that the tension in cords PB and PC equals 30 N and 50 N, respectively.

3-45. A force of $F = 200$ N holds the crate having a mass of $m = 25$ kg in equilibrium. If the tension along struts AC and AB is to be $T = 400$ N each, determine the coordinates $(0, y, z)$ of point A. Set $a = 1$ m and $b = 0.8$ m.

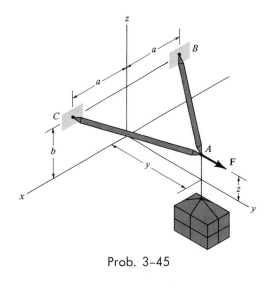

Prob. 3-45

3-45a. Solve Prob. 3-45 if $F = 150$ lb, the crate weighs $W = 300$ lb, $T = 500$ lb, $a = 4$ ft, and $b = 3$ ft.

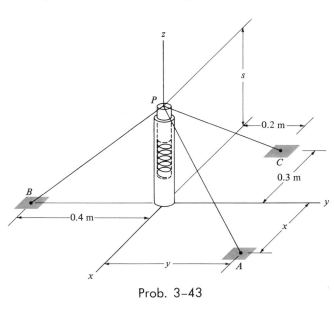

Prob. 3-43

*** 3-44.** Determine the magnitude and direction of $\mathbf{F}_1$ required to keep the concurrent force system in equilibrium.

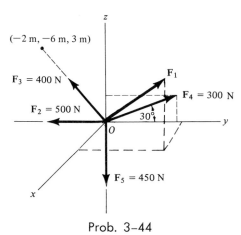

Prob. 3-44

81

Equivalent Force Systems

4-1. Effect of a Force on a Rigid Body

Rigid Body. A *rigid body* can be considered as a combination of a large number of particles in which all the particles remain at a fixed distance from one another before and after applying a load. This "model" of a physical body eliminates the need for any experimental testing, since it is assumed that the body does not deform and consequently the body's material properties do not have to be considered in the analysis of forces acting on the body. Because of this, the rigid-body assumption is an important idealization in the study of mechanics. In solving problems, however, one should clearly be aware of the limitations of the assumption. Since many of the principles of mechanics are related to the shape of the body, it is important to recognize problems where applied loadings might in reality cause large deformations. If the loads severely deform a body such that the orientation of the loading *after* it is applied is *not known,* the rigid-body assumption cannot be used. However, if the deformed shape of the body is *known,* so that the final position of the applied loads can be determined with sufficient accuracy, then assuming the body to be rigid may be justified. In many cases the deformations occurring in structures, machines, mechanisms, and so on, are relatively small, and the rigid-body assumption is suitable for analysis. Thus, with the exception of springs, the principles of mechanics as discussed in this book will be based on the assumption that all materials are rigid.

Unlike a particle, the size and shape of a rigid body become important in the force analysis of the body, since the applied force system may not necessarily be concurrent. Clearly, the two forces $\mathbf{F}_1$ and $\mathbf{F}_2$ acting on the

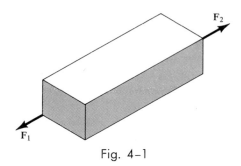

Fig. 4-1

block in Fig. 4-1 are not a system of concurrent forces; rather, they are noncollinear yet parallel to each other.

In the previous chapter, it was shown that the condition for the equilibrium of a particle or a concurrent force system simply requires that the resultant of the force system be equal to zero. In the next chapter, it will be shown that such a restriction is necessary but not sufficient for the equilibrium of a rigid body. A further restriction must be made with regard to the nonconcurrency of the applied force system, giving rise to the concept of *moment*. In this chapter, a formal definition of a moment will be developed and ways of finding the moments caused by concentrated forces and distributed loadings about points and axes will be discussed. Also, methods of reducing nonconcurrent concentrated force systems and distributed loadings to equivalent, yet simpler, systems will be considered. Before doing any of this, however, it is first necessary to introduce the cross product of vector multiplication.

4-2. Cross Product

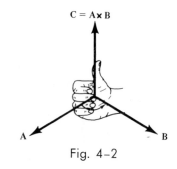

Fig. 4-2

The vector cross product of two vectors $\mathbf{A}$ and $\mathbf{B}$ yields the resultant vector $\mathbf{C}$, which is written

$$\mathbf{C} = \mathbf{A} \times \mathbf{B}$$

and is read "$\mathbf{C}$ equals $\mathbf{A}$ cross $\mathbf{B}$."

Magnitude. The *magnitude* of vector $\mathbf{C}$ is defined as the product of the magnitudes of the two vectors $\mathbf{A}$ and $\mathbf{B}$ and the sine of the angle θ made between their tails ($0° \leq \theta \leq 180°$). Thus, $C = AB \sin \theta$.

Direction. Vector $\mathbf{C}$ has a *direction* that is perpendicular to the plane containing the two vectors $\mathbf{A}$ and $\mathbf{B}$ such that $\mathbf{A}$, $\mathbf{B}$, and $\mathbf{C}$ form a *right-handed system;* i.e., the direction of $\mathbf{C}$ is specified by the right-hand rule, curling the fingers of the right hand from vector $\mathbf{A}$ (cross) to vector $\mathbf{B}$. The thumb then points in the direction of $\mathbf{C}$, as shown in Fig. 4-2.

Knowing both the magnitude and direction of $\mathbf{C}$, we can write

$$\mathbf{C} = \mathbf{A} \times \mathbf{B} = (AB \sin \theta)\mathbf{u}_C \qquad (4-1)$$

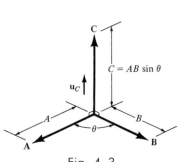

Fig. 4-3

where the scalar $AB \sin \theta$ defines the *magnitude* of $\mathbf{C}$ and the unit vector $\mathbf{u}_C$ defines the *direction* of $\mathbf{C}$. The terms of Eq. 4-1 are illustrated graphically in Fig. 4-3.

If the vector cross product $\mathbf{A} \times \mathbf{B} = \mathbf{0}$, then $AB \sin \theta \, \mathbf{u}_C = \mathbf{0}$. Provided $A \neq 0$ and $B \neq 0$, it is necessary that $\sin \theta = 0$, so that $\theta = 0°$ or $\theta = 180°$. This occurs if $\mathbf{A}$ is *parallel* to $\mathbf{B}$. In a similar manner,

$$\mathbf{A} \times \mathbf{A} = \mathbf{0} \qquad (4-2)$$

Laws of Operation

1. The commutative law is *not* valid, i.e.,

$$\mathbf{A} \times \mathbf{B} \neq \mathbf{B} \times \mathbf{A}$$

Rather,

$$\mathbf{A} \times \mathbf{B} = -\mathbf{B} \times \mathbf{A} \qquad (4\text{--}3)$$

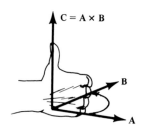

This is shown in Fig. 4–4 by using the right-hand rule. The cross product $\mathbf{B} \times \mathbf{A}$ yields a vector that acts in the opposite direction to $\mathbf{C}$, i.e., $\mathbf{B} \times \mathbf{A} = -\mathbf{C}$.

2. Multiplication by a scalar:

$$m(\mathbf{A} \times \mathbf{B}) = (m\mathbf{A}) \times \mathbf{B} = \mathbf{A} \times (m\mathbf{B}) = (\mathbf{A} \times \mathbf{B})m \qquad (4\text{--}4)$$

This property is easily shown, since the magnitude of the resultant vector ($|m|AB \sin \theta$) and its direction are the same in each case.

3. The distributive law:

$$\mathbf{A} \times (\mathbf{B} + \mathbf{D}) = (\mathbf{A} \times \mathbf{B}) + (\mathbf{A} \times \mathbf{D}) \qquad (4\text{--}5)$$

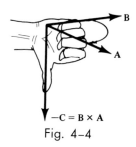

Fig. 4–4

The proof of this identity is left as an exercise (see Prob. 4–4). It is important to note that *proper order* of the cross products in Eq. 4–5 must be maintained, since they are not commutative.

Cartesian Vector Formulation. Equation 4–1 may be used to find the cross product of each of the Cartesian unit vectors. For example, to find $\mathbf{i} \times \mathbf{j}$, the *magnitude* of the resultant vector is $(i)(j)(\sin 90°) = (1)(1)(1) = 1$, and its *direction* is determined using the right-hand rule. As shown in Fig. 4–5, the resultant vector points in the $+\mathbf{k}$ direction. Thus, $\mathbf{i} \times \mathbf{j} = (1)\mathbf{k}$. In a similar manner,

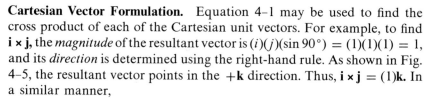

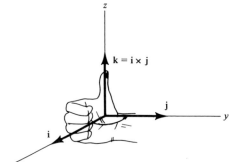

Fig. 4–5

These results should *not* be memorized; rather, it should be clearly understood how each is obtained by using the right-hand rule and the definition of the cross product. A simple scheme shown in Fig. 4–6 is helpful for obtaining the same results when the need arises. If the circle is constructed as shown, then "crossing" two unit vectors in a *clockwise* fashion around the circle yields the *positive* third unit vector, e.g., $\mathbf{k} \times \mathbf{i} = \mathbf{j}$. Moving *counterclockwise,* a *negative* unit vector is obtained, e.g., $\mathbf{i} \times \mathbf{k} = -\mathbf{j}$.

Consider now the cross product of two general vectors $\mathbf{A}$ and $\mathbf{B}$, which are expressed in Cartesian vector form as $\mathbf{A} = A_x\mathbf{i} + A_y\mathbf{j} + A_z\mathbf{k}$ and $\mathbf{B} = B_x\mathbf{i} + B_y\mathbf{j} + B_z\mathbf{k}$. Then

$$\mathbf{A} \times \mathbf{B} = (A_x\mathbf{i} + A_y\mathbf{j} + A_z\mathbf{k}) \times (B_x\mathbf{i} + B_y\mathbf{j} + B_z\mathbf{k})$$

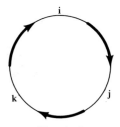

Fig. 4–6

Using the distributive law, Eq. 4–5, and the property of scalar multiplication, Eq. 4–4, we have

$$\mathbf{A} \times \mathbf{B} = A_x B_x (\mathbf{i} \times \mathbf{i}) + A_x B_y (\mathbf{i} \times \mathbf{j}) + A_x B_z (\mathbf{i} \times \mathbf{k})$$
$$+ A_y B_x (\mathbf{j} \times \mathbf{i}) + A_y B_y (\mathbf{j} \times \mathbf{j}) + A_y B_z (\mathbf{j} \times \mathbf{k})$$
$$+ A_z B_x (\mathbf{k} \times \mathbf{i}) + A_z B_y (\mathbf{k} \times \mathbf{j}) + A_z B_z (\mathbf{k} \times \mathbf{k})$$

Carrying out the cross-product operations and combining terms yields

$$\mathbf{A} \times \mathbf{B} = (A_y B_z - A_z B_y)\mathbf{i} - (A_x B_z - A_z B_x)\mathbf{j} + (A_x B_y - A_y B_x)\mathbf{k} \quad (4\text{–}6)$$

This equation may also be written in a more compact determinant form as

$$\mathbf{A} \times \mathbf{B} = \begin{vmatrix} \mathbf{i} & \mathbf{j} & \mathbf{k} \\ A_x & A_y & A_z \\ B_x & B_y & B_z \end{vmatrix} \quad (4\text{–}7)$$

Thus, to find the cross product of any two Cartesian vectors **A** and **B**, it is necessary to expand a determinant whose first row of elements are the unit vectors **i, j,** and **k** and whose second and third rows represent the x, y, z components of the two vectors **A** and **B**, respectively.*

---

*A determinant having three rows and three columns can be expanded using three minors, each of which is multiplied by one of the three terms in the first row. There are four elements in each minor, e.g.,

$$\begin{vmatrix} A_{11} & A_{12} \\ A_{21} & A_{22} \end{vmatrix}$$

By *definition,* this notation represents the terms $(A_{11}A_{22} - A_{12}A_{21})$, which is simply the product of the two elements of the arrow slanting downward to the right $(A_{11}A_{22})$ *minus* the product of the two elements intersected by the arrow slanting downward to the left $(A_{12}A_{21})$. For a 3×3 determinant, such as Eq. 4–7, the three minors can be generated in accordance with the following scheme:

For element **i**:
$$\begin{vmatrix} \mathbf{i} & \mathbf{j} & \mathbf{k} \\ A_x & A_y & A_z \\ B_x & B_y & B_z \end{vmatrix} = \mathbf{i}(A_y B_z - A_z B_y)$$

For element **j**:
$$\begin{vmatrix} \mathbf{i} & \mathbf{j} & \mathbf{k} \\ A_x & A_y & A_z \\ B_x & B_y & B_z \end{vmatrix} = -\mathbf{j}(A_x B_z - A_z B_x)$$

For element **k**:
$$\begin{vmatrix} \mathbf{i} & \mathbf{j} & \mathbf{k} \\ A_x & A_y & A_z \\ B_x & B_y & B_z \end{vmatrix} = \mathbf{k}(A_x B_y - A_y B_x)$$

Adding the results, and noting that the **j** element *must include the minus sign,* yields the expanded form of **A** × **B** given by Eq. 4–6.

4-3. Moment of a Force

Scalar Formulation. The concepts dealing with the moment of a force can be illustrated by a simple example. Consider force **F** applied at the end of a pipe wrench, Fig. 4–7. One effect of this force is to cause the pipe to rotate about the *aa* axis. Experience teaches that to produce the most efficient turning, **F** should be applied at right angles to the handle, and the distance *d* should be made as large as possible. This rotational effect of a force about the *aa* axis is called the *moment of a force* or simply the *moment* M_O. Note in particular, that the *aa* axis is perpendicular to a plane containing both **F** and *d*, Fig. 4–7, and that the *aa* axis intersects this plane at point *O*.

In the general case, force **F** and point *O* are located in space such that the line of action of **F** and the point are contained within a shaded plane as shown in Fig. 4–8a. The moment of **F** about an axis passing through *O* is defined as a *vector quantity*, M_O.

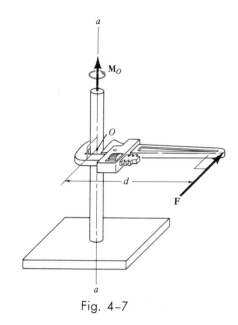

Fig. 4–7

Magnitude. The magnitude of M_O is

$$M_O = Fd \qquad (4\text{–}8)$$

where *d* is referred to as the *moment arm* or perpendicular distance from the axis at point *O* to the line of action of the force. Units of moment magnitude consist of force times distance, e.g., N · m.

Direction. The *direction* of M_O will be specified by using the right-hand rule. The fingers of the right hand are curled with the sense of rotation as caused by the force acting about the axis; the *thumb* then *points* in the direction of the moment vector along the *moment axis,* which is *perpendicular* to the shaded plane containing **F** and *d*, Fig. 4–8a. By this definition, the moment M_O can be considered as a *sliding vector* and therefore acts at any point along the moment axis.

In three dimensions, M_O is illustrated by a regular vector with a curl on it to *distinguish* it from a force vector, Fig. 4–8a. Many problems in mechanics, however, involve coplanar-force systems that may be conveniently viewed in two dimensions. For example, a two-dimensional view of Fig. 4–8a is given in Fig. 4–8b. Here M_O is simply represented by the (counterclockwise) curl, which indicates the action of **F**. This curl is used to show the *sense of rotation* caused by **F**. Using the right-hand rule, however, realize that the *direction* of the moment vector in Fig. 4–8b is indicated by the thumb and thus points *out* of the page, since the fingers follow the curl.

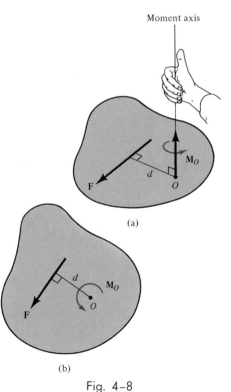

Fig. 4–8

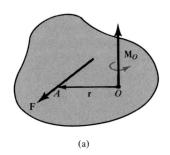

(a)

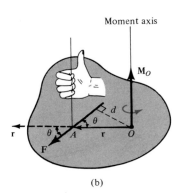

Moment axis

(b)

Fig. 4–9

Vector Formulation. The moment of a force **F** about an axis passing through point O may be expressed in terms of the vector cross product. With reference to Fig. 4–9a, this expression is written as

$$\mathbf{M}_O = \mathbf{r} \times \mathbf{F} \tag{4–9}$$

where **r** represents a position vector drawn from O to *any point A* lying on the line of action of **F.** It will now be shown that the moment vector $\mathbf{M}_O$, when determined by Eq. 4–9, has the proper magnitude and direction.

Magnitude. From the definition of the vector cross product, Eq. 4–1, the *magnitude* of $\mathbf{M}_O$ is

$$M_O = rF \sin \theta = F(r \sin \theta) = Fd$$

Since it is necessary that the angle θ be defined between the *tails* of **r** and **F**, then **r** must be treated as a sliding vector such that its tail is placed at point A, Fig. 4–9b. Note then that the moment arm d is defined as $d = r \sin \theta$. Hence, $M_O = Fd$, which is the same as Eq. 4–8.

Direction. The direction of $\mathbf{M}_O$ is determined by the right-hand rule in accordance with the definition of the vector cross product. Thus, extending **r** to the dashed position and curling the right-hand fingers from **r** towards **F,** "**r** cross **F**," the thumb is directed upward or perpendicular to the plane, in the *same direction* as defined by the moment of the force, $\mathbf{M}_O$, Fig. 4–9b. Since the cross product is not commutative, it is important that the *proper order* of vector multiplication be maintained in Eq. 4–9.

Cartesian Vector Formulation. Consider the rigid body of *arbitrary shape* shown in Fig. 4–10.* If a force **F** is applied to point A on the body, the moment of this force about point O is $\mathbf{M}_O = \mathbf{r} \times \mathbf{F}$, where **r** extends from O to A (or to any other point lying on the line of action of **F**). Expressing **r** and **F** in Cartesian vector form as $\mathbf{r} = r_x\mathbf{i} + r_y\mathbf{j} + r_z\mathbf{k}$ and $\mathbf{F} = F_x\mathbf{i} + F_y\mathbf{j} + F_z\mathbf{k}$, then from Eq. 4–7 we have

$$\mathbf{M}_O = \mathbf{r} \times \mathbf{F} = \begin{vmatrix} \mathbf{i} & \mathbf{j} & \mathbf{k} \\ r_x & r_y & r_z \\ F_x & F_y & F_z \end{vmatrix} \tag{4–10}$$

*The shaded object shown in Fig. 4–10 represents a rigid body having an *arbitrary shape*. Similarly, the object shown in Fig. 4–27 represents the plane view of a rigid body. Such objects are simply *graphical representations* used to explain the theory. However, it should be kept in mind that the arbitrary shape of these objects may in reality represent an airplane, beam, bridge, stone, etc.

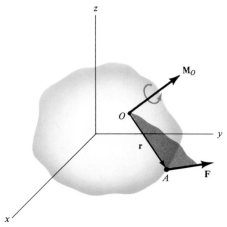

Fig. 4–10

By definition, $\mathbf{M}_O$ will be *perpendicular* to the shaded plane containing vectors $\mathbf{r}$ and $\mathbf{F}$, as shown in Fig. 4–10.

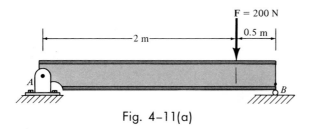

Fig. 4–11(a)

Example 4-1

Determine the moment of force $\mathbf{F}$ about points A and B of the beam shown in Fig. 4–11a.

Solution
(*Scalar Analysis*).   The moment arm $d = 2$ m, since it is the *perpendicular distance* from A to the line of action of $\mathbf{F}$. Hence, the magnitude of the moment of $\mathbf{F}$ about A is

$$M_A = Fd = 200 \text{ N}(2 \text{ m}) = 400 \text{ N} \cdot \text{m} \qquad Ans.$$

By the right-hand rule, the moment is directed *into* the page, Fig. 4–11b, since the force *tends* to rotate the beam in a clockwise direction about an axis passing through A. (In reality, this rotation is *prevented* by the roller constraint at B.) In a similar manner, the magnitude of moment at B is

$$M_B = Fd = (200 \text{ N})(0.5 \text{ m}) = 100 \text{ N} \cdot \text{m} \qquad Ans.$$

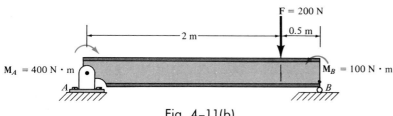

Fig. 4–11(b)

This moment is counterclockwise and hence directed out of the page. The three-dimensional views of the moment vectors are shown in Fig. 4–11c.

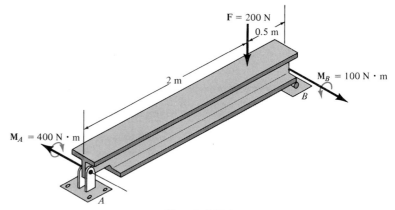

Fig. 4–11(c)

Example 4–2

Determine the moment of the 800-N force acting on the frame in Fig. 4–12 about points A, B, C, and D.

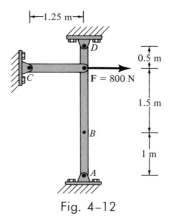

Fig. 4–12

Solution
(Scalar Analysis).   In general, $M = Fd$, where d is the moment arm or *perpendicular distance* from the *point* on the moment axis to the *line of action* of the force. Hence,

$M_A = 800 \text{ N}(2.5 \text{ m}) = 2000 \text{ N} \cdot \text{m} \downarrow$ *Ans.*

$M_B = 800 \text{ N}(1.5 \text{ m}) = 1200 \text{ N} \cdot \text{m} \downarrow$ *Ans.*

$M_C = 800 \text{ N}(0) = 0$ (line of action of **F** passes through C)   *Ans.*

$M_D = 800 \text{ N}(0.5 \text{ m}) = 400 \text{ N} \cdot \text{m} \uparrow$ *Ans.*

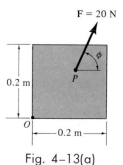

F = 20 N

0.2 m

0.2 m

Fig. 4-13(a)

Example 4-3

Determine the point of application P and the direction for a 20-N force that lies in the plane of the square plate shown in Fig. 4–13a, so that this force creates the greatest counterclockwise moment about point O. What is this moment?

Solution

(*Scalar Analysis*).   Since the maximum moment created by the force is required, the force must act on the plate at a distance *farthest* from point O. As shown in Fig. 4–13b, the point of application of **F** must therefore be at the diagonal corner. In order to produce *counterclockwise* rotation of the plate, **F** must act at an angle $45° < \phi < 225°$. The greatest moment is produced when the line of action of **F** is *perpendicular* to d_P, i.e., $\phi = 135°$, Fig. 4–13c. The maximum moment is, therefore,

$$M_O = Fd_P = (20 \text{ N})(0.2 \sqrt{2} \text{ m}) = 5.66 \text{ N} \cdot \text{m} \qquad Ans.$$

By the right-hand rule, $\mathbf{M}_O$ is directed out of the page.

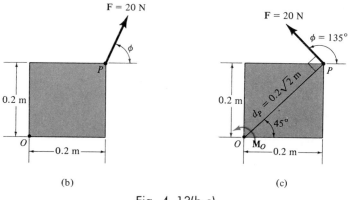

(b)

(c)

Fig. 4-13(b,c)

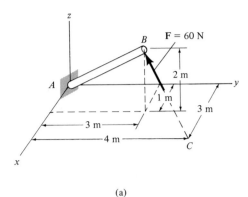

(a)

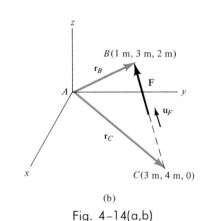

(b)

Fig. 4-14(a,b)

Example 4-4

The pole in Fig. 4–14a is subjected to a 60-N force that is directed from C to B. Determine completely the moment created by this force about point A.

Solution

(***Vector Analysis***). Problems in three dimensions, such as this one, where the moment arm is difficult to determine, can conveniently be solved using Cartesian vectors. As shown in Fig. 4–14b, either one of two position vectors can be used for the solution, since $\mathbf{M}_A = \mathbf{r}_B \times \mathbf{F}$ or $\mathbf{M}_A = \mathbf{r}_C \times \mathbf{F}$. The position vectors are represented as

$$\mathbf{r}_B = \{1\mathbf{i} + 3\mathbf{j} + 2\mathbf{k}\} \text{ m} \quad \text{and} \quad \mathbf{r}_C = \{3\mathbf{i} + 4\mathbf{j}\} \text{ m}$$

Since force $\mathbf{F}$ has a magnitude of 60 N and a direction specified by the unit vector $\mathbf{u}_F$, directed from C to B, we have

$$\mathbf{F} = (60 \text{ N})\mathbf{u}_F = (60)\left[\frac{(1-3)\mathbf{i} + (3-4)\mathbf{j} + (2-0)\mathbf{k}}{\sqrt{(-2)^2 + (-1)^2 + (2)^2}}\right]$$

$$= \{-40\mathbf{i} - 20\mathbf{j} + 40\mathbf{k}\} \text{ N}$$

Substituting into the determinant formulation, Eq. 4–7, and following the scheme for determinant expansion as stated in the footnote on page 86, we háve

$$\mathbf{M}_A = \mathbf{r}_B \times \mathbf{F} = \begin{vmatrix} \mathbf{i} & \mathbf{j} & \mathbf{k} \\ 1 & 3 & 2 \\ -40 & -20 & 40 \end{vmatrix}$$

$$= [3(40) - 2(-20)]\mathbf{i} - [1(40) - 2(-40)]\mathbf{j} + [1(-20) - 3(-40)]\mathbf{k}$$

or

$$\mathbf{M}_A = \mathbf{r}_C \times \mathbf{F} = \begin{vmatrix} \mathbf{i} & \mathbf{j} & \mathbf{k} \\ 3 & 4 & 0 \\ -40 & -20 & 40 \end{vmatrix}$$

$$= [4(40) - 0(-20)]\mathbf{i} - [3(40) - 0(-40)]\mathbf{j} + [3(-20) - 4(-40)]\mathbf{k}$$

In both cases,

$$\mathbf{M}_A = \{160\mathbf{i} - 120\mathbf{j} + 100\mathbf{k}\} \text{ N} \cdot \text{m}$$

The *magnitude* of $\mathbf{M}_A$ is therefore

$$M_A = \sqrt{(160)^2 + (-120)^2 + (100)^2} = 223.6 \text{ N} \cdot \text{m} \qquad Ans.$$

A unit vector acting in the direction of $\mathbf{M}_A$ is

$$\mathbf{u}_{M_A} = \frac{\mathbf{M}_A}{M_A} = 0.716\mathbf{i} - 0.537\mathbf{j} + 0.447\mathbf{k}$$

The components of $\mathbf{u}_{M_A}$ are the *direction cosines* of $\mathbf{M}_A$. Hence,

$$\alpha = \cos^{-1}(0.716) = 44.3° \qquad \textit{Ans.}$$
$$\beta = \cos^{-1}(-0.537) = 122.5° \qquad \textit{Ans.}$$
$$\gamma = \cos^{-1}(0.447) = 63.4° \qquad \textit{Ans.}$$

As shown in Fig. 4–14c, $\mathbf{M}_A$ acts perpendicular to the shaded plane containing vectors $\mathbf{F}$, $\mathbf{r}_B$, and $\mathbf{r}_C$. Had this problem been worked using a scalar approach, where $M_A = Fd$, notice the difficulty that might arise in obtaining the moment arm d, Fig. 4–14c, and establishing the coordinate direction angles α, β, and γ for $\mathbf{M}_A$.

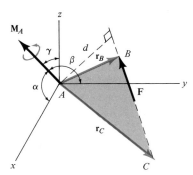

Fig. 4–14(c)

4-4. Varignon's Theorem

A concept often used in mechanics is *Varignon's theorem,* which is sometimes referred to as the *principle of moments.* This theorem states that *the sum of the moments of all the forces of a concurrent force system about a given point is equal to the moment created by the resultant force of the system about the point.* To prove this, consider the system of three concurrent forces shown in Fig. 4–15. The moment of all three forces about point O is

$$\mathbf{M}_R = \mathbf{r} \times \mathbf{F}_1 + \mathbf{r} \times \mathbf{F}_2 + \mathbf{r} \times \mathbf{F}_3 = \mathbf{r} \times (\mathbf{F}_1 + \mathbf{F}_2 + \mathbf{F}_3)$$

Since the resultant force at P is $\mathbf{F}_R = \mathbf{F}_1 + \mathbf{F}_2 + \mathbf{F}_3$, one can also write

$$\mathbf{M}_R = \mathbf{r} \times \mathbf{F}_R$$

This property was originally developed by the French mathematician Varignon (1654–1722). It has important applications to the solution of problems and proofs of theorems that follow, since it allows us to consider the moments of a force's components rather than the force itself.

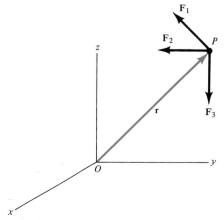

Fig. 4–15

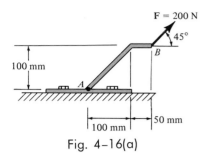

F = 200 N

45°

B

100 mm

A

50 mm

100 mm

Fig. 4-16(a)

Example 4-5

A 200-N force acts on the bracket shown in Fig. 4–16a. Determine the moment of the force about point A.

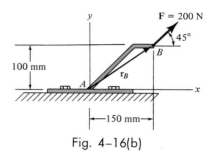

y

F = 200 N

45°

B

100 mm

r_B

A

x

—150 mm—

Fig. 4-16(b)

Solution I
(**Vector Analysis**).   Using a Cartesian vector approach, the force and position vector shown in Fig. 4–16b can be represented as

$$r_B = \{150i + 100j\} \text{ mm} = \{0.15i + 0.10j\} \text{ m}$$
$$F = \{200 \cos 45°i + 200 \sin 45°j\} \text{ N}$$

Hence, the required moment becomes

$$M_A = r_B \times F = \begin{vmatrix} i & j & k \\ 0.15 & 0.10 & 0 \\ 200 \cos 45° & 200 \sin 45° & 0 \end{vmatrix}$$

$$= 0i - 0j + [(0.15)(200 \sin 45°) - (0.10)(200 \cos 45°)]k \qquad (1)$$

$$= \{7.07k\} \text{ N} \cdot \text{m} \qquad \qquad Ans.$$

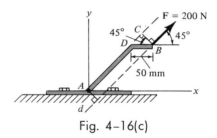

y

F = 200 N

45°   C

D   45°

B

50 mm

A

x

d

Fig. 4-16(c)

Solution II
(**Scalar Analysis**).   The moment arm d can be computed by trigonometry, using the construction shown in Fig. 4–16c. From triangle BCD,

$$CB = d = 50 \cos 45° = 35.35 \text{ mm} = 0.035\,35 \text{ m}$$

Thus,

$$M_A = Fd = 200 \text{ N}(0.035 \ 35 \text{ m}) = 7.07 \text{ N} \cdot \text{m}$$

According to the right-hand rule, $\mathbf{M}_A$ is directed in the $+\mathbf{k}$ direction since the force tends to rotate *counterclockwise* about point A. Hence,

$$\mathbf{M}_A = \{7.07\mathbf{k}\} \text{ N} \cdot \text{m} \qquad\qquad Ans.$$

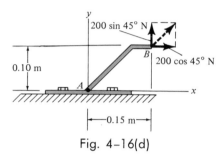

Fig. 4-16(d)

Solution III
(*Scalar Analysis Using Varignon's Theorem*).   The 200-N force may be resolved into x and y components, as shown in Fig. 4–16*d*. In accordance with Varignon's theorem, the moment computed about point A is equivalent to the sum of the moments produced by the two force components. Assuming counterclockwise rotation as positive, i.e., $+\mathbf{k}$ direction, we can add the moment of each force algebraically, since they both act along the same moment axis (perpendicular to the x, y plane, passing through A).

$$\zeta + M_A = (200 \sin 45°)(0.15) - (200 \cos 45°)(0.10) \qquad (2)$$
$$= 7.07 \text{ N} \cdot \text{m} \, \jmath$$

Thus,

$$\mathbf{M}_A = \{7.07\mathbf{k}\} \text{ N} \cdot \text{m} \qquad\qquad Ans.$$

By comparison, it is seen that solution III provides a more *convenient method* for analysis than solution II since the moment arm for each component force is easier to establish. Hence, *this method is generally recommended for solving problems involving a system of coplanar forces.* Although the Cartesian vector analysis, solution I, *automatically* accounts for the sign of the moment of each component force, Eq. (1), in two dimensions, these directions can easily be established *directly* from the diagram, Fig. 4–16*d*, Eq. (2). Hence, there really is no need to set up and evaluate a determinant. As a result, *Cartesian vector analysis is generally recommended only for solving three-dimensional problems* where the moment arms and force components may be more difficult to determine.

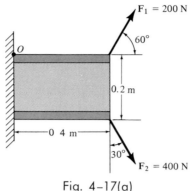

Fig. 4–17(a)

Example 4-6

Two coplanar forces act on the beam shown in Fig. 4–17a. Determine the moment of each force about point O.

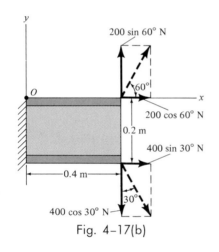

Fig. 4–17(b)

Solution
(*Scalar Analysis Using Varignon's Theorem*).   Each force is resolved into its x and y components as shown in Fig. 4–17b, and the moments of the components are computed about O. Assuming that positive moments act counterclockwise ($+\mathbf{k}$ direction) then for $\mathbf{F}_1$,

$$\curvearrowright +(M_O)_1 = 200 \cos 60°(0) + 200 \sin 60°(0.4)$$
$$= 69.3 \text{ N} \cdot \text{m} \curvearrowleft \qquad \qquad Ans.$$

For $\mathbf{F}_2$,

$$\curvearrowright +(M_O)_2 = 400 \sin 30°(0.2) - 400 \cos 30°(0.4)$$
$$= -98.6 \text{ N} \cdot \text{m} \curvearrowright \qquad \qquad Ans.$$

As stated in the previous example, the above method provides the most direct solution to this problem, since the moment arms for the *force components* are easily obtained. It generally involves more thought and calculation to use trigonometry to obtain the moment arm of each force or to set up and evaluate two determinants using Cartesian vectors.

Problems

4-1. Determine the resulting vector for each of the following cross products: (a) $(2\mathbf{i} + 4\mathbf{k}) \times 3\mathbf{j}$; (b) $5\mathbf{j} \times (3\mathbf{j} + 4\mathbf{k})$; (c) $(-10\mathbf{i} - 2\mathbf{k}) \times (3\mathbf{j} - 1\mathbf{k})$.

4-2. Given $\mathbf{A} = 3\mathbf{i} + 3\mathbf{j} - 1\mathbf{k}$, $\mathbf{B} = -2\mathbf{i} + 2\mathbf{k}$, and $\mathbf{C} = 2\mathbf{i} - 4\mathbf{j} + 3\mathbf{k}$, find: (a) $(\mathbf{A} \times \mathbf{B}) \times \mathbf{C}$; (b) $\mathbf{A} \times (\mathbf{B} \times \mathbf{C})$.

4-3. Determine a unit vector perpendicular to the plane containing the two vectors $\mathbf{A} = 4\mathbf{i} - 3\mathbf{j}$ and $\mathbf{B} = 5\mathbf{k}$.

*4-4. If $\mathbf{A}$, $\mathbf{B}$, and $\mathbf{D}$ are given vectors, prove the distributive law for the vector cross product, i.e., $\mathbf{A} \times (\mathbf{B} + \mathbf{D}) = (\mathbf{A} \times \mathbf{B}) + (\mathbf{A} \times \mathbf{D})$. *Suggestion:* Express each vector in Cartesian vector form and carry out the vector operations.

4–5. Determine the cross product $\mathbf{r}_1 \times \mathbf{r}_2$ of the two position vectors shown in the figure. Set $a = 400$ mm, $b = 300$ mm, and $c = 150$ mm.

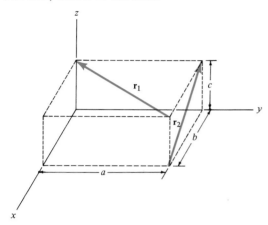

Prob. 4–5

4–5a. Solve Prob. 4–5 if $a = 0.8$ ft, $b = 0.4$ ft, and $c = 0.3$ ft.

4–6. A force $\mathbf{F}$ having a magnitude of $F = 100$ N acts along the diagonal of the parallelepiped. Using Cartesian vectors, determine the moment of $\mathbf{F}$ about point A where $\mathbf{M}_A = \mathbf{r}_B \times \mathbf{F}$ and $\mathbf{M}_A = \mathbf{r}_C \times \mathbf{F}$.

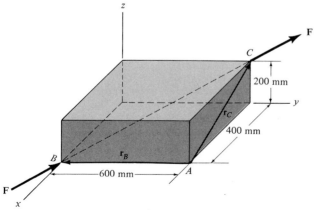

Prob. 4–6

4–7. The 400-N force acts on the end of the pipe at B. Determine: (a) the moment of this force about point A;

(b) the magnitude and direction of a horizontal force, applied at C, which produces the same moment.

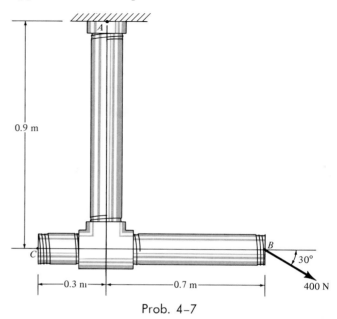

Prob. 4–7

* **4–8.** A force of 80 N acts on the handle of the paper cutter at A. Determine the moment created by this force about the hinge at O, if $\theta = 60°$. At what angle θ should the force be applied so that the moment it creates about point O is a maximum (clockwise)?

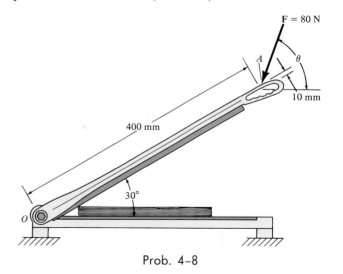

Prob. 4–8

4–9. The key at B prevents the arm A from slipping around the fixed shaft s. Determine the moment that the 800-N force creates about B.

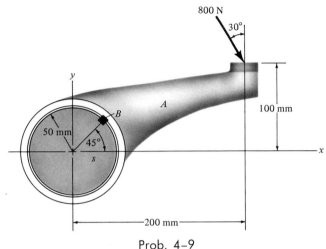

Prob. 4–9

4–10. The pole supports a traffic light that has a mass of $m = 10$ kg. Using Cartesian vectors, determine the moment of the weight of the traffic light about the base of the pole at A. Set $d = 3$ m and $h = 4$ m.

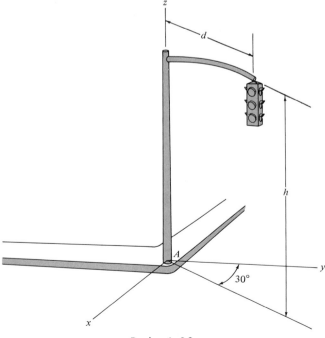

Prob. 4–10

4–10a. Solve Prob. 4–10 if the traffic light weighs $W = 20$ lb, $d = 12$ ft, and $h = 15$ ft.

4–11. Cable BC exerts a force of $F = 100$ N on the flag pole at B. Determine the moment created by this force about the base A of the pole.

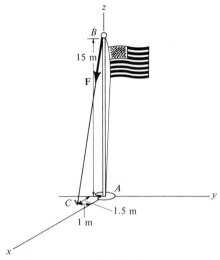

Prob. 4–11

* **4–12.** Determine the angle θ at which the 500-N force must act at A so that the moment of this force about point B is equal to zero.

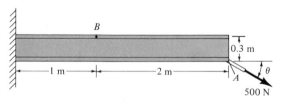

Prob. 4–12

4-13. Determine the moment of the 10-kN force about point O. Solve the problem using both a scalar analysis and a vector analysis.

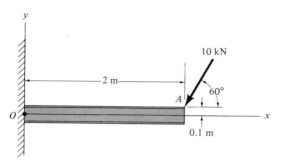

Prob. 4-13

4-14. A force of 40 N is applied to the handle of the wrench. Determine the moment of this force about point O. Solve the problem using both a scalar analysis and a vector analysis.

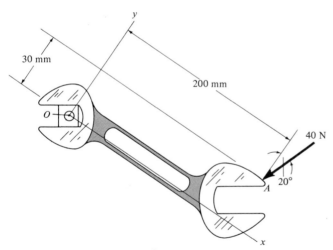

Prob. 4-14

4-15. Determine the moment created by the force $F = \{50i + 100j - 50k\}$ N acting at D, about each of the joints at B and C. Set $a = 1.25$ m, $b = 0.75$ m, and $c = 0.3$ m.

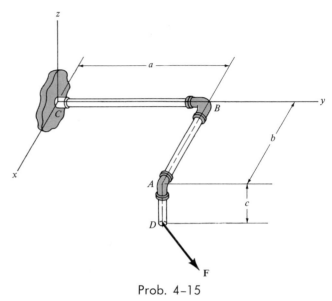

Prob. 4-15

4-15a. Solve Prob. 4-15 if $F = \{10i + 30j - 15k\}$ lb, and $a = 2$ ft, $b = 1.75$ ft, $c = 0.5$ ft.

*4-16. Determine the magnitude of the force F that should be applied at the end of the lever such that this force creates a clockwise moment of 15 N·m about point O when $\theta = 30°$.

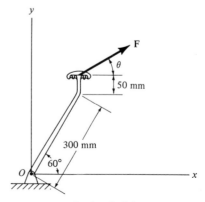

Prob. 4-16

99

4-17. If the force **F** applied at the end of the lever in Prob. 4–16 has a magnitude of 100 N, determine the angle θ so that the force develops a clockwise moment at O of 20 N · m.

4-18. A 20-N horizontal force is applied perpendicular to the handle of the socket wrench. Determine the magnitude and direction of the moment created by this force about point O.

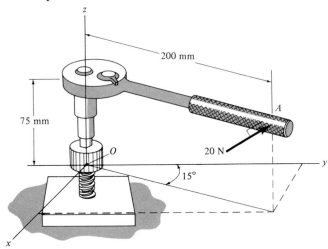

Prob. 4–18

4-19. If $\mathbf{F} = \{60\mathbf{i} + 40\mathbf{j} + 20\mathbf{k}\}$ N, determine the magnitude and direction of the moment of **F** about point A.

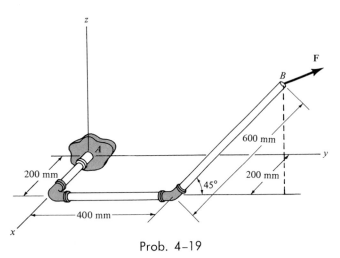

Prob. 4–19

*4-20. If each of the four traffic lights has a mass of $m = 6$ kg, determine the resultant moment created by the weight of all of them at the base of the post A. Set $a = 0.5$ m, $b = 2.5$ m, $c = 3$ m, $d = 0.75$ m, and $h = 3.5$ m.

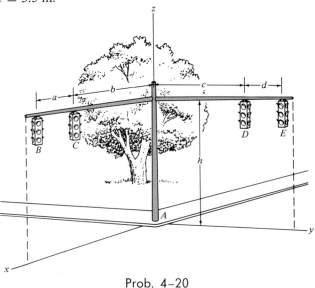

Prob. 4–20

*4-20a. Solve Prob. 4–10 if each of the four traffic lights has a weight of $W = 25$ lb, and $a = 1.5$ ft, $b = 8$ ft, $c = 9$ ft, $d = 1.75$ ft, $h = 15$ ft.

4-21. The crane can be adjusted for any angle $0° \leqslant \theta \leqslant 90°$ and any extension $0 \leqslant x \leqslant 5$ m. For a suspended mass of 200 kg, determine the moment developed at A as a function of x and θ. What values of both x and θ develop the maximum possible moment at A? Compute this moment. Neglect the size of the pulley at B.

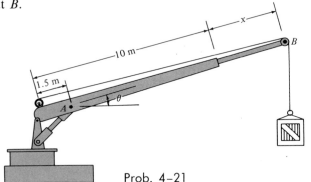

Prob. 4–21

4-22. The 5-m-long boom AB lies in the y-z plane. If a tension force of $F = 500$ N acts in cable BC, determine the moment of this force about point A. What is the shortest distance from point A to the cable?

4-23. Using a scalar analysis, determine the magnitude of force $\mathbf{F}_2$ that must be applied perpendicular to the handle so that the resultant moment of both $\mathbf{F}_1$ and $\mathbf{F}_2$ at O is zero.

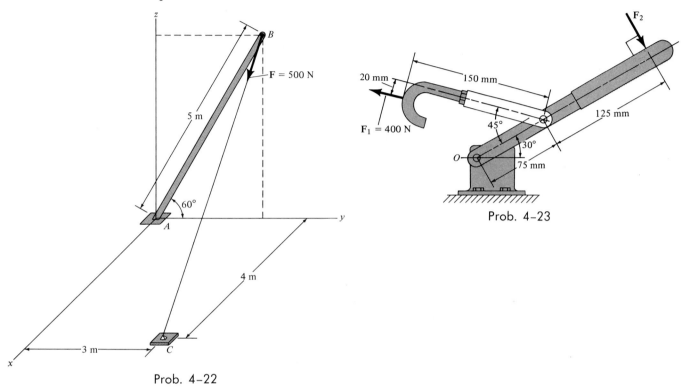

Prob. 4-23

Prob. 4-22

4-5. Moment of a Couple

A *couple* is defined as two parallel forces that have the same magnitude, opposite direction, and are separated by a perpendicular distance d, Fig. 4-18. Since the resultant force of the two forces composing the couple is zero, the effect of a couple is to produce a pure *moment*, or tendency of rotation in a specified direction.

The moment produced by a couple is equivalent to the moment of each of its two forces, computed about any arbitrary point O in space. To show

Fig. 4-18

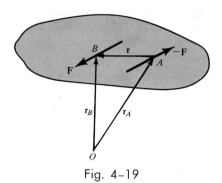

Fig. 4–19

this, consider position vectors $\mathbf{r}_A$ and $\mathbf{r}_B$, directed from O to points A and B lying on the line of action of $-\mathbf{F}$ and $\mathbf{F}$, Fig. 4–19. The couple moment computed about O is therefore,

$$\mathbf{M}_O = \mathbf{r}_A \times (-\mathbf{F}) + \mathbf{r}_B \times (\mathbf{F})$$
$$= (\mathbf{r}_B - \mathbf{r}_A) \times \mathbf{F}$$

By the triangle law of vector addition, $\mathbf{r}_A + \mathbf{r} = \mathbf{r}_B$ or $\mathbf{r} = \mathbf{r}_B - \mathbf{r}_A$, so that

$$\mathbf{M}_O = \mathbf{r} \times \mathbf{F} \qquad (4\text{–}11)$$

This result indicates that a couple is a *free vector* since $\mathbf{M}_O$ depends *only* upon the position vector directed *between* the forces and *not* the position vectors $\mathbf{r}_A$ or $\mathbf{r}_B$, directed from point O to the force. Understandably, this concept is unlike the moment of a force, which requires a definite point (or axis) about which moments are computed.

Scalar Formulation. The moment of a couple, $\mathbf{M}_C$, Fig. 4–20, is defined as having a *magnitude* of

$$M_C = Fd \qquad (4\text{–}12)$$

where F is the magnitude of one of the forces and d is the perpendicular distance or moment arm between the forces. The *direction* of the couple moment is determined by the right-hand rule, where the thumb indicates the direction when the fingers are curled with the sense of rotation caused by the forces. In all cases, $\mathbf{M}_C$ acts perpendicular to the plane containing the two forces.

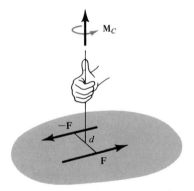

Fig. 4–20

Vector Formulation. The moment of a couple can be expressed by the vector cross product using Eq. 4–11, which can be written as

$$\mathbf{M}_C = \mathbf{r} \times \mathbf{F} \qquad (4\text{–}13)$$

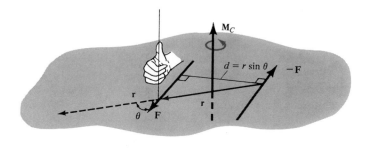

(a)

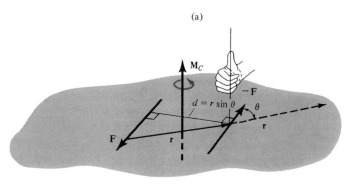

(b)

Fig. 4–21

Here **r** represents a position vector, drawn from *any point* on the line of action of one force to *any point* on the line of action of the other force. As shown in Figs. 4–21*a* and 4–21*b*, **F** in Eq. 4–13 represents that force to which **r** is *directed*. In each case, the *magnitude* of the couple is

$$M_C = rF \sin \theta = F(r \sin \theta) = Fd$$

The angle θ must be constructed between the tails of **r** and **F**. Thus, **r** is treated as a sliding vector and extended to the dashed position, Fig. 4–21. Hence, the moment arm $d = r \sin \theta$, so that $M_C = Fd$, which is the same as Eq. 4–12. The *direction* of **M**$_C$ is defined by the right-hand rule applied to the cross product. Drawing the fingers from **r** to **F** (**r** cross **F**) the thumb points in the direction of **M**$_C$.

Hence, by comparison the scalar and vector formulations of **M**$_C$ give the same results. For application, however, *it is suggested that vector analysis be used only for solving three-dimensional problems where the components of the force or couple are difficult to determine.* Most often, *the moments of coplanar couples are easily obtained in two dimensions by using scalars.*

Equivalent Couples.   Two couples are said to be equivalent if they produce the same moment vector. Since the moment produced by a couple is

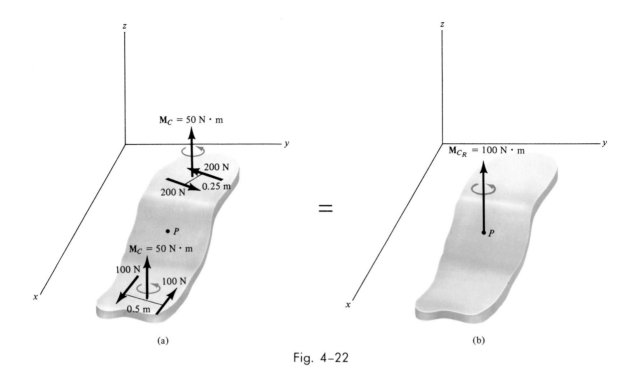

(a)

(b)

Fig. 4–22

always perpendicular to the plane containing the couple forces, it is therefore necessary that the forces of equal couples lie either in the same plane or in corresponding planes that are *parallel* to one another. In this way, the line of action of each couple moment will be the same, that is, perpendicular to the parallel planes. For example, the two couples shown in Fig. 4–22a are equivalent. One couple is produced by a pair of 100-N forces separated by a distance of $d = 0.5$ m, and the other is produced by a pair of 200-N forces separated by a distance of 0.25 m. Since the planes

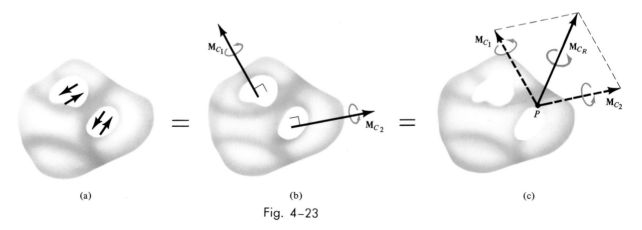

(a)   (b)   (c)

Fig. 4–23

in which the forces act are parallel to the x-y plane, the moment produced by each of the couples may be expressed as $\mathbf{M}_C = \{50\mathbf{k}\}\,\mathrm{N} \cdot \mathrm{m}$.

Resultant Couple. Since couple moments are free vectors, both couple moments in Fig. 4–22a may be applied at any point P, and added vectorially. As shown in Fig. 4–22b, the resultant of both couples is, therefore, $\mathbf{M}_{C_R} = 50\mathbf{k} + 50\mathbf{k} = \{100\mathbf{k}\}\,\mathrm{N} \cdot \mathrm{m}$.

The moments of two couples acting on nonparallel planes are also added using the rules of vector addition. For example, the two couples acting on different planes of the rigid body in Fig. 4–23a may be replaced by their corresponding moments $\mathbf{M}_{C_1}$ and $\mathbf{M}_{C_2}$, shown in Fig. 4–23b. These free vectors may be moved to the *arbitrary point P* and added to obtain their resultant vector sum $\mathbf{M}_{C_R} = \mathbf{M}_{C_1} + \mathbf{M}_{C_2}$, shown in Fig. 4–23c. Hence, the effect of the two couples in Fig. 4–23a imparts a "total twist" or moment $\mathbf{M}_{C_R}$ to the body acting about an axis that is parallel to the direction of $\mathbf{M}_{C_R}$.

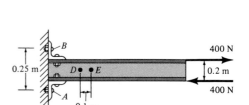

(a)

Example 4-7

A couple acts at the end of the beam shown in Fig. 4–24a. Replace the couple by an equivalent one having a pair of forces that act through (a) points A and B; (b) points D and E.

Solution

(*Scalar Analysis*). The couple has a magnitude of $M_C = Fd = 400(0.2) = 80\,\mathrm{N} \cdot \mathrm{m}$ and a direction that is into the page since the forces tend to rotate clockwise. $\mathbf{M}_C$ is a free vector so that it can be placed at any point on the beam, Fig. 4–24b.

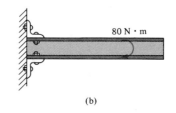

(b)

Part (a). To preserve the direction of $\mathbf{M}_C$, *horizontal* forces acting through points A and B must be directed as shown in Fig. 4–24c. The magnitude of each force is

$$M_C = Fd$$
$$80 = F(0.25)$$
$$F = 320\,\mathrm{N} \qquad\qquad Ans.$$

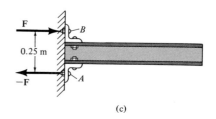

(c)

Part (b): To generate the required clockwise rotation, forces acting through points D and E must be *vertical* and directed as shown in Fig. 4–24d. The magnitude of each force is

$$M_C = Pd$$
$$80 = P(0.1)$$
$$P = 800\,\mathrm{N} \qquad\qquad Ans.$$

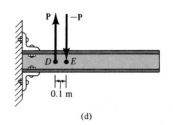

(d)

Fig. 4–24

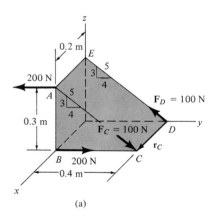

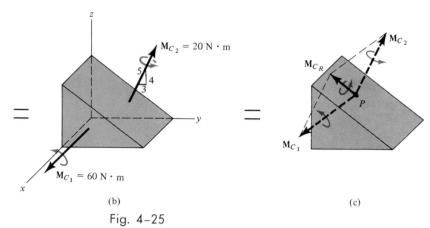

(a)

(b)

(c)

Fig. 4–25

Example 4–8

Replace the two couples acting on the triangular block in Fig. 4–25a by a single resultant couple.

Solution

The couple $\mathbf{M}_{C_1}$, caused by the forces at A and B, can easily be determined from a scalar formulation.

$$M_{C_1} = Fd = 200(0.3) = 60 \text{ N} \cdot \text{m}$$

By the right-hand rule, $\mathbf{M}_{C_1}$ acts in the $+\mathbf{i}$ direction, Fig. 4–25b. Hence,

$$\mathbf{M}_{C_1} = \{60\mathbf{i}\} \text{ N} \cdot \text{m}$$

Vector analysis will be used to determine $\mathbf{M}_{C_2}$, caused by forces at C and D. If $\mathbf{F}_C$ is considered for the calculation, then from Fig. 4–25a, $\mathbf{M}_{C_2} = \mathbf{r}_C \times \mathbf{F}_C$, where $\mathbf{r}_C$ extends from the line of action of $\mathbf{F}_D$ to the line of action of $\mathbf{F}_C$, Fig. 4–25a. We have

$$\begin{aligned}\mathbf{M}_{C_2} = \mathbf{r}_C \times \mathbf{F}_C &= (0.2\mathbf{i}) \times [100(\tfrac{4}{5})\mathbf{j} - 100(\tfrac{3}{5})\mathbf{k}] \\ &= (0.2\mathbf{i}) \times [80\mathbf{j} - 60\mathbf{k}] = 16(\mathbf{i} \times \mathbf{j}) - 12(\mathbf{i} \times \mathbf{k}) \\ &= \{12\mathbf{j} + 16\mathbf{k}\} \text{ N} \cdot \text{m}\end{aligned}$$

The *same result* can also be obtained by using a position vector which extends from D to A, E to C, or E to A since in all cases these points lie on the line of action of each force. Another possibility is to use $\mathbf{F}_D$ and $\mathbf{r}_D = -\mathbf{r}_C$, in which case

$$\begin{aligned}\mathbf{M}_{C_2} = \mathbf{r}_D \times \mathbf{F}_D &= (-0.2\mathbf{i}) \times [-100(\tfrac{4}{5})\mathbf{j} + 100(\tfrac{3}{5})\mathbf{k}] \\ &= \{12\mathbf{j} + 16\mathbf{k}\} \text{ N} \cdot \text{m}\end{aligned}$$

Try to obtain $\mathbf{M}_{C_2}$ by using the scalar formulation, Fig. 4–25b, and then express the result in terms of its $\mathbf{j}$ and $\mathbf{k}$ components.

Since $\mathbf{M}_{C_1}$ and $\mathbf{M}_{C_2}$ are free vectors, they may be moved to some arbitrary point P on the block and added vectorially, Fig. 4–25c. The resultant couple moment becomes

$$\mathbf{M}_{C_R} = \mathbf{M}_{C_1} + \mathbf{M}_{C_2} = \{60\mathbf{i} + 12\mathbf{j} + 16\mathbf{k}\} \, \text{N} \cdot \text{m} \qquad Ans.$$

Hence, if the block is free to rotate, it would turn about an axis defined by the coordinate direction angles of $\mathbf{M}_{C_R}$.

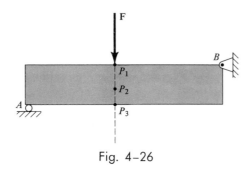

Fig. 4–26

4–6. The Principle of Transmissibility

The *principle of transmissibility* is an important concept often used in mechanics for studying the action of a force on a rigid body. This principle states that the conditions of equilibrium or the motion of the body remain *unchanged* when a force $\mathbf{F}$, acting at a given point on the body, is applied to another point *lying on the line of action of the force. In other words*, $\mathbf{F}$ can be considered as a *sliding vector.*

To illustrate this concept, consider the beam shown in Fig. 4–26, which is subjected to a force at point P_1. Rather than applying $\mathbf{F}$ at the beam's top surface, it can be applied from the bottom of the beam at P_3, or by drilling a hole in the beam to P_2, it can be applied at this point. In any case, since P_1, P_2, and P_3 lie on the line of action of $\mathbf{F}$, the principle of transmissibility states that $\mathbf{F}$ may be applied at *any* of these three points and the *external reactive forces* developed at the supports A and B will remain the same. The *internal forces* developed in the beam, however, will depend upon the location of $\mathbf{F}$. For example, if $\mathbf{F}$ acts at P_1, the internal forces in the beam have a high intensity around P_1; whereas if $\mathbf{F}$ acts at P_3, the effect of $\mathbf{F}$ on generating internal forces at P_1 will be less.

4–7. Resolution of a Force into a Force and a Couple

Before the methods of rigid-body equilibrium can be developed, it is first necessary to study what effects are caused by moving a force from one point to another on a rigid body.

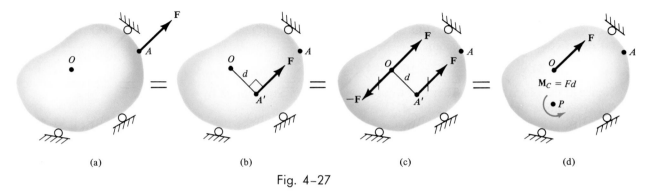

(a)　　　(b)　　　(c)　　　(d)

Fig. 4–27

Scalar Analysis. Consider moving the force **F** acting at point A on the constrained rigid body in Fig. 4–27a, to the arbitrary point O, which does *not* lie along the line of action of **F**. Using the principle of transmissibility, **F** may first be applied at point A', which lies at a perpendicular distance d from O to the line of action of the force, Fig. 4–27b. Applying equal but opposite forces **F** and $-$**F** at O, as shown in Fig. 4–27c, in no way alters the external effects on the body; however, the two forces indicated by a slash across them form a couple which has a magnitude $M_C = Fd$ and tends to rotate the body in a counterclockwise direction. Since the couple is a free vector, it may be applied at *any point P* on the body, as shown in Fig. 4–27d. In addition to this couple, **F** now acts at point O. By using this construction procedure, an equivalent system has been maintained between each of the diagrams in Fig. 4–27, as indicated by the "equal signs." In other words, when **F** acts at A, Fig. 4–27a, it will produce the *same reactions* at the three roller supports as when **F** is applied at O and a couple moment $\mathbf{M}_C$ is also applied to the body, Fig. 4–27d. From the construction, *note that the magnitude and direction of the couple moment can also be determined by taking the moment of* **F** *about O when the force is located at its original point A (or point A'). The line of action of* $\mathbf{M}_C$ *is thus perpendicular to the plane containing* **F** *and d.*

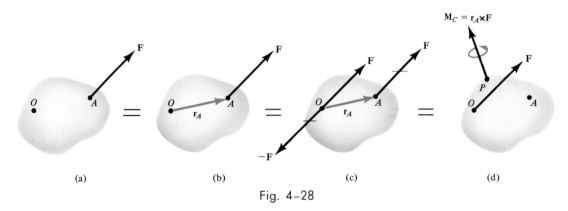

(a)　　　(b)　　　(c)　　　(d)

Fig. 4–28

Vector Analysis. For the general case, where the rigid body is viewed in three dimensions, it is often convenient to formulate the above procedure using vectors. The constructions shown in Fig. 4–28 are identical to those used for the scalar analysis, Fig. 4–27. With reference to Fig. 4–28, **F** *is moved from point A to point O and a couple moment* **M**$_C$ *is added,* where **M**$_C$ = **r**$_A$ × **F**. The position vector **r**$_A$ is drawn from *O* to *any point A* lying along the line of action of the force. According to the right-hand rule, the moment vector is perpendicular to the plane containing **F** and **r**$_A$. Furthermore, since **M**$_C$ is a free vector, it can act at *any point P* on the body.

PROCEDURE FOR ANALYSIS

The above concepts regarding the movement of a force to any point on a body may be summarized by the following two statements:

1. If the force is to be moved to a *point O located on its line of action,* by the principle of transmissibility, simply move the force to the point.
2. If the force is to be moved to a *point O that is not located on its line of action,* an equivalent system is maintained when the force is moved to point *O* and a couple moment is placed on the body. The magnitude and direction of the couple moment are determined by finding the moment of the force about point *O*.

As a physical illustration of these concepts, consider holding the end of a stick of negligible weight, Fig. 4–29. If a vertical force **F** is applied at the other end, and the stick is held in the vertical position, Fig. 4–29*a*, then, by the principle of transmissibility, the same force **F** is felt at the grip, Fig. 4–29*b*. When the stick is held in the horizontal position, Fig. 4–29*c*, the force has the effect of producing *both* a downward force **F** at the grip, and a clockwise twist, Fig. 4–29*d*. The twist can be thought of as being caused by a *couple* that is produced when **F** is moved to the grip. The couple moment has the *same* magnitude and direction as that caused by computing the moment of **F** about the grip, Fig. 4–29*c*.

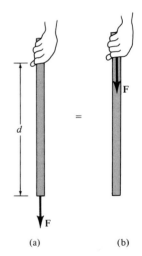

(a)   (b)

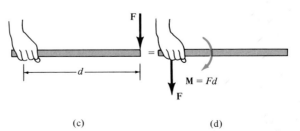

(c)   (d)

Fig. 4–29

  Example 4–9

Replace the force **F** acting on the end of the short beam segment shown in Fig. 4–30a by an equivalent force and couple system acting at point O.

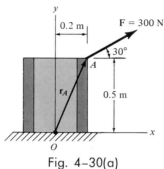

Fig. 4–30(a)

Solution I
(***Vector Analysis***). When **F** is moved to point O, a couple must be added. This couple is determined by taking the moment of **F** about point O, i.e., from Fig. 4–30a, $\mathbf{M}_O = \mathbf{r}_A \times \mathbf{F}$, where

$$\mathbf{r}_A = \{0.2\mathbf{i} + 0.5\mathbf{j}\} \text{ m}$$
$$\mathbf{F} = 300 \cos 30°\mathbf{i} + 300 \sin 30°\mathbf{j} = \{260\mathbf{i} + 150\mathbf{j}\} \text{ N} \qquad Ans.$$

Hence,

$$\mathbf{M}_O = \mathbf{r}_A \times \mathbf{F} = \begin{vmatrix} \mathbf{i} & \mathbf{j} & \mathbf{k} \\ 0.2 & 0.5 & 0 \\ 260 & 150 & 0 \end{vmatrix} = \{-100\mathbf{k}\} \text{ N} \cdot \text{m} \qquad Ans.$$

The results are shown in Fig. 4–30b.

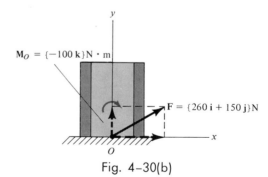

Fig. 4–30(b)

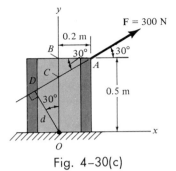

Fig. 4–30(c)

Solution II

(*Scalar Analysis*).   The moment arm d from O to $\mathbf{F}$ can be calculated by trigonometry, Fig. 4–30c. From right triangles ABC and CDO, we have

$$BC = 0.2 \tan 30° = 0.115 \text{ m}$$
$$CO = 0.5 - 0.115 = 0.385 \text{ m}$$
$$d = 0.385 \cos 30° = 0.333 \text{ m}$$

Hence, $\mathbf{F}$ is moved to O as shown in Fig. 4–30b and a couple $\mathbf{M}_O$ is added. The magnitude of $\mathbf{M}_O$ is

$$M_O = Fd = 300(0.333) = 100 \text{ N} \cdot \text{m}$$

and the direction is defined by the right-hand rule, i.e., $-\mathbf{k}$. Hence,

$$\mathbf{M}_O = \{-100\mathbf{k}\} \text{ N} \cdot \text{m} \qquad\qquad Ans.$$

The result is shown in Fig. 4–30b.

Solution III

(*Scalar Analysis Using Varignon's Theorem*).   Force $\mathbf{F}$ can be resolved into its x and y components, Fig. 4–30d, and each component can be moved to point O. The magnitude of the moment of these force components about O is easily obtained by using scalars. Their direction is obtained by the right-hand rule. From the data shown in Fig. 4–30d, assuming positive moments to be counterclockwise, i.e., $+\mathbf{k}$ direction, we have

$$\zeta + M_O = 150(0.2) - 260(0.5) = -100 \text{ N} \cdot \text{m}$$
$$\mathbf{M}_O = \{-100\mathbf{k}\} \text{ N} \cdot \text{m} \qquad\qquad Ans.$$

When the force components are added, the resultant $\mathbf{F}$ is obtained, Fig. 4–30b.

By comparison, solution III involves the least computations and is recommended when solving coplanar problems such as this one.

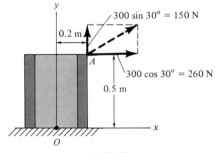

Fig. 4–30(d)

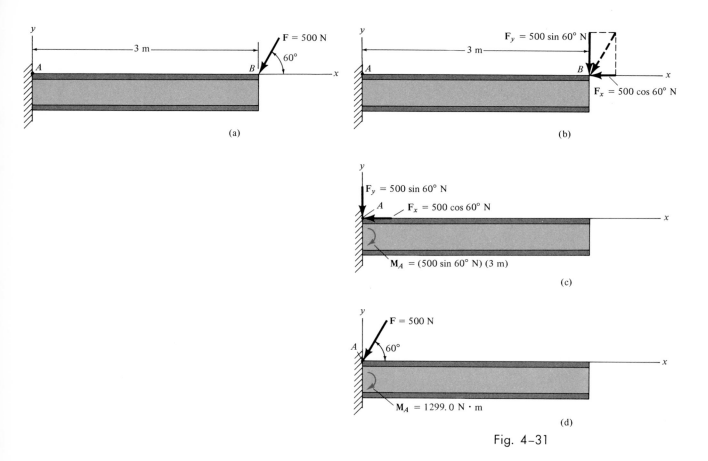

Fig. 4–31

Example 4–10

Replace the force **F**, acting at the end of the beam shown in Fig. 4–31a, by an equivalent force and couple system acting at point *A*.

Solution

Force **F** is resolved into components $\mathbf{F}_x$ and $\mathbf{F}_y$, Fig. 4–31b, and each component is moved separately to point *A*. Since the line of action of $\mathbf{F}_x$ passes through *A*, this force is simply moved to the point—principle of transmissibility, Fig. 4–31c. $\mathbf{F}_y$ is moved to *A* and a couple $\mathbf{M}_A$ is added to the beam. The moment of $\mathbf{F}_y$ about point *A* has a magnitude of

$$M_A = (500 \sin 60°)(3) = 1299.0 \text{ N} \cdot \text{m} \qquad Ans.$$

Since $\mathbf{F}_y$ tends to rotate the beam clockwise about *A*, by the right-hand rule, $\mathbf{M}_A$ acts into the page, i.e., in the $-\mathbf{k}$ direction. Adding $\mathbf{F}_x$ and $\mathbf{F}_y$ yields **F**. The results are shown in Fig. 4–31d.

Problems

*4-24. Determine the resultant couple acting on the beam.

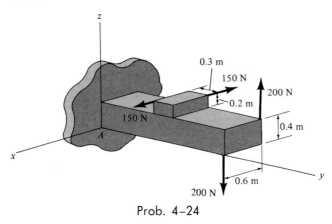

Prob. 4-24

4-25. If the resultant couple of the two couples acting on the fire hydrant is $M_R = \{-15i + 30j\}\ N \cdot m$, determine the force magnitude P. Set $F = 75\ N$, $a = 200\ mm$, and $b = 150\ mm$.

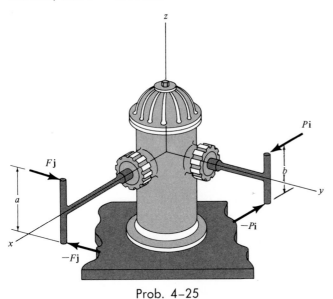

Prob. 4-25

4-25a. Solve Prob. 4-25 if $M_R = \{-22.5i + 40j\}\ lb \cdot ft$, $F = 18\ lb$, $a = 1.25\ ft$, and $b = 0.75\ ft$.

4-26. Express the moment of the couple acting on the pipe in Cartesian vector form. What is the magnitude of the couple moment?

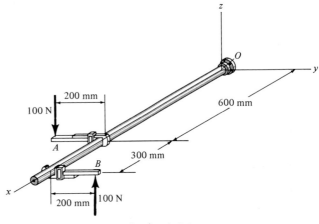

Prob. 4-26

4-27. The cord passing over the two small pegs A and B of the board is subjected to a tension of 100 N. Determine the required tension P acting on the cord that passes over the pegs C and D so that the resultant couple produced by the two couples is 15 N · m acting clockwise. The circular arc ACB has a radius of 300 mm as shown.

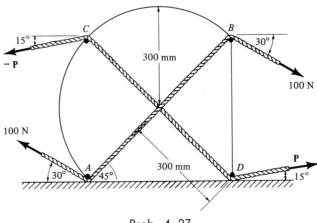

Prob. 4-27

*4-28. When the engine of the plane is running, the vertical reaction that the ground exerts at A is measured by means of a scale as 5 kN. When the engine is turned off, however, the vertical reactions at A and B are 3.75 kN each. The difference in readings at A is caused by a couple acting on the propeller when the engine is running. This couple tends to overturn the plane counterclockwise, which is opposite to the clockwise rotation of the propeller. Determine the magnitude of this couple and the magnitude of the vertical force exerted at B when the engine is running.

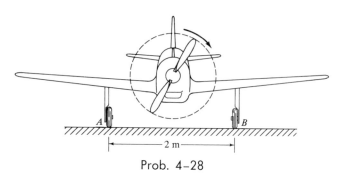

Prob. 4–28

4-29. If the resultant couple of the three couples acting on the triangular block is to be zero, determine the magnitude of forces **F** and **P**.

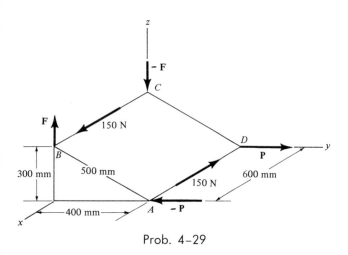

Prob. 4–29

4-30. A clockwise couple $M = 50\ \mathrm{N \cdot m}$ is resisted by the shaft of the electric motor. If $d = 150\ \mathrm{mm}$, deter-

mine the magnitude of the reactive forces **R** and $-\mathbf{R}$ which act at supports A and B so that the resultant of the two couples is zero.

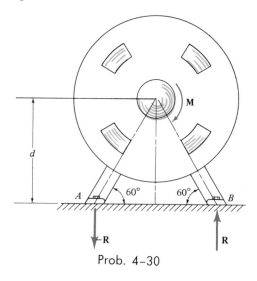

Prob. 4–30

4-30a. Solve Prob. 4–30 if $M = 60\ \mathrm{lb \cdot ft}$ and $d = 1.25\ \mathrm{ft}$.

4-31. Determine the resultant couple of the two couples that act on the shaft.

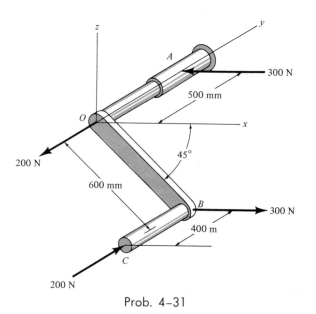

Prob. 4–31

*4-32. Replace the force **F**, having a magnitude of $F = 250$ N and acting at B, by an equivalent force and couple acting at A.

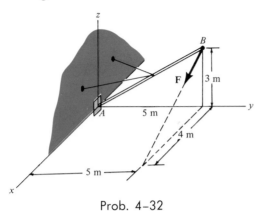

Prob. 4-32

4-33. Replace the force $\mathbf{F} = \{60\mathbf{i} - 70\mathbf{j} - 30\mathbf{k}\}$ N acting at the end of the beam by an equivalent force and couple system at point O.

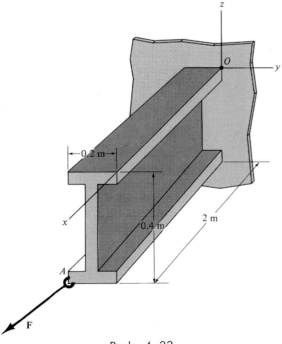

Prob. 4-33

4-34. Replace the force acting on the rigid frame by an equivalent force and couple system: (a) at point O; (b) at point B.

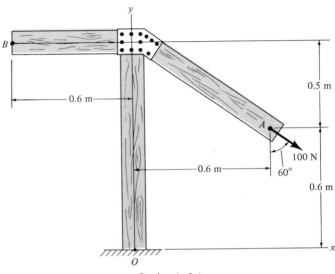

Prob. 4-34

4-35. The wheel is subjected to a force of $F = 100$ N. If $r_o = 400$ mm, $r_i = 250$ mm, determine an equivalent force and couple system at point A; repeat for point B.

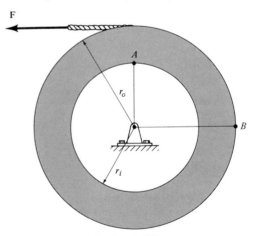

Prob. 4-35

4-35a. Solve Prob. 4-35 if $F = 60$ lb. $r_o = 2$ ft, and $r_i = 1.60$ ft.

*4-36. A force of 1.2 kN acts at the end of the beam. Replace this force by an equivalent force and couple system: (a) at point A; (b) at point O.

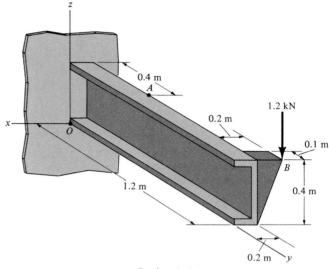

Prob. 4–36

4–8. Simplification of a Force and Couple System

It is always possible to reduce a system of forces and couples, acting on a rigid body, to a *single* resultant force acting at an arbitrary point O and a *single* resultant couple. To illustrate how this can be done, consider a rigid body of general shape and subjected to the loading shown in Fig. 4–32a. The couple moments M_{C_1} and M_{C_2} can simply be moved to point O since they are free vectors, Fig. 4–32c. Forces F_1, F_2, and F_3 are sliding vectors, and since O does *not* lie on the line of action of any of these forces, Fig. 4–32b, each force must be placed at O in accordance with the procedure outlined in the previous section. For example, when F_1 is applied at O, a corresponding couple $M_1 = r_1 \times F_1$ must *also* be applied to the body. Moving the other forces in the same manner, an *equivalent system* of concurrent forces and couples applied at O is obtained as shown in Fig. 4–32c. Using the law of vector addition, this sytem may be summed to a single resultant force and resultant couple, where

$$F_R = \Sigma F = F_1 + F_2 + F_3$$
$$M_{R_O} = \Sigma M_O = M_{C_1} + M_{C_2} + (r_1 \times F_1) + (r_2 \times F_2) + (r_3 \times F_3)$$

The results are shown in Fig. 4–32d. Note that the magnitude and direction of F_R are *independent* of the location of point O. However, M_{R_O} *depends* upon this location, since the moments of the forces are computed using position vectors r_1, r_2, and r_3.

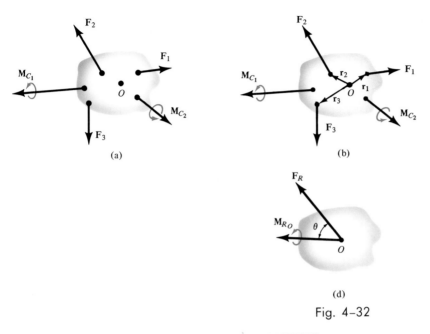
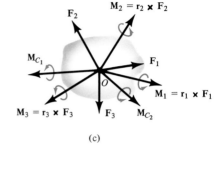

(a)

(b)

(c)

(d)

Fig. 4–32

PROCEDURE FOR ANALYSIS

The above method for simplifying any force and couple system to an equivalent single resultant force and resultant couple may be generalized by applying the following two steps:

Step 1: The resultant force $\mathbf{F}_R$ is equivalent to the vector sum of all the forces in the system, $\Sigma\mathbf{F}$; i.e.,

$$\mathbf{F}_R = \Sigma\mathbf{F} \qquad (4\text{–}14)$$

Step 2: The resultant couple $\mathbf{M}_{R_O}$ about point O is equivalent to the vector sum of all the couples in the system *plus* the vector sum of the moments about point O of all the forces in the system. This can be expressed mathematically as

$$\mathbf{M}_{R_O} = \Sigma\mathbf{M}_O \qquad (4\text{–}15)$$

In general, these equations should be applied in Cartesian vector form for solving three-dimensional problems for which the force components and the moment arms are difficult to determine; otherwise, a scalar analysis should be used.

Further simplification of $\mathbf{F}_R$ and $\mathbf{M}_{R_O}$ is possible, and this will be discussed in the next section. The following examples numerically illustrate applications of Eq. 4–14 and 4–15 using both a scalar and vector analysis.

Replace the forces acting on the pipe shown in Fig. 4–33*a* by an equivalent single force and couple system acting at point *A*.

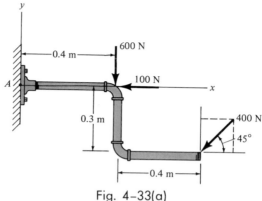

Fig. 4–33(a)

Solution

(*Scalar Analysis*).   Since this problem involves a coplanar-force system, the solution is easily obtained using a scalar analysis. Varignon's theorem will be applied to the 400-N force, whereby the moments of its two rectangular components will be considered.

Step 1: Applying Eq. 4–14 in the *x* and *y* directions, the resultant force has *x* and *y* components of

$$\xrightarrow{+} F_{R_x} = \Sigma F_x; \quad F_{R_x} = -100 - 400 \cos 45° = -382.8 \text{ N} \leftarrow$$
$$+\uparrow F_{R_y} = \Sigma F_y; \quad F_{R_y} = -600 - 400 \sin 45° = -882.8 \text{ N} \downarrow$$

As shown in Fig. 4–33*b*, $\mathbf{F}_R$ has a magnitude of

$$F_R = \sqrt{(F_{R_x})^2 + (F_{R_y})^2} = \sqrt{(382.8)^2 + (882.8)^2} = 962.2 \text{ N} \quad Ans.$$

and a direction defined by

$$\theta = \tan^{-1}\left(\frac{F_{R_y}}{F_{R_x}}\right) = \tan^{-1}\left(\frac{882.8}{382.8}\right) = 66.6° \quad Ans.$$

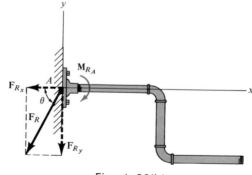

Fig. 4–33(b)

Step 2: The resultant couple moment $\mathbf{M}_{R_A}$ is determined by applying Eq. 4–15 about point A. Noting that the moment of each force in the system is perpendicular to the plane of the forces and assuming that positive moments act counterclockwise, i.e., in the $+\mathbf{k}$ direction, we have

$$\zeta + M_{R_A} = \Sigma M_A;$$
$$M_{R_A} = 100(0) - 600(0.4) - (400 \sin 45°)(0.8) - (400 \cos 45°)(0.3)$$
$$= -551 \text{ N} \cdot \text{m}$$

The negative sign indicates that $\mathbf{M}_{R_A}$ acts clockwise as shown in Fig. 4–33b.

Example 4-12

A beam is subjected to a couple moment $\mathbf{M}_C$ and forces $\mathbf{F}_1$ and $\mathbf{F}_2$ as shown in Fig. 4–34a. Replace this system by an equivalent single resultant force and couple acting at O.

Solution
(***Vector Analysis***). In order to simplify the problem of determining the force components and the moment arms, a Cartesian vector analysis will be used for the solution. Expressing the forces and couple as Cartesian vectors, we have

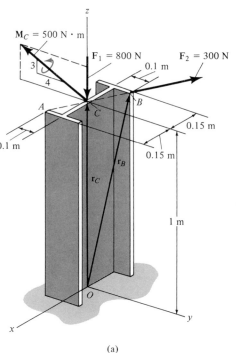

(a)

$$\mathbf{F}_1 = \{-800\mathbf{k}\} \text{ N}$$

$$\mathbf{F}_2 = (300 \text{ N})\mathbf{u}_{AB} = (300 \text{ N})\left(\frac{\mathbf{r}_{AB}}{r_{AB}}\right)$$

$$= 300\left[\frac{-0.3\mathbf{i} + 0.2\mathbf{j}}{\sqrt{(-0.3)^2 + (0.2)^2}}\right] = \{-249.6\mathbf{i} + 166.4\mathbf{j}\} \text{ N}$$

$$\mathbf{M}_C = -500(\tfrac{4}{5})\mathbf{j} + 500(\tfrac{3}{5})\mathbf{k} = \{-400\mathbf{j} + 300\mathbf{k}\} \text{ N} \cdot \text{m}$$

Step 1: Applying Eq. 4–14, yields

$$\mathbf{F}_R = \Sigma \mathbf{F}; \qquad \mathbf{F}_R = \mathbf{F}_1 + \mathbf{F}_2 = -800\mathbf{k} - 249.6\mathbf{i} + 166.4\mathbf{j}$$
$$= \{-249.6\mathbf{i} + 166.4\mathbf{j} - 800\mathbf{k}\} \text{ N} \qquad \qquad Ans.$$

Step 2: Applying Eq. 4–15,

$$\mathbf{M}_R = \Sigma \mathbf{M}_O; \quad \mathbf{M}_R = \mathbf{M}_C + \mathbf{r}_C \times \mathbf{F}_1 + \mathbf{r}_B \times \mathbf{F}_2 = (-400\mathbf{j} + 300\mathbf{k})$$

$$+ (1\mathbf{k}) \times (-800\mathbf{k}) + \begin{vmatrix} \mathbf{i} & \mathbf{j} & \mathbf{k} \\ -0.15 & 0.1 & 1 \\ -249.6 & 166.4 & 0 \end{vmatrix}$$

$$= \{-166.4\mathbf{i} - 649.6\mathbf{j} + 300\mathbf{k}\} \text{ N} \cdot \text{m} \qquad Ans.$$

These results are shown in Fig. 4–34b.

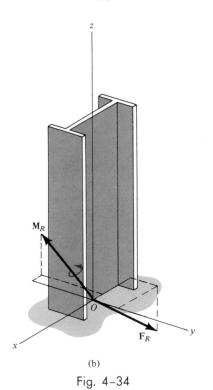

(b)

Fig. 4–34

4-9. Further Simplification of a Force and Couple System

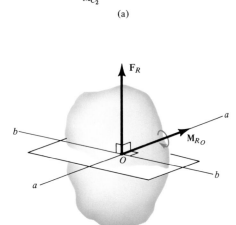

(a)

Simplification to a Single Resultant Force. Consider now a special case for which the system of forces and couples acting on a rigid body, Fig. 4-35a, reduces at point O to a resultant force $\mathbf{F}_R = \Sigma\mathbf{F}$ and resultant couple $\mathbf{M}_{R_O} = \Sigma\mathbf{M}_O$, which are *perpendicular* to one another, Fig. 4-35b. If this occurs, $\mathbf{M}_{R_O}$ can always be *eliminated* by moving $\mathbf{F}_R$ to a point P located on axis bb, which is *perpendicular* to both $\mathbf{F}_R$ and $\mathbf{M}_{R_O}$, Fig. 4-35c. Point P is chosen such that the moment arm d satisfies the scalar equation $M_{R_O} = F_R d$ or $d = M_{R_O}/F_R$. Furthermore, P is located, in this case, to the *left* of O to preserve the correct direction of $\mathbf{M}_{R_O}$ when $\mathbf{F}_R$ is moved *back* to O, Fig. 4-35b. (By the right-hand rule, $\mathbf{F}_R$, acting through P in Fig. 4-35c, tends to rotate the body about the aa axis, which is in the same direction as $\mathbf{M}_{R_O}$.)

If a system of forces is either concurrent, coplanar, or parallel, it can always be simplified, as in the above case, to a single resultant force $\mathbf{F}_R$ acting through a unique point P. This is because in each of these cases, $\mathbf{F}_R$ and $\mathbf{M}_{R_O}$ will always be perpendicular, when the force system is simplified at any point O, Fig. 4-35b.

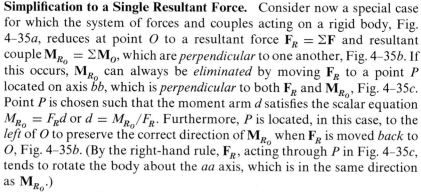

(b)

Fig. 4-35

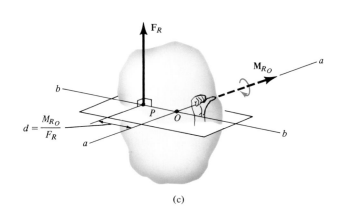

(c)

Concurrent force systems have been treated in detail in Chapter 3. Obviously, all the forces act at a point for which there is no resultant couple, so the point P is automatically specified, Fig. 4-36.

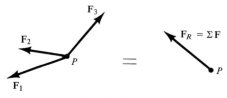

Fig. 4-36

Coplanar force systems, such as shown in Fig. 4–37a, can be simplified to a single resultant force, because when each force in the system is moved to any point O in the x-y plane, it produces a couple moment that is *perpendicular* to the plane, i.e., in the $\pm\mathbf{k}$ direction. The resultant moment $\mathbf{M}_{R_O} = \Sigma\mathbf{r} \times \mathbf{F}$ is thus perpendicular to the resultant force $\mathbf{F}_R$, Fig. 4–37b.

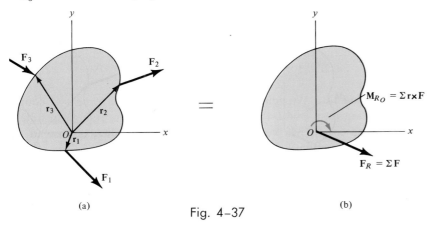

(a)

Fig. 4–37

(b)

Parallel force systems, such as shown in Fig. 4–38a, can be simplified to a single resultant force, because when each (vertical) force is moved to any point O in the x-y plane, it produces a couple that has moment components only about the x and y axes. The resultant moment $\mathbf{M}_{R_O} = \Sigma\mathbf{r} \times \mathbf{F}$ is thus perpendicular to the resultant force $\mathbf{F}_R$, Fig. 4–38b.

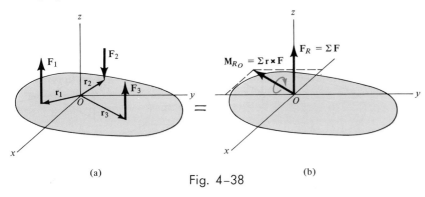

(a)

Fig. 4–38

(b)

PROCEDURE FOR ANALYSIS

The techniques used to reduce coplanar and parallel force systems to a single resultant force follow the general procedure outlined in the previous section. The simplification requires the following two steps:

Step 1: The resultant force $\mathbf{F}_R$ equals the sum of all the forces of the system, Eq. 4–14, i.e.,

$$\mathbf{F}_R = \Sigma\mathbf{F}$$

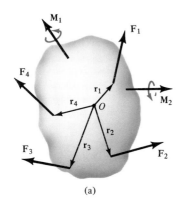

(a)

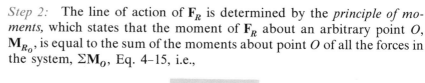

Step 2: The line of action of $\mathbf{F}_R$ is determined by the *principle of moments*, which states that the moment of $\mathbf{F}_R$ about an arbitrary point O, $\mathbf{M}_{R_O}$, is equal to the sum of the moments about point O of all the forces in the system, $\Sigma\mathbf{M}_O$, Eq. 4–15, i.e.,

$$\mathbf{M}_{R_O} = \Sigma\mathbf{M}_O$$

Most often a scalar analysis can be used to apply these equations, since the force components and the moment arms are easily determined for either coplanar or parallel force systems.

Simplification to a Wrench. In the general case, the force and couple system acting on a body, Fig. 4–39a, will simplify to a single resultant force $\mathbf{F}_R$ and couple $\mathbf{M}_{R_O}$ at O, which are *not* perpendicular. Instead, $\mathbf{F}_R$ will act at some angle θ from $\mathbf{M}_{R_O}$, Fig. 4–39b. As shown in Fig. 4–39c, however, $\mathbf{M}_{R_O}$ may be resolved into two components: one perpendicular, $\mathbf{M}_\perp$, and the other parallel, $\mathbf{M}_\parallel$, to the line of action of $\mathbf{F}_R$. Since the component $\mathbf{M}_\perp$ is *perpendicular* to $\mathbf{F}_R$, $\mathbf{M}_\perp$ may be *eliminated* by moving $\mathbf{F}_R$ to point P, as shown in Fig. 4–39d. This point lies on axis bb, which is perpendicular to both $\mathbf{M}_{R_O}$ and $\mathbf{F}_R$. The distance from P to O is $d = M_\perp/F_R$. Furthermore, when $\mathbf{F}_R$ is applied at P, the moment of $\mathbf{F}_R$ tending to cause rotation of the body *about O* is in the *same direction* as $\mathbf{M}_\perp$, Fig. 4–39d. Finally, since $\mathbf{M}_\parallel$ is a free vector, it may be moved to P so that it coincides with $\mathbf{F}_R$, Fig. 4–39e. This combination of a *collinear* force and couple is called a *wrench*. The *axis of the wrench* has the same line of action as the force. Hence, the wrench tends to cause both a translation along and a rotation about this axis. Comparing Fig. 4–39a to Fig. 4–39e, it is seen that a general force and couple system acting on a body can be simplified to a wrench. The axis of the wrench, and a point through which this axis passes, are unique and can always be determined.

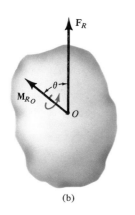

(b)

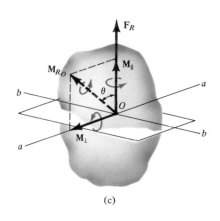

(c)

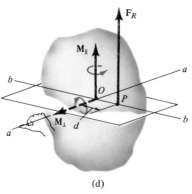

(d)

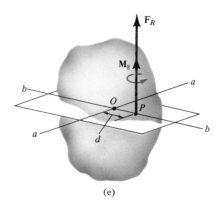

(e)

Fig. 4–39

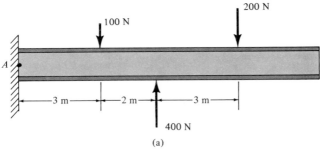

(a)

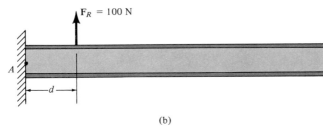

(b)

Fig. 4–40

Example 4–13

Replace the system of forces acting on the beam shown in Fig. 4–40a by an equivalent single resultant force. Specify the location of the force measured from the left end of the beam.

Solution

(*Scalar Analysis*). Problems involving a system of coplanar forces, such as this, are conveniently solved by using a scalar analysis.

Step 1:   If "positive" forces are assumed to act upward, then from Fig. 4–40a the force resultant F_R is

$$+\uparrow F_R = \Sigma F; \qquad F_R = -100 + 400 - 200 = 100 \text{ N}(\uparrow) \qquad \textit{Ans.}$$

Step 2:   The principle of moments will be applied with reference to point A. Considering counterclockwise rotations as positive, i.e., positive moment vectors are directed out of the page, then from Figs. 4–40a and 4–40b, we require

$$\zeta + M_{R_A} = \Sigma M_A;$$
$$100 \text{ N}(d) = -(100 \text{ N})(3 \text{ m}) + (400 \text{ N})(5 \text{ m}) - (200 \text{ N})(8 \text{ m})$$
$$(100)d = 100$$
$$d = 1 \text{ m} \qquad \textit{Ans.}$$

Using a clockwise sign convention would yield the same result. Since d is positive, F_R acts to the right of A as assumed in Fig. 4–40b.

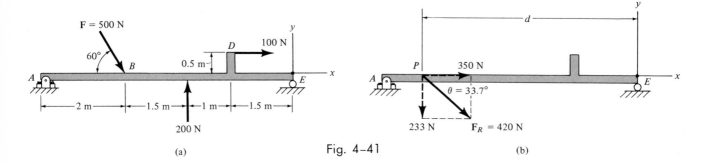

Fig. 4–41

(a)   (b)

Example 4–14

The beam AE in Fig. 4–41a is subjected to a system of coplanar forces. Determine the magnitude, the direction, and location on the beam of a single resultant force which is equivalent to the given system of forces.

Solution
(**Scalar Analysis**). *Step 1:* The origin of coordinates is located at point E as shown in Fig. 4–41a. Resolving the 500-N force into x and y components, and summing the force components, yields

$$\xrightarrow{+} F_{R_x} = \Sigma F_x; \quad F_{R_x} = 500 \cos 60° + 100 = 350 \text{ N}(\rightarrow)$$
$$+\uparrow F_{R_y} = \Sigma F_y; \quad F_{R_y} = -500 \sin 60° + 200 = -233 \text{ N}(\downarrow)$$

The magnitude and direction of the resultant force shown in Fig. 4–41b is therefore

$$F_R = \sqrt{(350)^2 + (-233)^2} = 420 \text{ N} \qquad\qquad Ans.$$
$$\theta = \tan^{-1}\left(\frac{233}{350}\right) = 33.7° \qquad\qquad Ans.$$

Step 2: The principle of moments is applied about point E. Since P lies on the x axis, $\mathbf{F}_{R_x}$ (350 N) does not create a moment about E (principle of transmissibility), only $\mathbf{F}_{R_y}$ does. Hence, from Figs. 4–41a and 4–41b, we require that

$$\lrcorner + M_{R_E} = \Sigma M_E;$$
$$233(d) = (500 \sin 60°)(4) + (50 \cos 60°)(0) - (100)(0.5) - (200)(2.5)$$
$$d = \frac{1182}{233} = 5.07 \text{ m} \qquad\qquad Ans.$$

Example 4–15

The slab in Fig. 4–42a is subjected to a series of four parallel forces. Determine the magnitude and direction of a single resultant force equivalent to the given force system and locate its point of application on the slab.

124

Solution I

(*Vector Analysis*). *Step 1:* From Fig. 4–42b, the resultant force is

$$\mathbf{F}_R = \Sigma \mathbf{F}; \qquad \mathbf{F}_R = -600\mathbf{k} + 100\mathbf{k} - 400\mathbf{k} - 500\mathbf{k}$$
$$= \{-1400\mathbf{k}\}\ \text{N} \qquad\qquad\qquad Ans.$$

Step 2: Using the principle of moments, choosing the origin as a reference for computing moments, and assuming that $\mathbf{F}_R$ acts at a point $P(x, y)$, Fig. 4–42c, we have

$$\Sigma \mathbf{M}_{R_O} = \Sigma \mathbf{M}_O;$$
$$\mathbf{r}_P \times \mathbf{F}_R = [\mathbf{r}_A \times (-600\mathbf{k})] + [\mathbf{r}_B \times (100\mathbf{k})] + [\mathbf{r}_C \times (-400\mathbf{k})]$$

or

$$(x\mathbf{i} + y\mathbf{j}) \times (-1400\mathbf{k}) = [8\mathbf{i} \times (-600\mathbf{k})] + [(6\mathbf{i} + 5\mathbf{j}) \times (100\mathbf{k})]$$
$$+ [(10\mathbf{j}) \times (-400\mathbf{k})]$$
$$-1400y\mathbf{i} + 1400x\mathbf{j} = -3500\mathbf{i} + 4200\mathbf{j}$$

The solution of this equation is obtained by equating the respective $\mathbf{i}$ and $\mathbf{j}$ components, which yields the two scalar equations

$$-1400y = -3500; \qquad\qquad y = 2.50\ \text{m} \qquad Ans.$$
$$1400x = 4200; \qquad\qquad x = 3.00\ \text{m} \qquad Ans.$$

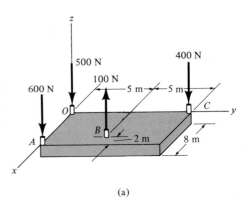

(a)

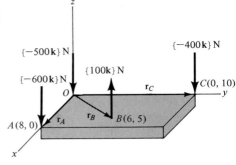

(b)

Solution II

(*Scalar Analysis*). *Step 1:* From Fig. 4–42a, the resultant force is

$$+\downarrow F_R = \Sigma F; \qquad F_R = 600 - 100 + 400 + 500$$
$$= 1400\ \text{N}(\downarrow) \qquad\qquad Ans.$$

Step 2: By the principle of moments, the moment about the x axis of the resultant force, Fig. 4–42c, is equal to the sum of the moments about the x axis created by each force in the system, Fig. 4–42a. The moment arms for each force are determined from the y coordinates. Using the right-hand rule, where positive moments act in the $+\mathbf{i}$ direction, we have

$$M_{R_x} = \Sigma M_x; \quad -(1400)y = 600(0) + (100)(5) - (400)(10) + 500(0)$$
$$-1400y = -3500 \qquad y = 2.50\ \text{m} \qquad Ans.$$

In a similar manner, assuming that positive moments act in the $+\mathbf{j}$ direction, a moment equation can be written about the y axis using moment arms defined by the x coordinates of each force.

$$M_{R_y} = \Sigma M_y; \quad 1400x = 600(8) - (100)(6) + 400(0) + 500(0)$$
$$1400x = 4200 \qquad x = 3.00\ \text{m} \qquad Ans.$$

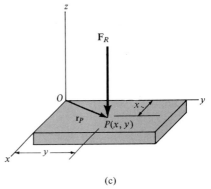

(c)

Fig. 4–42

Example 4-16

Three parallel forces act on the rim of the tube shown in Fig. 4–43*a*. If it is required that the resultant force $\mathbf{F}_R$ of the system have a line of action that coincides with the central *z* axis of the tube, determine the magnitude of $\mathbf{F}_C$ and its location θ on the rim. What is the magnitude of the resultant force $\mathbf{F}_R$?

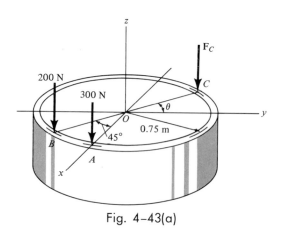

Fig. 4–43(a)

Solution

(***Vector Analysis***). *Step 1:*   From Fig. 4–43*b*, the force resultant $\mathbf{F}_R$ is

$$\mathbf{F}_R = \Sigma\mathbf{F}; \qquad -F_R\mathbf{k} = -300\mathbf{k} - 200\mathbf{k} - F_C\mathbf{k}$$
$$F_R = 500 + F_C \tag{1}$$

Step 2:   The location of $\mathbf{F}_C$ is determined from the principle of moments applied at point *O*. Since $\mathbf{F}_R$ passes through *O*, Fig. 4–43*b*, the moment of $\mathbf{F}_R$ about *O* is zero. Hence, we require that

$$\mathbf{M}_{R_O} = \Sigma\mathbf{M}_O; \qquad \mathbf{0} = \mathbf{r}_A \times (-300\mathbf{k}) + \mathbf{r}_B \times (-200\mathbf{k}) + \mathbf{r}_C \times (-F_C\mathbf{k})$$

or

$$\mathbf{0} = (0.75\mathbf{i}) \times (-300\mathbf{k}) + (0.75 \cos 45°\mathbf{i} - 0.75 \sin 45°\mathbf{j}) \times (-200\mathbf{k})$$
$$+ (-0.75 \sin \theta \, \mathbf{i} + 0.75 \cos \theta \, \mathbf{j}) \times (-F_C\mathbf{k})$$

Evaluating the cross products, we have

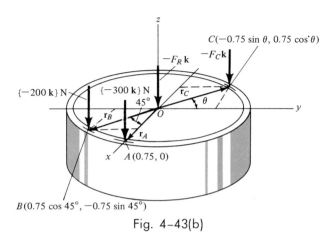

C(-0.75 sin θ, 0.75 cos'θ)

Fig. 4-43(b)

$$0 = \begin{vmatrix} \mathbf{i} & \mathbf{j} & \mathbf{k} \\ 0.75 & 0 & 0 \\ 0 & 0 & -300 \end{vmatrix} + \begin{vmatrix} \mathbf{i} & \mathbf{j} & \mathbf{k} \\ 0.530 & -0.530 & 0 \\ 0 & 0 & -200 \end{vmatrix}$$

$$+ \begin{vmatrix} \mathbf{i} & \mathbf{j} & \mathbf{k} \\ -0.75 \sin\theta & 0.75 \cos\theta & 0 \\ 0 & 0 & -F_C \end{vmatrix}$$

$$= 225\mathbf{j} + 106\mathbf{i} + 106\mathbf{j} - 0.75 \cos\theta(F_C)\mathbf{i} - 0.75 \sin\theta(F_C)\mathbf{j}$$

In order to satisfy this equation the respective **i** and **j** components must be equal to zero, i.e.,

$$0 = 106 - 0.75 \cos\theta(F_C) \tag{2}$$
$$0 = 331 - 0.75 \sin\theta(F_C) \tag{3}$$

Solving Eq. (2) for F_C and substituting into Eq. (3) yields

$$\frac{\sin\theta}{\cos\theta} = \tan\theta = \frac{331}{106}$$

$$\theta = \tan^{-1}\left(\frac{331}{106}\right) = 72.2° \qquad\qquad Ans.$$

Using this value of θ in Eq. (2) or (3) and solving for F_C, we have

$$F_C = 463.4 \text{ N} \qquad\qquad Ans.$$

The magnitude of $\mathbf{F}_R$ is obtained from Eq. (1), which yields

$$F_R = 963.4 \text{ N} \qquad\qquad Ans.$$

As a review, solve this problem using a scalar analysis, similar to that given in Example 4–15 and show that one obtains Eqs. (2) and (3) by applying the principle of moments about the x and y axis, respectively.

Problems

4-37. Replace the force and couple acting at the end of the beam by an equivalent force and couple system at point *A*.

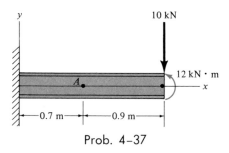

Prob. 4–37

4-38. Replace the loading system acting on the beam by an equivalent force and couple system at point *O*.

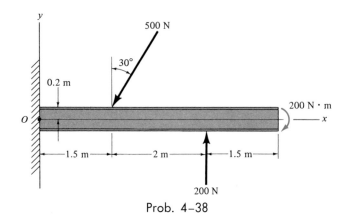

Prob. 4–38

4-39. Replace the system of three forces shown acting on the beam's cross section by an equivalent force and couple system acting at point *A*.

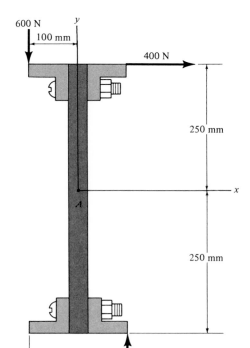

Prob. 4–39

*4–40. Replace the three forces $F_1 = 800$ N, $F_2 = 3$ kN, and $F_3 = 600$ N, acting on the column, by an equivalent resultant force and couple system at point O. Set $d = 0.4$ m and $a = 0.75$ m.

4–41. Replace the loading acting at the end of the cantilever shaft by an equivalent force and couple system at point O.

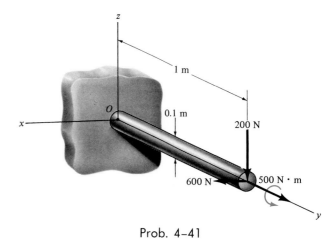

Prob. 4–41

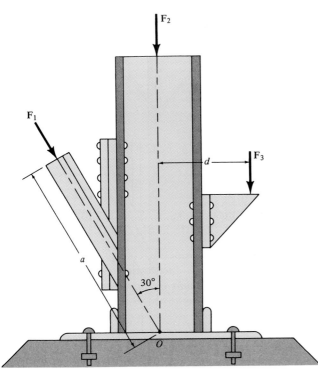

Prob. 4–40

*4–40a. Solve Prob. 4–40 if $F_1 = 200$ lb, $F_2 = 800$ lb, $F_3 = 300$ lb, $d = 1.75$ ft, and $a = 2$ ft.

4–42. The ropes AB and CD each exert tension forces of $F_D = F_B = 100$ N on the wing of the plane. Replace these forces by an equivalent force and couple system at point O.

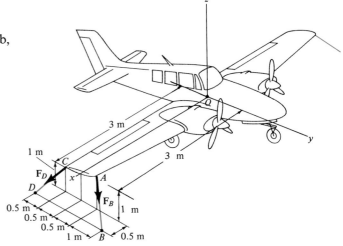

Prob. 4–42

4-43. Replace the force system acting on the plate by an equivalent single resultant force. Specify the location of this force.

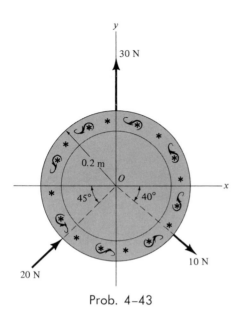

Prob. 4-43

*4-44. The system of parallel forces acts on the top surface of the *Warren girder*. Determine the resultant force of the system and specify its location, measured from the 1 kN force acting at end *A*.

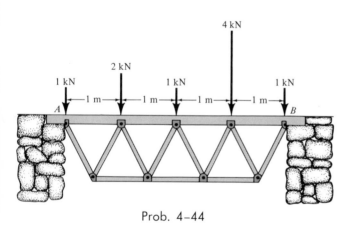

Prob. 4-44

4-45. The system of four parallel forces, $F_1 = 1$ kN, $F_2 = 2$ kN, $F_3 = 1.5$ kN, and $F_4 = 0.75$ kN, acts on the

roof truss. If $a = 1.25$ m, determine the resultant force and specify its location along *AB*, measured from point *A*.

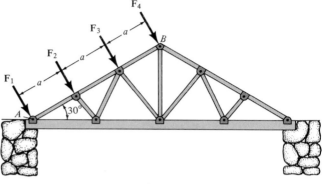

Prob. 4-45

4-45a. Solve Prob. 4-45 if $F_1 = 150$ lb, $F_2 = 375$ lb, $F_3 = 250$ lb, $F_4 = 100$ lb, and $a = 4$ ft.

4-46. The reinforced concrete column supports the four parallel forces shown. Determine the magnitudes of forces F_C and F_D acting at C and D so that the resultant force of the system acts through the midpoint O of the column.

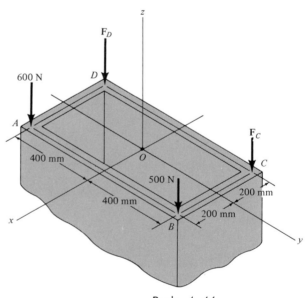

Prob. 4-46

4-47. Three parallel column forces act on the circular slab. Determine the resultant force, and specify its location (x, y) on the slab.

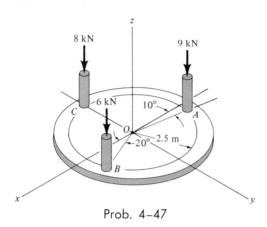

Prob. 4–47

4-49. The pole is stabilized in the vertical position by tension forces developed in three guy cables. If cables AB and CD sustain a force of 800 N and 600 N, respectively, determine the required location $P(x, y)$ of cable AP and the magnitude of force $\mathbf{T}$ developed in it so that the tension in all three cables creates a single resultant force of $\mathbf{F}_R = \{-216\mathbf{i} - 1800\mathbf{k}\}$ N acting at O.

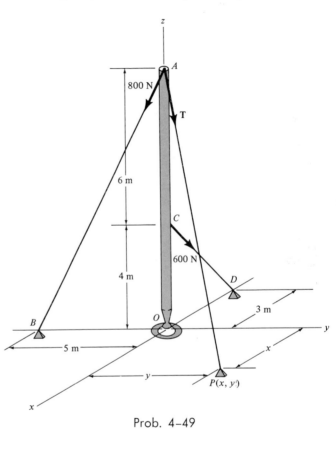

Prob. 4–49

***4-48.** Replace the couple and force acting on the pipe assembly by an equivalent force and couple system at point O.

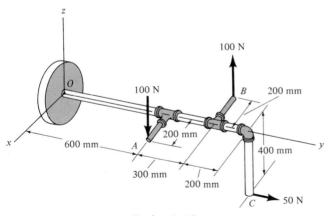

Prob. 4–48

4-50. Replace the two wrenches $F_1 = 300$ N, $M_1 = 100$ N $\cdot$ m, and $F_2 = 200$ N, $M_2 = 180$ N $\cdot$ m, and the force $F_3 = 100$ N, acting on the pipe assembly, by an equivalent force and couple system at point O. Set $a = 0.5$ m, $b = 0.6$ m, and $c = 0.8$ m.

4-51. The pipe assembly is subjected to the action of a wrench at B and a couple at A. Simplify this system to a single resultant wrench and specify the location of the wrench along the axis of pipe CD measured from point C.

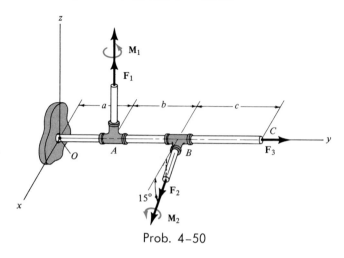

Prob. 4-50

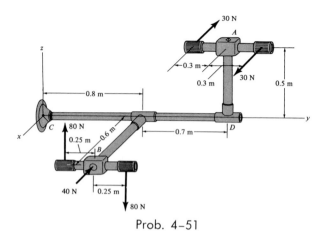

Prob. 4-51

4-50a. Solve Prob. 4-50 if $F_1 = 60$ lb, $M_1 = 40$ lb $\cdot$ ft, $F_2 = 80$ lb, $M_2 = 30$ lb $\cdot$ ft, $F_3 = 25$ lb, $a = 1$ ft, $b = 1.25$ ft, and $c = 1.50$ ft.

4-10. Reduction of a Simple Distributed Loading

In many situations a very large surface area of a body may be subjected to *distributed loadings,* such as those caused by wind, fluids, or simply the weight of material supported over the body's surface. The magnitude of the *intensity* of these loadings at each point is defined as a force per unit area or *pressure* which can be measured in units of pascals (Pa), where 1 pascal equals the pressure created by a force of 1 newton acting over a surface area of 1 m²; i.e., 1 Pa = 1 N/m². This unit is rather small for general engineering use, so prefixes are often used to represent more "realistic" magnitudes.*

Distributed Loading in Three Dimensions. The direction of the intensity of a distributed load is indicated by arrows shown on a *load-intensity diagram.* For example, the load distribution or pressure acting on the plate

---

*Although strictly not part of the SI system of units, the *bar* (b), which is 10^5 Pa, is sometimes used for pressure measurements.

in Fig. 4–44a is described by the loading function $p = p(x, y)$ Pa. Knowing this function, it is possible to determine the *magnitude* of force $d\mathbf{F}$ acting on the differential area dA m² of the plate, located at the arbitrary point (x, y). This force magnitude is simply $dF = [p(x, y)\text{N}/\text{m}^2](dA \text{ m}^2) = dV$ N, where dV represents the colored differential *volume element* shown in Fig. 4–44a.

The entire loading on the plate is therefore represented as a system of *parallel forces* infinite in number and each acting on separate differential areas dA. Using the methods of the previous section, this system of parallel forces can be simplified to a single resultant force $\mathbf{F}_R$ acting through a unique point $(\bar{x}, \bar{y})$, Fig. 4–44b.

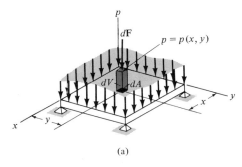

(a)

Magnitude of Resultant Force. To determine the *magnitude* of $\mathbf{F}_R$, it is necessary to sum each of the differential forces $d\mathbf{F}$ or volume elements dV, acting over the plate's *entire surface area A*. Applying Eq. 4–14 this sum may be expressed mathematically as an integral,

$$F_R = \Sigma F; \qquad F_R = \int_A p(x, y) \, dA = \int_V dV = V \qquad (4\text{–}16)$$

The result indicates that the *magnitude of the resultant force is equal to the total volume under the distributed-loading diagram.*

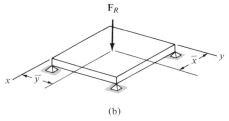

(b)

Fig. 4–44

Location of Resultant Force. The location, $(\bar{x}, \bar{y})$ of $\mathbf{F}_R$ is determined by using the principle of moments. For example, to determine $\bar{x}$, it is necessary that the moment produced by $\mathbf{F}_R$ about the y axis, $M_{R_y} = \bar{x}F_R$ (Fig. 4–44b), equal the sum of the moments of all forces $d\mathbf{F}$ about the y axis. Since $d\mathbf{F}$ produces a moment that has a magnitude of $dM_y = x \, dF = x \, dV$ (Fig. 4–44a), then applying Eq. 4–15 to the loading on the entire plate, we have

$$M_{R_y} = \Sigma M_y; \qquad \bar{x}F_R = \int_V x \, dV \qquad (4\text{–}17)$$

Solving for $\bar{x}$, using Eq. 4–16, we can write

$$\bar{x} = \frac{\displaystyle\int_V x \, dV}{\displaystyle\int_V dV} \qquad (4\text{–}18)$$

In a similar manner, applying the principle of moments about the x axis, $M_{R_x} = \Sigma M_x$, yields

$$\bar{y} = \frac{\displaystyle\int_V y \, dV}{\displaystyle\int_V dV} \qquad (4\text{–}19)$$

133

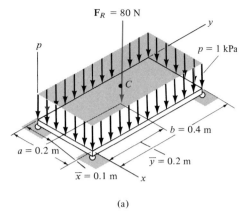

(a)

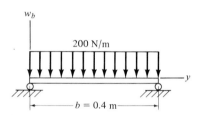

(b)

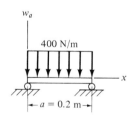

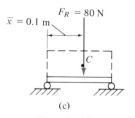

(c)

Fig. 4–45

In Chapter 9, it is shown that $(\bar{x}, \bar{y})$, as expressed by Eqs. 4–18 and 4–19, represents the x and y coordinates for the geometric center or *centroid* of the *volume* under the distributed-loading diagram $p(x, y)$. In other words, *the resultant force has a line of action that passes through the centroid of the volume defined by the distributed-loading diagram $p(x, y)$*.

A detailed treatment for locating the centroid of volumes by integration is given in Chapter 9. In many cases, however, the distributed-loading diagram is of a simple geometric form so that the results of integration can be *tabulated*. A representative tabulation that can be used to determine the centroids of common geometric shapes is given in Appendix B. Hence, if the volume shape of the loading diagram is contained in Appendix B, both the volume and centroid can be obtained directly from this table without the need for integration.

To illustrate the techniques used to reduce a distributed loading to a single resultant force, consider the rectangular plate shown in Fig. 4–45a, which is subjected to a uniform load of $p = 1$ kPa. The *magnitude of the resultant force* of this distributed loading is equal to the *volume* of the rectangular "block,"

$$F_R = V = (1000 \text{ N/m}^2)(0.2 \text{ m})(0.4 \text{ m}) = 80 \text{ N}$$

The line of action of this force acts through the centroid C or geometric center of the "block" and therefore is located on the plate at point

$$\bar{x} = 0.1 \text{ m} \qquad \bar{y} = 0.2 \text{ m}$$

Distributed Loading in Two Dimensions. If the loading function is uniform along any axis it can be simplified geometrically and thereby represented as an *area* rather than a volume. For example, in Fig. 4–46a, the plate loading $p = p(x)$ Pa is uniform along the y axis and, therefore, multiplying by the *width* a of the plate, we obtain $w = (p(x) \text{ N/m}^2)(a \text{ m}) = w(x) \text{ N/m}$, shown in Fig. 4–46b. The loading function $w = w(x)$ is a measure of force per unit length rather than force per unit area, and consequently the load-intensity diagram for $w = w(x)$ represents a system of *coplanar parallel forces*, Fig. 4–46b. Each differential force $d\mathbf{F}$ acts over an element of length dx such that at any point x, $dF = (w(x) \text{ N/m})(dx \text{ m}) = dA \text{ N}$. In other words, the magnitude of $d\mathbf{F}$ is represented by the colored differential area dA under the loading curve.

Magnitude of Resultant Force. In accordance with Eq. 4–14, the magnitude of the resultant force $\mathbf{F}_R$ supported by the plate is determined by integration or summing of the areas dA over the entire plate length b, i.e.,

$$F_R = \Sigma F; \qquad F_R = \int_b w(x) \, dx = \int_A dA = A \qquad (4\text{–}20)$$

This result indicates that *the magnitude of the resultant force is equal to the area under the loading diagram*.

Location of Resultant Force. The location $\bar{x}$ of the line of action of $\mathbf{F}_R$ is determined by applying the principle of moments, which requires that the moment of $\mathbf{F}_R$ about point O, Fig. 4–46c, be equal to the sum of the moments of all differential forces $d\mathbf{F}$ about O, Fig. 4–46b. Since $d\mathbf{F}$ produces a moment of $x\,dF = x\,dA$ about O, then for the entire plate,

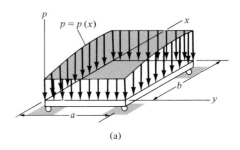

(a)

$$M_{R_O} = \Sigma M_O; \qquad \bar{x}F_R = \int_A x\,dA$$

Solving for $\bar{x}$, using Eq. 4–20, we can write

$$\bar{x} = \frac{\displaystyle\int_A x\,dA}{\displaystyle\int_A dA} \qquad (4\text{–}21)$$

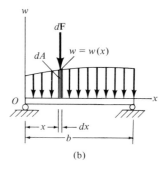

(b)

In Chapter 9, it is shown that this equation determines the $\bar{x}$ coordinate for the geometric center or centroid of the *area* under the distributed-loading diagram $w(x)$. It may therefore be concluded that *the resultant force has a line of action which passes through the centroid C of the area defined by the distributed-loading diagram* $w(x)$, Fig. 4–46c. Once $\bar{x}$ is determined, by symmetry, $\mathbf{F}_R$ passes through point $(\bar{x}, a/2)$ on the surface of the plate, Fig. 4–46d. Detailed treatment of the integration techniques for computing the centroids of areas is given in Chapter 9. In many cases, however, the distributed-loading diagram is of a simple geometric form, such as a rectangle, triangle, or trapezoid. The centroids for such common area shapes can be obtained directly from the tabulation in Appendix B.

The above concepts may be illustrated numerically by considering again the uniformly loaded plate of Fig. 4–45a. Since the loading is uniform in two directions, the loading function $p(x, y) = 1$ kPa may either be multiplied by the *width* of the plate to obtain the distribution along the length, i.e., $w_b = (1000 \text{ N/m}^2)(0.2 \text{ m}) = 200 \text{ N/m}$, Fig. 4–45b; or multiplied by the length to obtain the distribution along the width, i.e., $w_a = (1000 \text{ N/m}^2)(0.4 \text{ m}) = 400 \text{ N/m}$, Fig. 4–45c. In both cases the magnitude of $\mathbf{F}_R$ is equal to the *area* under the rectangular-loading diagram.

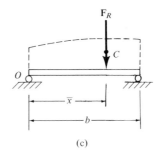

(c)

$$F_R = A = (200 \text{ N/m})(0.4 \text{ m}) = (400 \text{ N/m})(0.2 \text{ m}) = 80 \text{ N}$$

The *centroids* of each rectangle are at the midpoint of the area, so that

$$\bar{x} = 0.1 \text{ m} \qquad \bar{y} = 0.2 \text{ m}$$

These results are shown in Figs. 4–45b and 4–45c and compare with those of Fig. 4–45a.

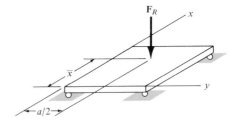

(d)

Fig. 4–46

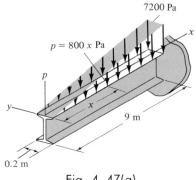

Fig. 4–47(a)

Example 4–17

A distributed loading of $p = 800x$ Pa acts over the top surface of the beam shown in Fig. 4–47a. Determine the magnitude and location of the resultant force.

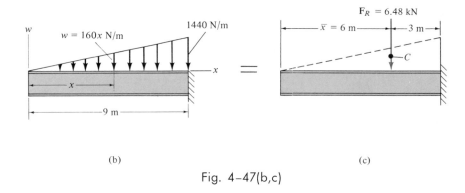

Fig. 4–47(b,c)

Solution

The loading function $p = 800x$ Pa indicates that the load intensity varies uniformly from $p = 0$ at $x = 0$ to $p = 7200$ Pa at $x = 9$ m. Since the intensity is uniform along the width of the beam, the loading may be viewed in two dimensions as shown in Fig. 4–47b. Here

$$w = (800x \text{ N/m}^2)(0.2 \text{ m})$$
$$= (160x) \text{ N/m}$$

At $x = 9$ m, note that $w = 1440$ N/m.

The *area A* of the *triangle* defined by the loading function $w(x)$ represents the *magnitude of the resultant force* $\mathbf{F}_R$. From Appendix B, $A = \frac{1}{2}bh$, so that

$$F_R = \tfrac{1}{2}(9 \text{ m})(1440 \text{ N/m}) = 6480 \text{ N} = 6.48 \text{ kN} \qquad \textit{Ans.}$$

The *line of action* of $\mathbf{F}_R$ passes through the *centroid* C of the triangle. From Appendix B, this point lies at a distance of one third of the length of the beam, measured from the right side. Hence,

$$\bar{x} = 9 \text{ m} - \tfrac{1}{3}(9 \text{ m}) = 6 \text{ m} \qquad\qquad Ans.$$

The results are shown in Fig. 4–47c.

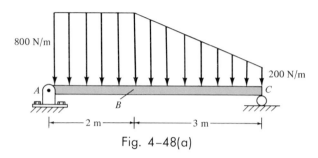

Fig. 4–48(a)

Example 4–18

Determine the magnitude and location of the resultant of the distributed load acting on the beam shown in Fig. 4–48a.

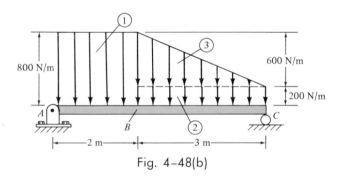

Fig. 4–48(b)

Solution

The distributed loading on the beam can be considered in three parts, Fig. 4–48b:
(1) a rectangular portion from A to B having an intensity of 800 N/m;
(2) a rectangular portion from B to C having an intensity of 200 N/m;
(3) a triangular portion from B to C having an intensity of $(800 \text{ N/m} - 200 \text{ N/m}) = 600 \text{ N/m}$.

The magnitude of the force represented by each piece is equal to its associated area, Fig. 4–48b. Hence,

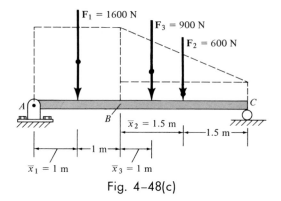

Fig. 4–48(c)

$$F_1 = 800 \text{ N/m}(2 \text{ m}) = 1600 \text{ N}$$
$$F_2 = 200 \text{ N/m}(3 \text{ m}) = 600 \text{ N}$$
$$F_3 = \tfrac{1}{2}(600 \text{ N/m})(3 \text{ m}) = 900 \text{ N}$$

The lines of action of each of these parallel forces act through the centroid of their associated areas, Fig. 4–48c, and therefore intersect the beam at

$$\bar{x}_1 = \tfrac{1}{2}(2 \text{ m}) = 1 \text{ m}$$
$$\bar{x}_2 = \tfrac{1}{2}(3 \text{ m}) = 1.5 \text{ m}$$
$$\bar{x}_3 = \tfrac{1}{3}(3 \text{ m}) = 1 \text{ m}$$

The magnitude of the resultant force $\mathbf{F}_R$ acting on the beam is thus

$$+\downarrow F_R = \Sigma F; \qquad F_R = 1600 + 900 + 600 = 3100 \text{ N} \qquad\qquad Ans.$$

Applying the principle of moments with reference to point A, Figs. 4–48c and 4–48d, we have

$$\curvearrowright + M_{R_A} = \Sigma M_A; \quad d(3100) = 1(1600) + (2 + 1.5)(600) + (2 + 1)(900)$$
$$d = 2.06 \text{ m} \qquad\qquad Ans.$$

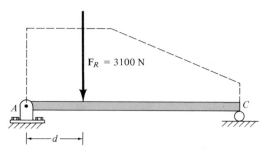

Fig. 4–48(d)

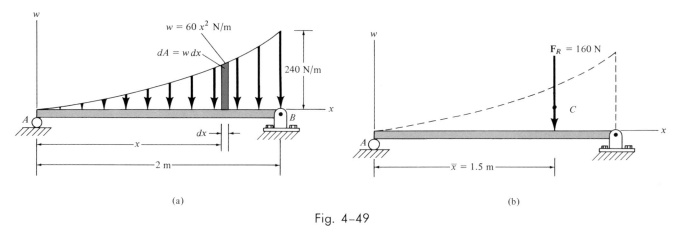

Fig. 4-49

Example 4-19

Determine the magnitude and location of the resultant force acting on the beam in Fig. 4-49a.

Solution

As shown in Fig. 4-49a, the colored differential area element $dA = w\,dx = 60x^2\,dx$ will be used for integration. Applying Eq. 4-20, by summing these elements from $x = 0$ to $x = 2$ m, the resultant force $\mathbf{F}_R$, Fig. 4-49b, is

$$F_R = \Sigma F;$$

$$F_R = \int_A dA = \int_0^2 60x^2\,dx = 60\left[\frac{x^3}{3}\right]_0^2 = 60\left[\frac{2^3}{3} - \frac{0^3}{3}\right]$$

$$= 160\text{ N} \qquad\qquad Ans.$$

Since the element of area dA is located an arbitrary distance x from A, Fig. 4-49a, the location $\bar{x}$ of $\mathbf{F}_R$, *measured from A*, Fig. 4-49b, is determined from Eq. 4-21.

$$\bar{x} = \frac{\int_A x\,dA}{\int_A dA} = \frac{\int_0^2 x(60\,x^2)\,dx}{160} = \frac{60\left[\frac{x^4}{4}\right]_0^2}{160} = \frac{60\left[\frac{2^4}{4} - \frac{0^4}{4}\right]}{160}$$

$$= 1.5\text{ m} \qquad\qquad Ans.$$

From the analysis it may be concluded that $\mathbf{F}_R$ is equal to the area under the distributed-load diagram, and the line of action of $\mathbf{F}_R$ passes through the centroid C of this area, Fig. 4-49b. These results may be checked by using Appendix B, where it is shown that for a semiparabolic area of height a, length b, and shape shown in Fig. 4-49a,

$$A = \frac{ab}{3} = \frac{240(2)}{3} = 160\text{ N} \quad\text{and}\quad \bar{x} = \frac{3}{4}b = \frac{3}{4}(2) = 1.5\text{ m}$$

139

Problems

*4–52. Determine the magnitude and direction of the resultant of the distributed load acting on the beam.

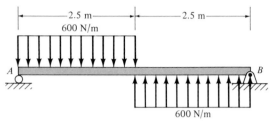

Prob. 4–52

4–53. Determine the magnitude and location of the resultant force acting on the beam.

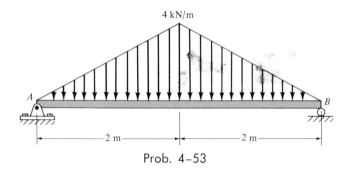

Prob. 4–53

4–54. Replace the loading system acting on the beam by a single resultant force. Specify the location of the resultant, measured from point A.

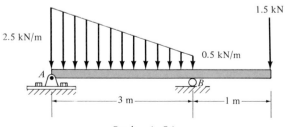

Prob. 4–54

4–55. The lifting force acting across the span of the wing on an airplane has a semiparabolic distribution that is symmetrical about the w axis. Replace this loading by an equivalent force and couple system at A. Set $w_o = 1.5$ kN/m and $L = 8$ m.

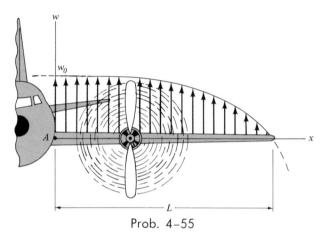

Prob. 4–55

4–55a. Solve Prob. 4–55 if $w_o = 1500$ lb/ft and $L = 52$ ft.

*4-56. The footing of the retaining wall is subjected to a soil loading that varies in a triangular fashion on the bottom and uniformly on the top. Replace this loading by an equivalent force and couple system at point O.

4-58. Determine the magnitude and location of the resultant force acting on the beam.

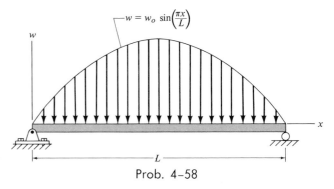

Prob. 4-58

4-59. The post supports two signs, each of which is subjected to a uniform wind load. Replace this loading by an equivalent force and couple system at point O.

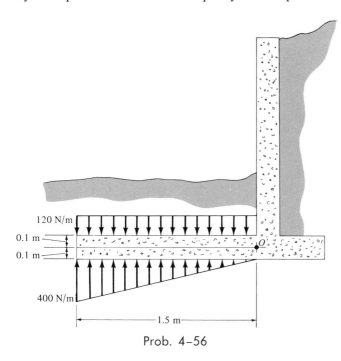

Prob. 4-56

4-57. Determine the equivalent force and couple system at point O on the cantilever beam.

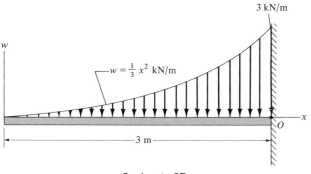

Prob. 4-57

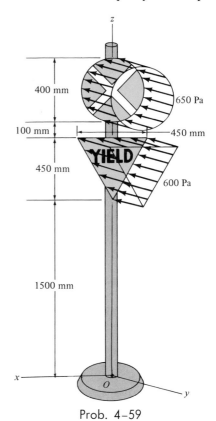

Prob. 4-59

*4-60. Wind blowing across a roof creates a triangular suction pressure that acts perpendicular to the roof on the leeward side. Determine the magnitude and location of the resultant force of this pressure distribution if $p_o = 100$ Pa, $l = 8$ m, and $a = 4.5$ m.

4-61. The wind pressure acting on a triangular sign is uniform. Replace this loading by an equivalent force and couple system at point O.

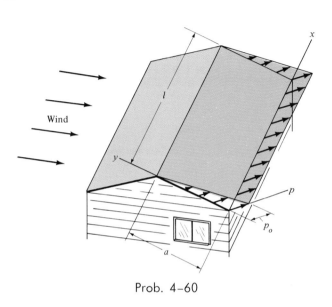

Prob. 4-60

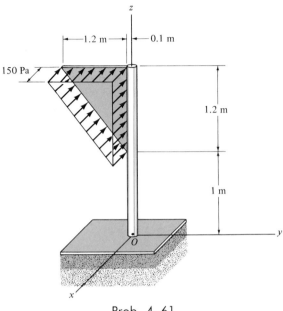

Prob. 4-61

*4-60a. Solve Prob. 4-60 if $p_o = 4$ lb/ft^2, $l = 30$ ft, and $a = 10$ ft.

4-62. Determine the length b of the uniform load and its position a on the beam such that the resultant force and resultant couple acting on the beam are zero.

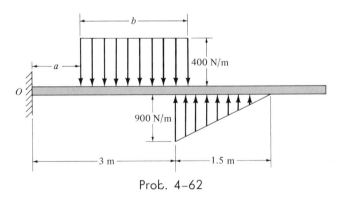

Prob. 4-62

4-63. The beam supports 40 sandbags, each having a mass of 50 kg. Determine the resultant weight of the bags, and specify its location on the beam, measured from point *A*. Each bag has a length of 0.75 m.

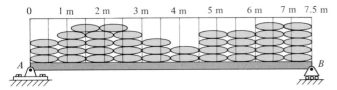

Prob. 4–63

4–11. Dot Product

The *dot product* of two vectors **A** and **B,** written **A · B,** read "**A** dot **B,**" is defined as the product of the magnitude of vectors **A** and **B** and the cosine of the angle between them. Expressed in equation form,

$$\mathbf{A} \cdot \mathbf{B} = AB \cos \theta \qquad (4\text{–}22)$$

where $0° \leqslant \theta \leqslant 180°$. The dot product is often referred to as the *scalar product* of vectors, since, unlike the cross product, the result is a *scalar* and not a vector. If **A · B** is *zero,* then $AB \cos \theta = 0$. Provided **A** and **B** are not equal to zero, then $\cos \theta = 0$, or $\theta = 90°$, in which case **A** is *perpendicular* to **B.**

Laws of Operation

1. Commutative law:

$$\mathbf{A} \cdot \mathbf{B} = \mathbf{B} \cdot \mathbf{A} \qquad (4\text{–}23)$$

2. Multiplication by a scalar:

$$m(\mathbf{A} \cdot \mathbf{B}) = (m\mathbf{A}) \cdot \mathbf{B} = \mathbf{A} \cdot (m\mathbf{B}) = (\mathbf{A} \cdot \mathbf{B})m \qquad (4\text{–}24)$$

3. Distributive law:

$$\mathbf{A} \cdot (\mathbf{B} + \mathbf{D}) = (\mathbf{A} \cdot \mathbf{B}) + (\mathbf{A} \cdot \mathbf{D}) \qquad (4\text{–}25)$$

Equations 4–23 and 4–24 can be easily proven by using the definition of the dot product, Eq. 4–22. The proof of Eq. 4–25 is left as an exercise (see Prob. 4–65).

Cartesian Vector Formulation.   Equation 4–22 may be used to find the dot product of each of the Cartesian unit vectors. For example, $\mathbf{i} \cdot \mathbf{i} = (1)(1) \cos 0° = 1$. In a similar manner,

$$\mathbf{i} \cdot \mathbf{i} = 1 \qquad \mathbf{i} \cdot \mathbf{j} = 0 \qquad \mathbf{i} \cdot \mathbf{k} = 0$$
$$\mathbf{j} \cdot \mathbf{j} = 1 \qquad \mathbf{j} \cdot \mathbf{k} = 0 \qquad \mathbf{j} \cdot \mathbf{i} = 0 \qquad (4\text{-}26)$$
$$\mathbf{k} \cdot \mathbf{k} = 1 \qquad \mathbf{k} \cdot \mathbf{i} = 0 \qquad \mathbf{k} \cdot \mathbf{j} = 0$$

These results should not be memorized; rather, it should be clearly understood how each is obtained using Eq. 4–22.

Consider now the dot product of two general vectors **A** and **B** which are expressed in Cartesian vector form as $\mathbf{A} = A_x\mathbf{i} + A_y\mathbf{j} + A_z\mathbf{k}$ and $\mathbf{B} = B_x\mathbf{i} + B_y\mathbf{j} + B_z\mathbf{k}$. Then

$$\mathbf{A} \cdot \mathbf{B} = (A_x\mathbf{i} + A_y\mathbf{j} + A_z\mathbf{k}) \cdot (B_x\mathbf{i} + B_y\mathbf{j} + B_z\mathbf{k})$$

Using the scalar and distributive laws of operation, Eqs. 4–24 and 4–25, we have

$$\mathbf{A} \cdot \mathbf{B} = A_x B_x(\mathbf{i} \cdot \mathbf{i}) + A_x B_y(\mathbf{i} \cdot \mathbf{j}) + A_x B_z(\mathbf{i} \cdot \mathbf{k})$$
$$+ A_y B_x(\mathbf{j} \cdot \mathbf{i}) + A_y B_y(\mathbf{j} \cdot \mathbf{j}) + A_y B_z(\mathbf{j} \cdot \mathbf{k})$$
$$+ A_z B_x(\mathbf{k} \cdot \mathbf{i}) + A_z B_y(\mathbf{k} \cdot \mathbf{j}) + A_z B_z(\mathbf{k} \cdot \mathbf{k})$$

In accordance with Eqs. 4–26, the final result becomes

$$\mathbf{A} \cdot \mathbf{B} = A_x B_x + A_y B_y + A_z B_z \qquad (4\text{-}27)$$

Thus, to determine the dot product of two Cartesian vectors, multiply the corresponding x, y, and z components together and sum their products.

Applications.   The dot product has two important applications in mechanics.

1. *Computing the angle formed between two vectors or intersecting lines.* Consider vectors **A** and **B** shown in Fig. 4–50. From Eq. 4–22, the angle θ between them is

$$\theta = \cos^{-1}\left(\frac{\mathbf{A} \cdot \mathbf{B}}{AB}\right) \qquad 0 \leqslant \theta \leqslant \pi$$

where $\mathbf{A} \cdot \mathbf{B}$ is given by Eq. 4–27

2. *Computing the projection of a vector along an axis.* In Fig. 4–51, the projection of vector **A** along the aa axis, for which the direction is specified by the unit vector $\mathbf{u}_a$, is defined as a scalar having a magnitude of $A_p = A \cos \theta$. Since the magnitude $u_a = 1$, using Eq. 4–22, A_p may be determined from

$$A_p = A \cos \theta = \mathbf{A} \cdot \mathbf{u}_a$$

The projected vector $\mathbf{A}_p$ is therefore represented as

$$\mathbf{A}_p = A \cos \theta\, \mathbf{u}_a = (\mathbf{A} \cdot \mathbf{u}_a)\mathbf{u}_a$$

The above two applications are illustrated numerically in the following example problems.

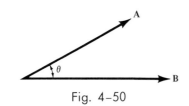

Fig. 4–50

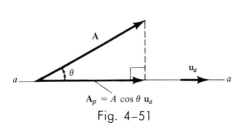

$A_p = A \cos \theta\, \mathbf{u}_a$

Fig. 4–51

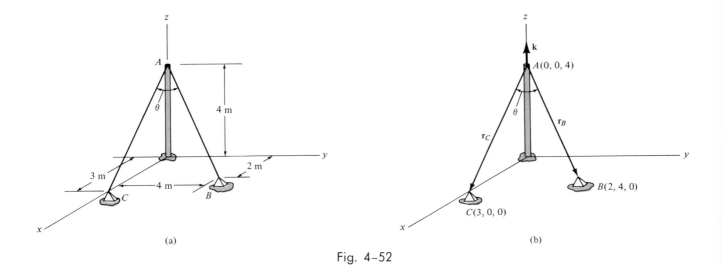

Fig. 4-52

Example 4-20

Two wires AB and AC are attached to the top of a pole as shown in Fig. 4-52a. Determine the angle θ formed between the wires at A.

Solution

From Fig. 4-52b and Eq. 4-22, $\mathbf{r}_B \cdot \mathbf{r}_C = r_B r_C \cos \theta$. Hence,

$$\theta = \cos^{-1} \left(\frac{\mathbf{r}_B \cdot \mathbf{r}_C}{r_B r_C} \right) \qquad (1)$$

Using the coordinates to formulate the position vectors, we have

$$\mathbf{r}_B = \{2\mathbf{i} + 4\mathbf{j} - 4\mathbf{k}\} \text{ m} \qquad r_B = \sqrt{(2)^2 + (4)^2 + (-4)^2} = 6 \text{ m}$$
$$\mathbf{r}_C = \{3\mathbf{i} - 4\mathbf{k}\} \text{ m} \qquad r_C = \sqrt{(3)^2 + (-4)^2} = 5 \text{ m}$$

Using Eq. 4-27 to obtain the dot product,

$$\mathbf{r}_B \cdot \mathbf{r}_C = (2\mathbf{i} + 4\mathbf{j} - 4\mathbf{k}) \cdot (3\mathbf{i} - 4\mathbf{k})$$
$$= 2(3) + (4)(0) + (-4)(-4) = 22 \text{ m}^2$$

Substituting into Eq. (1),

$$\theta = \cos^{-1} \left[\frac{22}{(6)(5)} \right] = 42.8° \qquad \qquad Ans.$$

Example 4-21

The rod AB shown in Fig. 4-53a, is subjected to a horizontal force of $\mathbf{F} = \{300\mathbf{j}\} \text{ N}$ acting at its end. Determine the projection of this force along the axis AB of the rod.

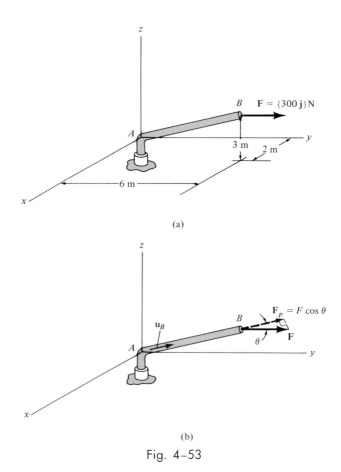

(a)

(b)

Fig. 4–53

Solution

The direction of the axis of the rod can be defined by the unit vector $\mathbf{u}_B$, where

$$\mathbf{u}_B = \frac{\mathbf{r}_B}{r_B} = \frac{2\mathbf{i} + 6\mathbf{j} + 3\mathbf{k}}{\sqrt{(2)^2 + (6)^2 + (3)^2}} = 0.286\mathbf{i} + 0.857\mathbf{j} + 0.429\mathbf{k}$$

As shown in Fig. 4–53b, the required projection of $\mathbf{F}$ along AB has a magnitude of $F_p = F\cos\theta$. Without obtaining θ, this magnitude is determined by using Eq. 4–22. Since $u_B = 1$, $\mathbf{F} \cdot \mathbf{u}_B = F\cos\theta$. Hence,

$$F\cos\theta = \mathbf{F} \cdot \mathbf{u}_B = (300\mathbf{j}) \cdot (0.286\mathbf{i} + 0.857\mathbf{j} + 0.429\mathbf{k})$$
$$= (0)(0.286) + (300)(0.857) + (0)(0.429) = 257.1 \text{ N}$$

Expressing $\mathbf{F}_p$ in Cartesian vector form, we have

$$\mathbf{F}_p = (F\cos\theta)\mathbf{u}_B = 257.1 \text{ N}(0.286\mathbf{i} + 0.857\mathbf{j} + 0.429\mathbf{k})$$
$$= \{73.5\mathbf{i} + 220.3\mathbf{j} + 110.3\mathbf{k}\} \text{ N} \qquad \textit{Ans.}$$

4-12. Moment of a Force About a Specified Axis

Recall that when the moment of a force is computed about a point, the moment and its axis are *always* perpendicular to the plane containing the force and the moment arm. In some problems, however, it is important to find the projection of this moment along another axis that passes through the point. If the force components or the appropriate moment arms are difficult to determine, vector analysis should be used for the computation. To show how this is done, consider a rigid body of general shape that is subjected to a force $\mathbf{F}$ acting at point A, Fig. 4–54. We wish to determine the effect of $\mathbf{F}$ in tending to rotate the body about the *aa* axis. This tendency for rotation is measured by the moment $\mathbf{M}_p$ which $\mathbf{F}$ produces about this axis. To determine $\mathbf{M}_p$, consider first computing the moment of $\mathbf{F}$ about an arbitrary point O that lies on the axis. Here $\mathbf{M}_O$ is expressed by the cross product $\mathbf{M}_O = \mathbf{r} \times \mathbf{F}$, where $\mathbf{r}$ is directed from O to A. Since $\mathbf{M}_O$ acts along the *bb* axis, which is perpendicular to the plane containing $\mathbf{r}$ and $\mathbf{F}$, the projection of $\mathbf{M}_O$ onto the *aa* axis is then represented by $\mathbf{M}_p$. The *magnitude* of $\mathbf{M}_p$ can be determined by the dot product $M_p = M_O \cos \theta = \mathbf{u}_a \cdot \mathbf{M}_O$, where $\mathbf{u}_a$ is a unit vector used to define the direction of the *aa* axis. Hence, as a general expression we have

$$M_p = \mathbf{u}_a \cdot (\mathbf{r} \times \mathbf{F}) \qquad (4\text{–}28)$$

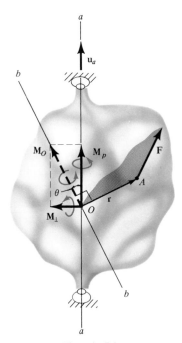

Fig. 4–54

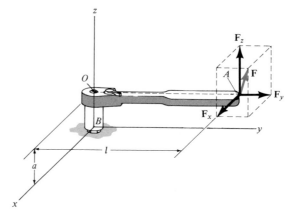

Fig. 4–55(a)

In vector algebra, this combination of dot and cross product yielding the scalar M_p is termed the *mixed triple product*. When the Cartesian components of each of the vectors in Eq. 4–28 are known, the mixed triple product may be written in determinant form as

$$M_p = (u_{a_x}\mathbf{i} + u_{a_y}\mathbf{j} + u_{a_z}\mathbf{k}) \cdot \begin{vmatrix} \mathbf{i} & \mathbf{j} & \mathbf{k} \\ r_x & r_y & r_z \\ F_x & F_y & F_z \end{vmatrix}$$

or simply

$$M_p = \mathbf{u}_a \cdot (\mathbf{r} \times \mathbf{F}) = \begin{vmatrix} u_{a_x} & u_{a_y} & u_{a_z} \\ r_x & r_y & r_z \\ F_x & F_y & F_z \end{vmatrix} \qquad (4\text{–}29)$$

where u_{a_x}, u_{a_y}, u_{a_z} represent the Cartesian components of the unit vector defining the direction of the *aa* axis

r_x, r_y, r_z represent the Cartesian components of the position vector drawn from any point O on the *aa* axis to any point A on the line of action of the force

F_x, F_y, F_z represent the Cartesian components of the force vector.

   The above concepts may be illustrated numerically using both a scalar analysis and a vector analysis by considering what effect the force **F,** acting on the end of the socket wrench shown in Fig. 4–55a has on turning the bolt *B*. The effective action of **F** in this regard depends only upon the moment that **F** creates about the *z* axis, since a rotation about this axis will cause the bolt to turn.

Scalar Analysis.   Separating **F** into its Cartesian components $\mathbf{F}_x$, $\mathbf{F}_y$, and $\mathbf{F}_z$ in Fig. 4–55*a*, it is seen that *only* component $\mathbf{F}_x$ causes a rotation of the

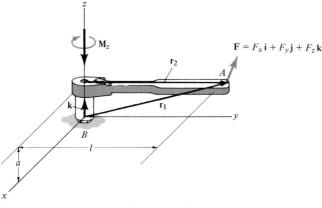

Fig. 4–55(b)

wrench about the z axis. (Since the line of action of $\mathbf{F}_y$ passes through a point lying on the z axis, and the line of action of $\mathbf{F}_z$ is parallel to the z axis, both of these forces create zero moment about the z axis.) The moment created by $\mathbf{F}_x$ has a magnitude $M_z = F_x l$ and a direction defined by the right-hand rule, i.e., in the $-\mathbf{k}$ direction. Thus, $\mathbf{M}_z = -F_x l\mathbf{k}.$

Vector Analysis. Applying Eq. 4–29 to the solution, the z axis is defined by the unit vector $\mathbf{u}_a = \mathbf{k}$, Fig. 4–55$b$. Both vectors $\mathbf{r}_1 = l\mathbf{j} + a\mathbf{k}$ and $\mathbf{r}_2 = l\mathbf{j}$ satisfy the requirements for $\mathbf{r}$ defined in Eq. 4–29. Why? Hence, using the scheme illustrated in the footnote on page 86 for expanding the determinant, we have

$$M_z = \mathbf{u}_a \cdot (\mathbf{r}_1 \times \mathbf{F}) = \begin{vmatrix} 0 & 0 & 1 \\ 0 & l & a \\ F_x & F_y & F_z \end{vmatrix}$$
$$= 0(lF_z - aF_y) - 0(0 - aF_x) + 1(0 - lF_x)$$
$$= -lF_x$$

or

$$M_z = \mathbf{u}_a \cdot (\mathbf{r}_2 \times \mathbf{F}) = \begin{vmatrix} 0 & 0 & 1 \\ 0 & l & 0 \\ F_x & F_y & F_z \end{vmatrix}$$
$$= 0(lF_z - 0) - 0(0 - 0) + 1(0 - lF_x)$$
$$= -lF_x$$

The negative sign indicates that $\mathbf{M}_z$ acts opposite to $\mathbf{u}_a = +\mathbf{k}$. Since both the magnitude and direction of $\mathbf{M}_z$ are known, $\mathbf{M}_z$ may be written in vector form as $\mathbf{M}_z = M_z(\mathbf{k}) = -F_x l\mathbf{k}.$

The following two examples further illustrate these concepts numerically.

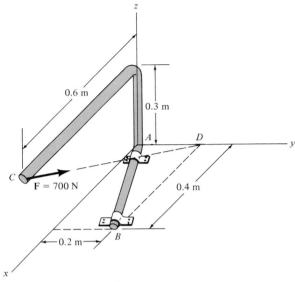

0.6 m

0.3 m

A

D

C

$F = 700$ N

0.4 m

0.2 m

B

z

y

x

Fig. 4–56(a)

Example 4–22

The rod shown in Fig. 4–56a is supported by two brackets at A and B. Determine the moment $\mathbf{M}_{AB}$ produced by force $\mathbf{F}$, which tends to rotate the rod about the AB axis.

Solution
(*Vector Analysis*). A vector analysis using Eq. 4–29 will be considered for the solution since the components of the force or the appropriate moment arm are difficult to determine. In general, the magnitude of $\mathbf{M}_{AB}$ is

$$M_{AB} = \mathbf{u}_B \cdot (\mathbf{r} \times \mathbf{F}) \tag{1}$$

Each of the terms in this equation will now be identified.

Unit vector $\mathbf{u}_B$ defines the direction of the AB axis of the rod, Fig. 4–56b, where

$$\mathbf{u}_B = \frac{\mathbf{r}_B}{r_B} = \frac{0.4\mathbf{i} + 0.2\mathbf{j}}{\sqrt{(0.4)^2 + (0.2)^2}} = 0.894\mathbf{i} + 0.447\mathbf{j}$$

Vector $\mathbf{r}$ in Eq. (1) is directed from *any point* on the AB axis to *any* point on the line of action of the force. For example, position vectors $\mathbf{r}_C$ and $\mathbf{r}_D$ are suitable, Fig. 4–56b. For simplicity, choose $\mathbf{r}_D$, where

$$\mathbf{r}_D = \{0.2\mathbf{j}\} \text{ m}$$

Expressing $\mathbf{F}$ as a Cartesian vector,

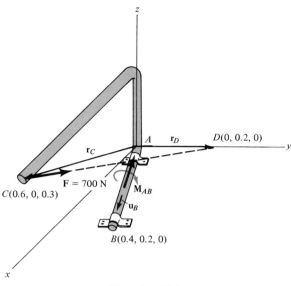

Fig. 4–56(b)

$$\mathbf{F} = F\mathbf{u}_{CD} = F\frac{\mathbf{r}_{CD}}{r_{CD}} = 700 \text{ N} \left[\frac{-0.6\mathbf{i} + 0.2\mathbf{j} - 0.3\mathbf{k}}{\sqrt{(-0.6)^2 + (0.2)^2 + (-0.3)^2}} \right]$$
$$= \{-600\mathbf{i} + 200\mathbf{j} - 300\mathbf{k}\} \text{ N}$$

Substituting these vectors into the determinant form of Eq. (1) and expanding, using the scheme listed in the footnote on page 86, we have

$$M_{AB} = \mathbf{u}_B \cdot (\mathbf{r}_D \times \mathbf{F}) = \begin{vmatrix} 0.894 & 0.447 & 0 \\ 0 & 0.2 & 0 \\ -600 & 200 & -300 \end{vmatrix}$$
$$= 0.894(0.2)(-300) + 0 + 0$$
$$= -53.64 \text{ N} \cdot \text{m}$$

Expressing $\mathbf{M}_{AB}$ as a vector,

$$\mathbf{M}_{AB} = M_{AB}\mathbf{u}_B = (-53.64 \text{ N} \cdot \text{m})(0.894\mathbf{i} + 0.447\mathbf{j})$$
$$= \{-48.0\mathbf{i} - 24.0\mathbf{j}\} \text{ N} \cdot \text{m} \qquad\qquad Ans.$$

The result is shown in Fig. 4–56b.

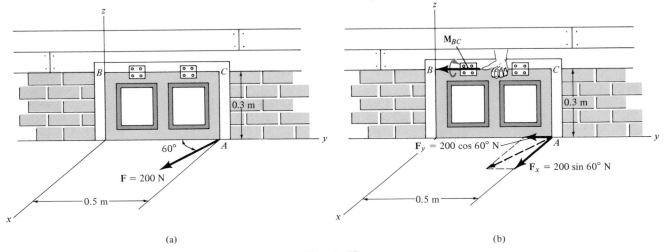

(a)

(b)

Fig. 4-57

Example 4-23

The window is subjected to a 200-N force at its corner A as shown in Fig. 4-57a. Compute the moment of this force about the window's hinged axis BC. The force lies in the x-y plane.

Solution
(*Scalar Analysis*).   Since the force components and the moment arm are easy to determine, a scalar analysis will be used for the solution. If the force is resolved into its rectangular components, Fig. 4-57b, it is noted that only $\mathbf{F}_x$ produces a moment about the BC axis. (The line of action of $\mathbf{F}_y$ is *parallel* to this axis, and hence the moment about BC is zero.) The moment arm extends from C to A. Hence, the magnitude of moment is

$$M_{BC} = F_x d_{CA} = (200 \sin 60°)(0.3) = 52.0 \text{ N} \cdot \text{m}$$

By the right-hand rule, Fig. 4-57b, $\mathbf{F}_x$ tends to rotate the window out and around so that the thumb is directed from C to B, i.e., in the $-\mathbf{j}$ direction. Thus,

$$\mathbf{M}_{BC} = \{-52.0\mathbf{j}\} \text{ N} \cdot \text{m} \qquad\qquad Ans.$$

As an exercise, try solving this problem using Eq. 4-29.

Problems

* 4-64. If $\mathbf{r}_1 = \{3\mathbf{i} + 2\mathbf{j}\}$ m and $\mathbf{r}_2 = \{1\mathbf{i} + 2\mathbf{j} + 1.5\mathbf{k}\}$ m, determine the angle made between the tails of the vectors.

4-65. Given the three vectors **A**, **B**, and **D**, show that $\mathbf{A} \cdot (\mathbf{B} + \mathbf{D}) = (\mathbf{A} \cdot \mathbf{B}) + (\mathbf{A} \cdot \mathbf{D})$. *Suggestion:* Express each vector in Cartesian vector form and carry out the vector operations.

4-66. Given the three nonzero vectors **A, B,** and **C,** show that if $\mathbf{A} \cdot (\mathbf{B} \times \mathbf{C}) = 0$, the three vectors *must* lie in the same plane.

4-67. Find the equation of a plane that contains the three points $A(2, 4, -1)$, $B(3, 1, 1)$, and $C(2, -2, 4)$. *Hint:* Construct position vectors between each of these points and some arbitrary point $D(x, y, z)$ lying in the plane, then use the result of Prob. 4-66.

*4-68. Determine the projection of the moment created by the 30-N force along the z axis of the socket wrench. The force lies in the x-y plane.

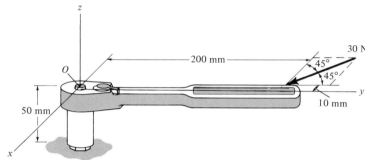

Prob. 4-68

4-69. The semicircular vent is subjected to a uniform wind pressure of 2 kPa. Compute the moment of this loading about the hinged *aa* axis of the vent.

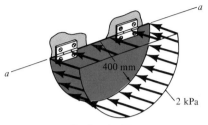

Prob. 4-69

4-70. A force $F = 80$ N acts vertically downward on the Z bracket. If $a = 100$ mm, $b = 300$ mm, and $c = 200$ mm, determine the moment of this force about the bolt axis (z axis), which is directed at 15° from the vertical.

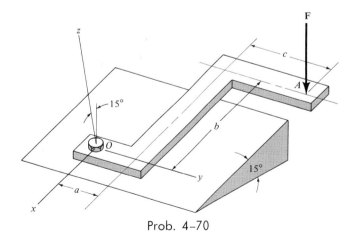

Prob. 4-70

4-70a. Solve Prob. 4-70 if $F = 40$ lb and $a = 2$ in., $b = 6$ in., and $c = 4$ in.

4-71. The bracket is acted upon by a 600-N force at A. Determine the projection of the moment of this force about the y' axis.

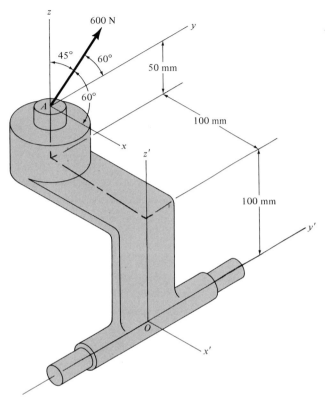

600 N

45° 60°

50 mm

60°

100 mm

100 mm

A

z'

x

y

y'

x'

O

Prob. 4-71

*4-72. Three forces act at the end of the cantilever beam. Determine the moment created by each of these forces about the x axis. Solve the problem using both a vector analysis and a scalar analysis.

Scaler

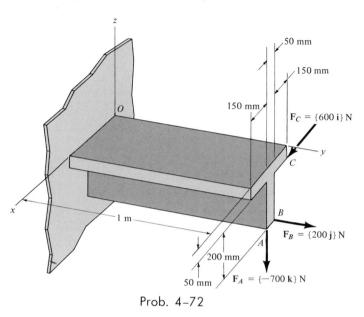

50 mm

150 mm

O

150 mm

$F_C = \{600\,\mathbf{i}\}$ N

y

C

1 m

B

$F_B = \{200\,\mathbf{j}\}$ N

A

x

200 mm

50 mm

$F_A = \{-700\,\mathbf{k}\}$ N

Prob. 4-72

4-73. If the force $F = 100$ N lies in the plane $DBEC$, which is parallel to the x-z plane, and makes an angle of $10°$ with the extended line DB as shown, determine the angle that $\mathbf{F}$ makes with the diagonal AB of the crate.

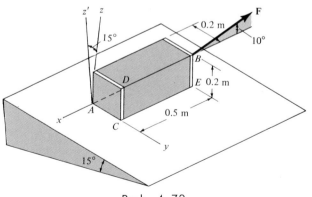

z' z

15°

0.2 m

F

10°

B

D

E 0.2 m

0.5 m

x

A

C

y

15°

Prob. 4-73

4-74. For the crate in Prob. 4-73, determine the angle that the line of action of **F** makes with the vertical z' axis.

4-75. The A frame is being hoisted into an upright position by the *vertical* towing force of $F = 800$ N. If $a = 4$ m, determine the moment of this force about the x' axis passing through points A and B when the frame is in the position shown.

4-75a. Solve Prob. 4-75 if $F = 125$ lb and $a = 8$ ft.

*4-76. The chain AB exerts a force of $F = 200$ N on the door at B. Determine the projection of the moment of this force acting along the hinged axis, x, of the door.

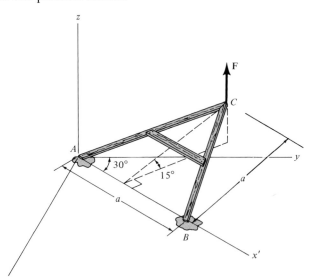

Prob. 4-75

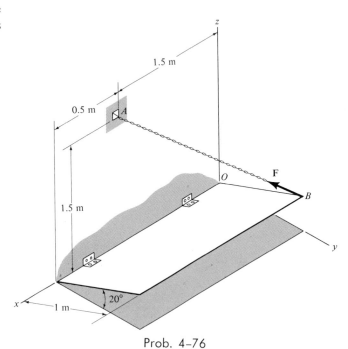

Prob. 4-76

Equilibrium of a Rigid Body

In this chapter the fundamental concepts of rigid-body equilibrium will be discussed. Recall that in Chapter 3 it was shown that the equilibrium of a rigid body subjected to a system of *concurrent forces* can be analyzed by using the methods of particle equilibrium. In the more general case, however, rigid-body equilibrium requires both a *balance of forces* to prevent the body from translating or moving along a straight or curved path, and a *balance of moments* to prevent the body from rotating.

Due to the symmetry of loading, many types of engineering problems can be solved by projecting all the forces acting on the body onto a single plane. Hence, in the first part of this chapter, the equilibrium of a body subjected to a *coplanar* or *two-dimensional force system* will be considered. Ordinarily, the geometry of such problems is not very complex, so a scalar solution is suitable for analysis. The more general discussion of rigid bodies subjected to *three-dimensional force systems* is given in the second part of this chapter. It will be seen that many of these types of problems can best be solved by using Cartesian vector analysis.

Coplanar Force Systems Acting on a Rigid Body

5-1. Free-Body Diagrams and Support Reactions

The free-body diagram concept, introduced in Chapter 3 in connection with problems involving the equilibrium of a particle, plays an important role in the solution of rigid-body equilibrium problems. Provided the free-body diagram is correctly drawn, the effects of all the forces and couples acting on the rigid body can be accounted for when the equations of equilibrium are applied. *For this reason, a comprehensive understanding as to how to draw a free-body diagram is of primary importance for solving problems in mechanics.* It is the preliminary step before applying the equations of equilibrium.

Support Reactions.   Before discussing how to draw the free-body diagram, it is necessary to consider the various types of *reactions* that occur at supports and points of contact between bodies. For two-dimensional problems, i.e., bodies subjected to coplanar force systems, the types of supports commonly encountered are given in Table 5–1. It is important to carefully study each of the symbols used to represent these supports and the types of reactions that occur. Notice that a support develops a *force* on its attached member if it *prevents translation* of the member, and it develops a *couple* if it *prevents rotation* of the member. For example, a member in contact with the smooth surface (6) is prevented from translating *only* in the downward direction, perpendicular or normal to the surface. Hence, the surface exerts only a *normal force* **F** on the member at the point of contact. The magnitude of this force represents *one unknown*. Since the member is free to rotate on the surface, a couple cannot be developed by the surface on the member at the point of contact. The pin or hinge support (8) prevents translation of the connecting member at its point of connection. Unlike the smooth surface, translation is prevented in any direction. Hence, a force **F** must be developed at the support such that it has *two unknowns*, its magnitude F and direction ϕ or, equivalently, the magnitudes of its two components $\mathbf{F}_x$ and $\mathbf{F}_y$. Since the connecting member is allowed to rotate freely in the plane about the pin, a pin support does not resist a couple acting perpendicular to this plane. The fixed support (9), however, prevents *both* planar translation and rotation of the connecting member at the point of connection. Therefore, this type of support exerts both a force and couple on the member. Note that the couple acts *perpendicular* to the plane of the page since rotation is prevented in the plane. Hence, there are *three unknowns* at a fixed support.

Table 5–1   Supports for Rigid Bodies Subjected to Two-Dimensional Force Systems

Types of Connection	Reaction	Number of Unknowns
(1) light cable		One unknown: magnitude of force **F**; this force prevents translation of the attached member in the direction of the cable. The cable remains in tension.
(2) weightless link	or	One unknown: magnitude of force **F**; this force prevents translation of the attached member in the θ direction.

Table 5-1 (Contd.)

Type of Connection	Reaction	Number of Unknowns
(3) roller		One unknown: magnitude of force **F**; this force prevents translation of the attached member normal to the tangent at the point of contact with the surface.
(4) roller or pin in confined smooth slot	or	One unknown: magnitude of force **F**; this force prevents translation of the attached member in either direction normal to the confined slot.
(5) rocker		One unknown: magnitude of force **F**; this force prevents translation of the attached member normal to the *tangent* at the point of contact with the surface.
(6) smooth contacting surface		One unknown: magnitude of force **F**; this force prevents translation of the member normal to the *tangent* at the point of contact with the surface.
(7) collar on smooth rod	or	One unknown: magnitude of force **F**; this force prevents translation of the attached member normal to the axis of the supporting rod.
(8) smooth pin or hinge	or	Two unknowns: the magnitudes of the two force components F_x and F_y, or the magnitude of the force **F** and its direction ϕ; note that ϕ and θ are not necessarily equal [usually not, unless the rod shown is a link as in (2)]. The force prevents translation of the attached member.
(9) fixed support	or	Three unknowns: the magnitudes of the couple moment **M** and two force components F_x and F_y, or the magnitudes of the couple moment **M** and force **F** and the direction of the force, ϕ. The force prevents translation of the attached member. The couple moment prevents rotation of the member about an axis perpendicular to the plane of the member.

In reality, all supports actually exert *distributed surface loads* on their contacting members. The concentrated forces and couples shown in Table 5–1 represent the *resultants* of this load distribution. This representation is, of course, an idealization. However, as explained in Sec. 2–4, it can be used here provided the surface area over which the distributed load acts is considerably *smaller* than the *total* surface area of the connecting member.

PROCEDURE FOR DRAWING A FREE-BODY DIAGRAM

To construct a free-body diagram for a rigid body or group of bodies, considered as a single system, the following three steps must be performed:

Step 1:   Imagine the body to be <u>isolated</u> from its surroundings, by drawing (sketching) its outlined shape.

Step 2:   Indicate on this sketch all the forces and couples that act on the body. Also include the dimensions of the body necessary for computing the moments of the forces. Forces and couples generally encountered are those due to (1) applied *external* loadings, (2) reactions occurring at the supports or at points of contact between bodies, and (3) the weight of the isolated body. Forces that are *internal* to the body, such as the forces of contact between any two particles of the body, are *not shown* on the free-body diagram of the entire body. These forces always occur in equal and opposite collinear pairs, and therefore their *net effect* on the body will be zero.

Step 3:   The forces and couples that are known should be labeled with their proper magnitude and direction. Letters are used to represent the magnitudes and direction angles of forces and couples that are *unknown*. In particular, if a force or couple has a known line of action but unknown magnitude, the arrowhead which defines the directional sense of the vector can be assumed. The correctness of the assumed direction will become apparent after solving the equilibrium equations for the unknown magnitude. By definition, the *magnitude* of a vector is *always positive,* so that if the solution yields a "negative" magnitude, the *minus sign* indicates that the vector's "arrowhead" direction is *opposite* to that which was originally assumed.

Before proceeding, carefully review this section; then attempt to draw the free-body diagrams for the following example problems using the above three-step procedure before "looking" at the solutions. Further practice in drawing free-body diagrams should be gained by solving the problems given at the end of this section.

Example 5-1

Draw the free-body diagram for the uniform beam shown in Fig. 5–1a. The beam has a mass of 100 kg.

Solution

The free-body diagram for the beam is shown in Fig. 5–1b. Since the support is a fixed wall, there are three reactions acting *on the beam* at A, denoted as $\mathbf{A}_x$, $\mathbf{A}_y$, and $\mathbf{M}_A$. The magnitudes of these vectors are *unknown* and their directions have all been *assumed*. (How does one obtain the *correct* directions for these unknown vectors?)

For convenience in computing the reactions at A, the distributed loading is reduced to a concentrated force. From the methods of Sec. 4–10, the resultant of the distributed loading is equal to the area under the triangular loading diagram, i.e., $\frac{1}{2}(6\text{ m})(400\text{ N/m}) = 1200$ N. This force acts through the centroid of the triangle, i.e., $\frac{1}{3}(6\text{ m}) = 2$ m from A. The weight of the beam, $W = (100)(9.81) = 981$ N, acts 3 m from A since the beam is uniform.

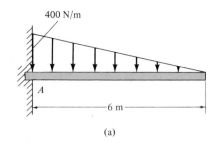

(a)

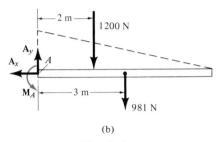

(b)

Fig. 5–1

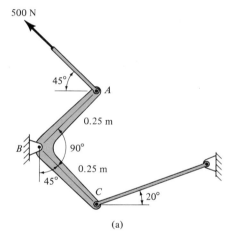

(a)

Example 5-2

Draw the free-body diagram for the bell crank ABC shown in Fig. 5–2a, and determine the number of unknown force components.

Solution

The free-body diagram is shown in Fig. 5–2b. The pin support at B exerts forces $\mathbf{B}_x$ and $\mathbf{B}_y$ *on the crank,* each component having a known line of action, but unknown magnitude. The link at C exerts a force $\mathbf{F}_C$ acting in the direction of the link and having an unknown magnitude. The dimensions of the crank are also labeled on the free-body diagram since this information will be useful in computing the moments of the forces. As usual, the directions of the three unknown forces have been assumed. The correct directions will become apparent after solving the equilibrium equations.

Fig. 5–2

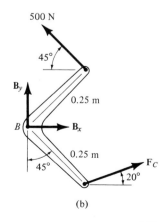

(b)

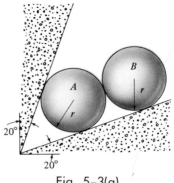

Fig. 5–3(a)

Example 5–3

Two smooth cylinders A and B, each having a mass of 2 kg, rest between the inclined planes shown in Fig. 5–3a. Draw the free-body diagrams for (a) cylinder A and B individually; (b) cylinders A and B combined.

Solution

Part (a). The free-body diagrams for A and B individually are shown in Fig. 5–3b. Note that the weight of each cylinder is represented as $W = 2(9.81) = 19.62$ N. Since all contacting surfaces are *smooth,* the reactive forces **P, R, F,** and **T** act in a direction *normal* to the tangent at their surfaces of contact. The magnitudes of these forces are unknown. In particular, the reactive force **R** acting *on A* represents the effect that B *exerts on A*. Using Newton's third law, the free-body diagram of B must include an *equal but opposite force* **R** which A *exerts on B*.

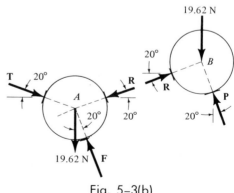

Fig. 5–3(b)

Part (b). The free-body diagram for both cylinders A and B combined is shown in Fig. 5–3c. The reactions at the surfaces of contact with the inclined planes must be the *same* as those listed on the separate free-body diagrams given in part (a). Since the contact force **R**, which acts between A and B, occurs as an equal and opposite collinear pair of forces, it is considered here as an *internal force* and hence is not shown on the "combined" free-body diagram in Fig. 5–3c.

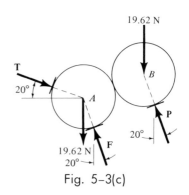

Fig. 5–3(c)

Example 5-4

The highway sign shown in Fig. 5–4a has a uniform mass of 100 kg with a center of gravity at G. In order to provide roadway clearance, it is pin supported at C and D and held over a traffic lane by means of a tie rod AB. Draw a free-body diagram of the sign and the supporting frame. Neglect the weight of the frame.

Solution

By observation, the frame and sign and the loading are all symmetrical about the vertical x-y plane, hence the problem may be analyzed as a system of *coplanar forces*. The free-body diagram is shown in Fig. 5–4b. The tension force **T** in the tie rod ("link") has a known line of action indicated by the 3-4-5 triangle. The force components C'_x and C'_y represent the horizontal and vertical reactions of *both* pins C and D, respectively. Consequently, after the solution for these reactions is obtained, *half* the magnitude is applied at C and half at D.

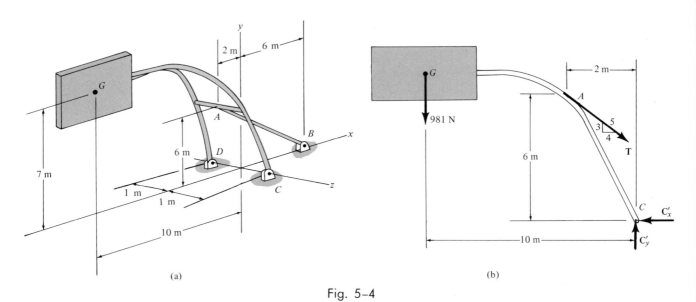

Fig. 5–4

163

Problems

Draw the free-body diagram in each of the following problems and determine the total number of unknown force and couple magnitudes and/or directions. Neglect the weight of the members unless otherwise stated.

5-1. The cylinder of mass 10 kg resting between the smooth inclined planes.

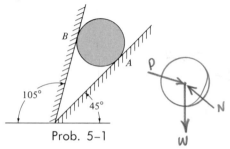

Prob. 5-1

5-2. The uniform 4-m-long beam pinned at A and supported by a cord at B; the mass of the beam is 100 kg.

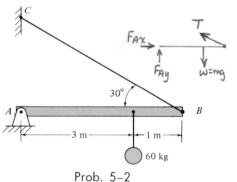

Prob. 5-2

5-3. The bell crank supported at A by a pin and subjected to cable forces at B and C.

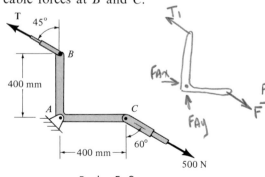

Prob. 5-3

*5-4. The uniform rod ABC supported by a pin at A and a short link BD.

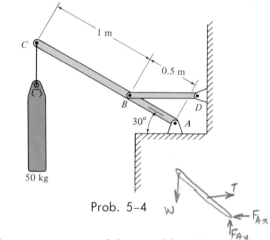

Prob. 5-4

5-5. The truss supported by a cable AB and pin C.

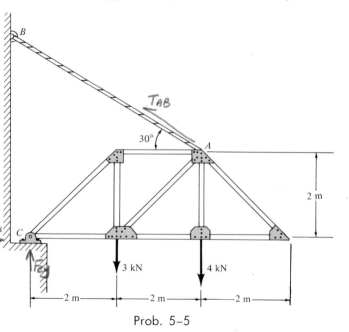

Prob. 5-5

164

5-6. The spanner wrench subjected to the 200-N force; the support at *A* acts as a pin and the surface of contact at *B* is smooth.

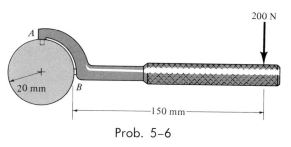

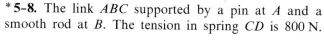

Prob. 5-6

5-7. The uniform beam that has a mass of 100 kg and is supported by making contact with the smooth surfaces at *A*, *B*, and *C*.

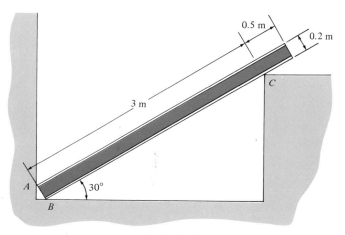

Prob. 5-7

*5-8. The link *ABC* supported by a pin at *A* and a smooth rod at *B*. The tension in spring *CD* is 800 N.

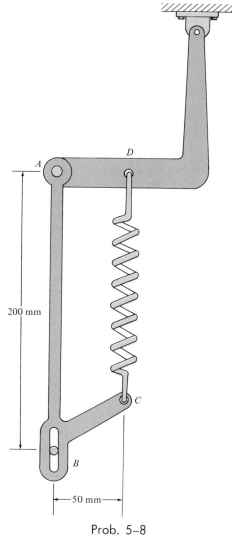

Prob. 5-8

5–9. The uniform link supported at A by a pin and at C by a cord; the spring is subjected to a tension of 300 N.

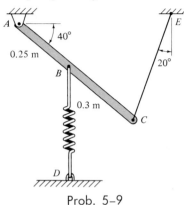

Prob. 5–9

5–10. The ratchet gear G that is pin-connected at its center A and contacts a smooth pawl P at B. A cable is wrapped around the hub of the gear and is subjected to a tension of 800 N.

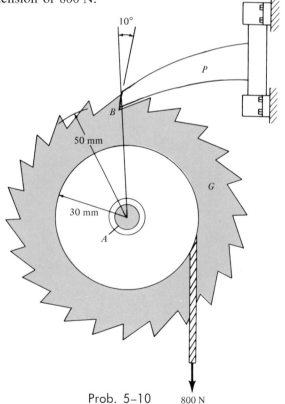

Prob. 5–10   800 N

5–11. The beam supported at A by a fixed support and at B by a roller.

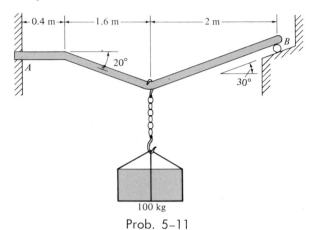

100 kg

Prob. 5–11

* **5–12.** The beam supported at A by a smooth roller and at B by a fixed support.

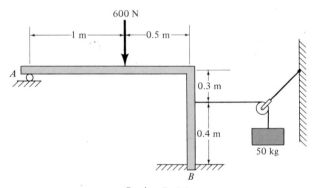

Prob. 5–12

It was shown in Sec. 4-9 that a system of *coplanar forces* acting on a body can always be simplified to a resultant force $\mathbf{F}_R = \Sigma\mathbf{F}$ (Eq. 4-14), acting through an arbitrary point O, and a resultant couple moment $\mathbf{M}_{R_O} = \Sigma\mathbf{M}_O$ (Eq. 4-15) which acts *perpendicular* to $\mathbf{F}_R$.

Vector Equations of Equilibrium. If $\mathbf{F}_R$ and $\mathbf{M}_{R_O}$ both equal zero, the body is said to be in *statical equilibrium* since physically this situation maintains a balance of both force and moment. Equations that express these conditions are

$$\Sigma\mathbf{F} = \mathbf{0}$$
$$\Sigma\mathbf{M}_O = \mathbf{0} \tag{5-1}$$

Here $\Sigma\mathbf{F}$ represents the sum of *all* the external forces acting on the body and $\Sigma\mathbf{M}_O$ is the sum of the moments of *all* these forces about *any point O* that lies in the plane of the forces.

Scalar Equations of Equilibrium. If the applied forces lie in the x-y plane, then from Eq. 5-1, the force summation can be expressed in Cartesian vector form as $\Sigma\mathbf{F} = \Sigma F_x\mathbf{i} + \Sigma F_y\mathbf{j}$, and the corresponding moments $\Sigma\mathbf{M}_O$ are directed either in the plus or minus $\mathbf{k}$ direction. Consequently, by setting the vector components equal to zero, the following *three* independent scalar equations can be generated:

$$\Sigma F_x = 0$$
$$\Sigma F_y = 0$$
$$\Sigma M_O = 0 \tag{5-2}$$

Here ΣF_x and ΣF_y represent, respectively, the algebraic sums of the x and y components of all the forces acting on the body, and ΣM_O represents the algebraic sum of the moments of all these force components about an axis perpendicular to the x-y plane and passing through point O.

Equations 5-2 provide the *necessary conditions* for equilibrium of a rigid body subjected to a coplanar force system. If the body is *properly supported* so that all the forces acting on the body can be determined from these equations, they are also *sufficient conditions* for equilibrium. In this book only those types of bodies that meet these conditions of proper support will be considered. Situations for which bodies are improperly supported, or have too many supports needed for equilibrium, are discussed further in Sec. 5-6.

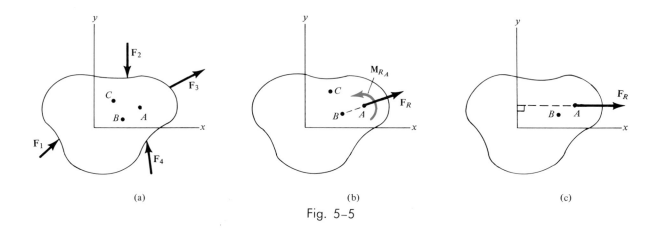

(a)   (b)   (c)

Fig. 5–5

Alternative Sets of Scalar Equilibrium Equations. Although Eqs. 5–2 are *most often* used for solving equilibrium problems involving coplanar-force systems, certain *alternative* sets of three independent equilibrium equations may also be used. One such set is

$$\Sigma F_y = 0$$
$$\Sigma M_A = 0 \qquad (5\text{–}3)$$
$$\Sigma M_B = 0$$

When using these equations it is required that the moment points A and B do *not* lie on a line that is *perpendicular* to the y axis. To prove that Eqs. 5–3 provide *necessary conditions* for equilibrium, consider the free-body diagram of an arbitrarily shaped body shown in Fig. 5–5a. Using the methods of Sec. 4–9, the system of forces acting on the free-body diagram may be replaced at point A by a single resultant force $\mathbf{F}_R = \Sigma \mathbf{F}$ and a resultant couple $\mathbf{M}_{R_A} = \Sigma \mathbf{M}_A$, Fig. 5–5b. For equilibrium, $\mathbf{F}_R = \mathbf{0}$ and $\mathbf{M}_{R_A} = \mathbf{0}$. Hence, if it is required that $\Sigma M_A = 0$, it is necessary that $\mathbf{M}_{R_A} = \mathbf{0}$. Furthermore, in order that $\mathbf{F}_R$ satisfy $\Sigma F_y = 0$, it must have no components along the y axis, and therefore its line of action must be perpendicular to the y axis, Fig. 5–5c. Finally, if it is required that $\Sigma M_B = 0$, where B does not lie on the line of action of $\mathbf{F}_R$, then $\mathbf{F}_R = \mathbf{0}$, and indeed the body shown in Fig. 5–5a must be in equilibrium.

In a similar manner, it can be shown that

$$\Sigma F_x = 0$$
$$\Sigma M_A = 0 \qquad (5\text{–}4)$$
$$\Sigma M_B = 0$$

form a complete set of three necessary equilibrium equations provided points A and B do *not* lie on a line *perpendicular* to the x axis.

A third alternative set of equilibrium equations is

168

$$\Sigma M_A = 0$$
$$\Sigma M_B = 0 \qquad\qquad (5\text{-}5)$$
$$\Sigma M_C = 0$$

Here it is necessary that points A, B, and C do not lie on the same line. To prove that these equations when satisfied ensure equilibrium, consider again the free-body diagram in Fig. 5-5b. If it is required that $\Sigma M_A = 0$, then the resultant couple $\mathbf{M}_{R_A} = \mathbf{0}$. $\Sigma M_B = 0$ is satisfied if the line of action of $\mathbf{F}_R$ passes through point B as shown, and finally, if $\Sigma M_C = 0$, where C does not lie on line AB, it is necessary that $\mathbf{F}_R = \mathbf{0}$ and the body in Fig. 5-5a must be in equilibrium.

If one *violates* the stated restriction for using any of the three sets of Eqs. 5-3, 5-4, or 5-5, one obtains equations that are not *linearly independent;* i.e., one equation will differ from another only by a constant multiple and therefore a simultaneous solution of equations cannot be determined.

PROCEDURE FOR ANALYSIS

The following two-step procedure should be used when solving coplanar-force equilibrium problems:

Step 1:   Draw a free-body diagram of the body (or group of bodies considered as a single system) as outlined in Sec. 5-1. Briefly this requires that all the known and unknown external force and couple magnitudes be labeled and the directions of these vectors specified. The "arrowhead" direction of a force or couple having an *unknown* magnitude but known line of action can be *assumed*. Dimensions of the body, necessary for computing the moments of forces, are also included on the free-body diagram.

Step 2:   Apply the equations of equilibrium: $\Sigma F_x = 0, \Sigma F_y = 0, \Sigma M_O = 0$ (or the alternative sets of Eqs. 5-3, 5-4, or 5-5). To *avoid* solving the equilibrium equations simultaneously, apply the moment equation $\Sigma M_O = 0$ about a point (O) that *intersects the lines of action of two of the three unknown forces.* In this way, the moment of these unknowns is *zero* about O, and one can obtain a *direct solution* for the third unknown. When applying the force equations $\Sigma F_x = 0$ and $\Sigma F_y = 0$, orient the x and y axes along lines that will provide the simplest reduction of the forces into their x and y components. If the solution of the equilibrium equations yields a *negative* magnitude for an unknown force or couple, it indicates that its "arrowhead" direction is *opposite* to that which was assumed on the free-body diagram.

The following example problems numerically illustrate this two-step procedure.

Example 5-5

Determine the horizontal and vertical components of reaction for the loaded beam shown in Fig. 5–6a. Neglect the weight of the beam in the calculations.

Solution I
(**Scalar Analysis**).   *Step 1:*   The free-body diagram of the beam is shown in Fig. 5–6b. The reaction at A is "vertical" since there is a smooth *roller* at this support. The *pin support* at B exerts an unknown force, represented by the two components $\mathbf{B}_x$ and $\mathbf{B}_y$. The directions of the three vectors $\mathbf{A}_y$, $\mathbf{B}_x$, and $\mathbf{B}_y$ have been assumed. For simplicity in applying the equilibrium equations, $\mathbf{F}_1$ is represented by its x and y components as shown.

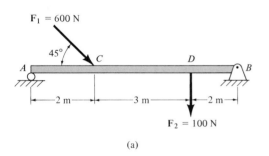

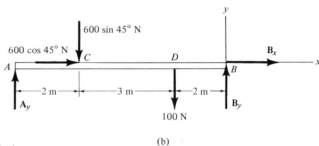

(a)   (b)

Fig. 5–6

Step 2:   Applying the equilibrium equation $\Sigma F_x = 0$ to the force system on the free-body diagram in Fig. 5–6b yields

$$\xrightarrow{+} \Sigma F_x = 0; \qquad 600 \cos 45° \text{ N} + B_x = 0$$
$$B_x = -424.3 \text{ N} \qquad\qquad Ans.$$

Since the result is a negative quantity, $\mathbf{B}_x$ has a direction *opposite* to that shown (assumed) in Fig. 5–6b; i.e., $\mathbf{B}_x$ acts to the left.

A direct solution for $\mathbf{A}_y$ can be obtained by applying the moment equation $\Sigma M_B = 0$ about point B. For the calculation, it should be apparent that forces $600 \cos 45°$ N, $\mathbf{B}_x$ and $\mathbf{B}_y$ create zero moment about B, since the line of action of each of these forces passes through this point. Assuming counterclockwise rotation about B to be positive ($+\mathbf{k}$ direction), Fig. 5–6b, we have

$$\zeta + \Sigma M_B = 0; \quad 100 \text{ N}(2 \text{ m}) + (600 \sin 45° \text{ N})(5 \text{ m}) - A_y(7 \text{ m}) = 0$$
$$A_y = 331.6 \text{ N} \qquad\qquad Ans.$$

Show that this same result would be obtained if moments had been assumed positive clockwise.

Summing forces in the y direction, using the result $A_y = 331.6$ N, we have

$+\uparrow\Sigma F_y = 0;\quad 331.6\text{ N} - 600\sin 45°\text{ N} - 100\text{ N} + B_y = 0$

$$B_y = 192.7\text{ N} \qquad Ans.$$

The magnitude of force $\mathbf{B}_y$ may *also* be obtained by summing moments about point A, i.e.,

$\zeta+\Sigma M_A = 0;\quad -(600\sin 45°\text{ N})(2\text{ m}) - 100\text{ N}(5\text{ m}) + B_y(7\text{ m}) = 0$

$$B_y = 192.7\text{ N} \qquad Ans.$$

The first solution for B_y is the least complicated; however, the second solution may be desirable, since (for accuracy) it does not depend upon prior knowledge of the magnitude of $\mathbf{A}_y$.

Solution II

(*Vector Analysis*). *After* the free-body diagram for the beam is drawn, the equations of equilibrium may be applied using the methods of vector algebra. With reference to the x and y axes shown in Fig. 5–6b, each of the forces acting on the free-body diagram must first be expressed in Cartesian vector form.

$$\mathbf{F}_A = A_y\mathbf{j}$$
$$\mathbf{F}_1 = \{600\cos 45°\,\mathbf{i} - 600\sin 45°\,\mathbf{j}\}\text{ N}$$
$$\mathbf{F}_2 = \{-100\mathbf{j}\}\text{ N}$$
$$\mathbf{F}_B = B_x\mathbf{i} + B_y\mathbf{j}$$

Thus,

$$\Sigma\mathbf{F} = \mathbf{0};\quad A_y\mathbf{j} + 600\cos 45°\,\mathbf{i} - 600\sin 45°\,\mathbf{j} - 100\mathbf{j} + B_x\mathbf{i} + B_y\mathbf{j} = \mathbf{0}$$

To satisfy this vector equation, the scalar components in the $\mathbf{i}$ and $\mathbf{j}$ directions, respectively, must be equated to zero. Hence,

$$\xrightarrow{+}\Sigma F_x = 0;\qquad 600\cos 45° + B_x = 0 \qquad (1)$$
$$+\uparrow\Sigma F_y = 0;\qquad A_y - 600\sin 45° - 100 + B_y = 0 \qquad (2)$$

Summing moments about point B requires the use of position vectors extending from B to points on the line of action of each force. Thus,

$$\Sigma\mathbf{M}_B = \mathbf{0};\qquad (\mathbf{r}_D \times \mathbf{F}_2) + (\mathbf{r}_C \times \mathbf{F}_1) + (\mathbf{r}_A \times \mathbf{F}_A) = \mathbf{0}$$
$$(-2\mathbf{i}) \times (-100\mathbf{j}) + (-5\mathbf{i}) \times (600\cos 45°\,\mathbf{i} - 600\sin 45°\,\mathbf{j})$$
$$+ (-7\mathbf{i}) \times (A_y\mathbf{j}) = \mathbf{0}$$

Each vector cross product yields a moment vector acting in the positive or negative $\mathbf{k}$ direction. After computing these cross products, the above equation reduces to

$$\zeta+\Sigma M_B = 0;\qquad 100(2) + 600\sin 45°\,(5) - A_y(7) = 0 \qquad (3)$$

By comparison, Eqs. (1) through (3) are identical to those obtained in solution I.

Although the vector approach *automatically* accounts for the direction of force and moment vectors in the equilibrium equations, the *scalar analysis* gives a more *direct solution*. For this reason, and because the x and y force components and moments are easily determined, equilibrium problems involving *coplanar-force systems* will be solved throughout this book using a scalar analysis.

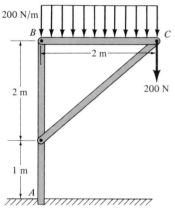

(a)

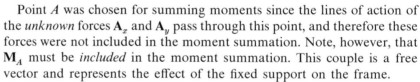

(b)

Fig. 5–7

Example 5–6

Determine the reactions at the fixed support A for the loaded frame shown in Fig. 5–7a.

Solution

Step 1: The free-body diagram for the frame is shown in Fig. 5–7b. There are three unknowns at the fixed support, represented by the magnitudes of A_x, A_y, and M_A. The distributed load is simplified to a resultant force equal to the area under the loading curve, i.e., $(2 \text{ m})(200 \text{ N/m}) = 400 \text{ N}$ (Sec. 4–10). This force acts through the centroid or geometric center of the loading curve, 1 m from B or C.

Step 2: Applying the equilibrium equations, we have

$$\xrightarrow{+}\Sigma F_x = 0; \qquad\qquad A_x = 0 \qquad\qquad Ans.$$
$$+\uparrow\Sigma F_y = 0; \qquad A_y - 400 \text{ N} - 200 \text{ N} = 0$$
$$A_y = 600 \text{ N} \qquad\qquad Ans.$$
$$\zeta+\Sigma M_A = 0; \quad M_A - 400 \text{ N}(1 \text{ m}) - 200 \text{ N}(2 \text{ m}) = 0$$
$$M_A = 800 \text{ N} \cdot \text{m} \qquad\qquad Ans.$$

Point A was chosen for summing moments since the lines of action of the *unknown* forces A_x and A_y pass through this point, and therefore these forces were not included in the moment summation. Note, however, that M_A must be *included* in the moment summation. This couple is a free vector and represents the effect of the fixed support on the frame.

Although only *three* independent equilibrium equations can be written for a rigid body, it is a good practice to *check* all calculations using a fourth equilibrium equation. The latter equation may be obtained from one of the other sets of equilibrium equations, 5–3, 5–4, or 5–5. For example, the above computations may be verified by summing moments about point C:

$$\zeta+\Sigma M_C = 0; \: 400 \text{ N}(1 \text{ m}) - 600 \text{ N}(2 \text{ m}) + 800 \text{ N} \cdot \text{m} \equiv 0$$
$$400 \text{ N} \cdot \text{m} - 1200 \text{ N} \cdot \text{m} + 800 \text{ N} \cdot \text{m} \equiv 0$$

Example 5–7

The uniform beam shown in Fig. 5–8a is subjected to a concentrated force and couple. If the beam is supported at A by a smooth wall and at

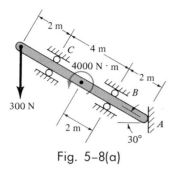

Fig. 5-8(a)

B and *C* by rollers, determine the reactions at these supports. Neglect the weight of the beam.

Solution

Step 1:   The free-body diagram for the beam is shown in Fig. 5-8*b*. All the support reactions act normal to the surface of contact since the contacting surfaces are smooth. The reactions at *B* and *C* are shown acting in the positive *y′* direction. This assumes that only the rollers located on the bottom of the beam are used for support. The 4000-N · m couple is a free vector, and it can therefore be shown acting at *any point* on the free-body diagram wihout affecting the computed results.

Step 2:   Using the *xy* coordinate system in Fig. 5-8*b* and applying the equations of equilibrium, we have

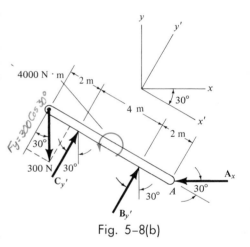

Fig. 5-8(b)

$$\xrightarrow{+}\Sigma F_x = 0; \qquad C_{y'} \sin 30° + B_{y'} \sin 30° - A_x = 0 \qquad (1)$$

$$+\uparrow\Sigma F_y = 0; \qquad -300 \text{ N} + C_{y'} \cos 30° + B_{y'} \cos 30° = 0 \qquad (2)$$

$$\zeta+\Sigma M_A = 0; \qquad -B_{y'}(2 \text{ m}) + 4000 \text{ N} \cdot \text{m} - C_{y'}(6 \text{ m})$$
$$+ (300 \cos 30° \text{ N})(8 \text{ m}) = 0 \quad (3)$$

When writing the moment equation, it should be noticed that the line of action of the force component 300 sin 30° N passes through point *A*, and therefore this force is not included in the moment equation.

Solving Eqs. (2) and (3) simultaneously, we obtain

$$B_{y'} = -1000.0 \text{ N} \qquad\qquad \textit{Ans.}$$
$$C_{y'} = 1346.4 \text{ N} \qquad\qquad \textit{Ans.}$$

Since $B_{y'}$ is a negative quantity, its direction is opposite to that shown on the free-body diagram in Fig. 5-8*b*. Therefore, the top roller at *B* serves as the support rather than the bottom one.

When the values of $B_{y'}$ and $C_{y'}$ are substituted into Eq. (1) we obtain

$$1346.4 \sin 30° \text{ N} - 1000.0 \sin 30° \text{ N} - A_x = 0$$
$$A_x = 173.2 \text{ N} \qquad\qquad \textit{Ans.}$$

Note that the negative sign of $B_{y'}$ must be *retained* when substituting into this equation, since the equations of equilibrium were *originally written* with the assumption that $\mathbf{B}_{y'}$ acts upward on the beam.

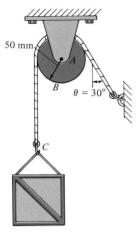

50 mm

$\theta = 30°$

B

A

C

Fig. 5-9(a)

Example 5-8

A 100-kg crate shown in Fig. 5-9a is held in equilibrium by means of a continuous cable and frictionless pulley. Determine the tension T in the cable and the horizontal and vertical components of reaction at the pin A.

Solution

Step 1: The free-body diagrams of the crate, cord, and pulley are shown in Fig. 5-9b. Note that the principle of action, equal but opposite reaction, must be carefully observed when drawing each of these diagrams. In particular, the cord exerts an unknown pressure distribution p along part of the pulley surface, whereas the pulley exerts an equal but opposite effect on the cord. For the solution, it is simpler to *combine* the free-body diagrams of the pulley and a portion of the cord, so that the pressure distribution becomes *internal* to the system and is therefore eliminated from the analysis, Fig. 5-9c.

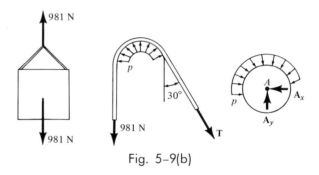

981 N

p

30°

981 N

T

A_x

p

A_y

Fig. 5-9(b)

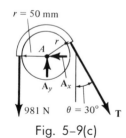

$r = 50$ mm

r

A

A_y A_x

981 N $\theta = 30°$ T

Fig. 5-9(c)

Step 2: Summing moments about point A to eliminate $\mathbf{A}_x$ and $\mathbf{A}_y$, Fig. 5-9c, we have

$$\zeta + \Sigma M_A = 0; \qquad 981 \text{ N}(50 \text{ mm}) - T(50 \text{ mm}) = 0$$
$$T = 981 \text{ N} \qquad\qquad Ans.$$

It is seen that the rope tension remains *constant* as it passes over the pulley. (This of course is true for *any angle* θ for which the rope is directed and for *any radius* r of the pulley.) Using the result for T, a force summation is applied to determine the reaction at pin A.

$$\xrightarrow{+} \Sigma F_x = 0; \qquad\qquad -A_x + 981 \sin 30° = 0$$
$$A_x = 490.5 \text{ N} \qquad\qquad Ans.$$

$$+\uparrow \Sigma F_y = 0; \qquad A_y - 981 - 981 \cos 30° = 0$$
$$A_y = 1830.5 \text{ N} \qquad\qquad Ans.$$

Example 5-9

The beam shown in Fig. 5–10*a* is pin-connected at *A* and rests against a smooth support at *B*. The cylinder has a mass of 100 kg and is suspended by a cord that is attached to the beam at *D* and passes over a smooth pulley at *C*. Compute the horizontal and vertical components of reaction at *A*.

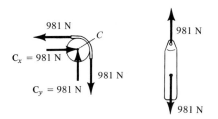

(a)

Solution I

Step 1: The free-body diagrams of the cylinder, beam, and pulley are shown in Fig. 5–10*b*. Since the cable is subjected to a uniform tension of 981 N, the pulley is in equilibrium if the beam exerts reactions $C_x = 981$ N and $C_y = 981$ N on the pulley as shown. In accordance with Newton's third law of motion, the pulley exerts equal but opposite reactions on the beam at *C*. The reaction N_B is perpendicular (or tangent) to the beam at *B*, since the support is smooth.

Step 2: Summing moments about *A*, we obtain a direct solution of N_B.

$$\zeta + \Sigma M_A = 0; \qquad -N_B(0.6) + 981(0.1) + 981(0.9) = 0$$
$$N_B = 1635 \text{ N}$$

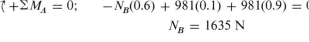

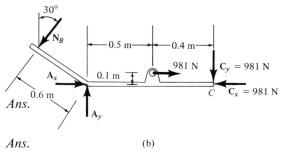

Using this result,

$$\xrightarrow{+}\Sigma F_x = 0; \qquad -1635 \sin 30° + A_x + 981 - 981 = 0$$
$$A_x = 817.5 \text{ N} \qquad \text{Ans.}$$

$$+\uparrow\Sigma F_y = 0; \qquad -1635 \cos 30° + A_y - 981 = 0$$
$$A_y = 2396.9 \qquad \text{Ans.}$$

(b)

Solution II

Step 1: An *easier* method for solving this problem consists of using a *single* free-body diagram of a "system" consisting of the beam, pulley, cord, and cylinder, Fig. 5–10*c*. Here the reactive forces at *C* and those of the rope (Fig. 5–10*b*) become *internal* to the system and are thereby eliminated from the analysis.

Applying the equations of equilibrium,

$$\zeta + \Sigma M_A = 0; \qquad -N_B(0.6) + 981(1.0) = 0$$
$$N_B = 1635 \text{ N}$$

$$\xrightarrow{+}\Sigma F_x = 0; \qquad -1635 \sin 30° + A_x = 0$$
$$A_x = 817.5 \text{ N} \qquad \text{Ans.}$$

$$+\uparrow\Sigma F_y = 0; \qquad -1635 \cos 30° + A_y - 981 = 0$$
$$A_y = 2396.9 \text{ N} \qquad \text{Ans.}$$

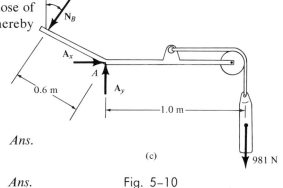

(c)

Fig. 5–10

175

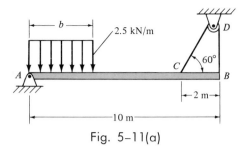

Fig. 5-11(a)

Example 5-10

The 100-kg uniform beam AB shown in Fig. 5-11a is supported at A by a pin and at B and C by a continuous cable which wraps around a frictionless pulley located at D. If a maximum tension force of 800 N can be developed in the cable before it breaks, determine the greatest length b of a uniform 2.5-kN/m distributed load that can be placed on the beam. The load is applied from the left support. What are the horizontal and vertical components of reaction at A just before the cable breaks?

Solution

Step 1: A free-body diagram of the beam is shown in Fig. 5-11b. Since the cable is continuous and passes over a frictionless pulley, the entire cable is subjected to its maximum tension of 800 N when the maximum loading is on the beam. Hence, the cable exerts an 800-N force at points C and B *on the beam* in the direction of the cable.

The distributed load is reduced to a concentrated force in accordance with the methods of Sec. 4–10. The magnitude of this force is equivalent to the area under the diagram, i.e., $(2500b)$ N. The force acts through the centroid of the area, a distance of $b/2$ from point A.

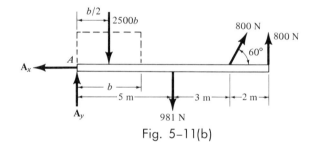

Fig. 5-11(b)

Step 2: By summing moments of the force system about point A (Fig. 5-11b) it is possible to obtain a direct solution for the dimension b. Why?

$$\zeta + \Sigma M_A = 0; \qquad -(2500b \text{ N}) \left(\frac{b}{2} \text{ m}\right) - 981 \text{ N}(5 \text{ m})$$
$$+ (800 \sin 60° \text{ N})(8 \text{ m}) + 800 \text{ N}(10 \text{ m}) = 0$$
$$b = 2.63 \text{ m} \qquad \textit{Ans.}$$

Using this result and summing forces in the x and y directions, we have

$$\xrightarrow{+} \Sigma F_x = 0; \qquad -A_x + 800 \cos 60° \text{ N} = 0$$
$$A_x = 400 \text{ N} \qquad \textit{Ans.}$$

$$+\uparrow \Sigma F_y = 0;$$
$$A_y - 2500 \text{ N/m}(2.63 \text{ m}) - 981 \text{ N} + 800 \sin 60° \text{ N} + 800 \text{ N} = 0$$
$$A_y = 6059.9 \text{ N} \qquad \textit{Ans.}$$

The solution to some equilibrium problems can be somewhat simplified if one is able to recognize members that are subjected to either two or three forces.

Two-Force Members.   When forces are applied at only two points on a member, the member is called a *two-force member*. Examples are shown in Fig. 5–12. In each case, the weight of the member is assumed to be negligible compared to the applied forces. *For both force and moment equilibrium, it is necessary that the resultant force* $\mathbf{F}_A$, *acting at one point A on the member, be of equal magnitude, opposite direction, and collinear with the resultant force* $\mathbf{F}_B$ *acting at point B*. Hence, the line of action of the force is specified from geometry, and only the force magnitude must be determined or stated.

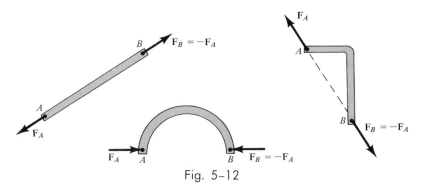

Fig. 5–12

Three-Force Members.   *If a member is subjected to three forces, then it is necessary that the forces be either concurrent or parallel if the member is to be in equilibrium.* To show this, consider the body in Fig. 5–13a and suppose that any two of the three forces acting on the body have lines of action that intersect at point O. To satisfy moment equilibrium about O, i.e., $\Sigma M_O = 0$, the third force must also pass through O, which then makes the force system *concurrent*. If two of the three forces are parallel, the point of concurrency, O, is considered to be at "infinity" and the third force must be parallel to the other two forces to intersect at this "point."

Since the three forces are concurrent at a point, the force equilibrium equations ($\Sigma F_x = 0$, $\Sigma F_y = 0$) can be solved by using either graphics or trigonometry. In this regard, it is necessary that the three forces, extended to the point of concurrency, form a closed triangle when they are added in a tip-to-tail fashion, Fig. 5–13b. The unknowns can then be determined by applying the law of sines or the law of cosines to the vector triangle.

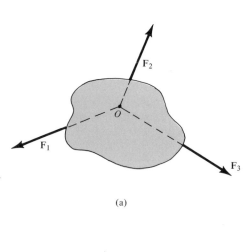

(a)

(b)

Fig. 5–13

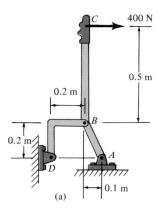

(a)

Example 5–11

The lever *ABC* is pin-supported at *A*, and connected to a short link *BD* as shown in Fig. 5–14*a*. If the weights of the members are negligible, determine the force developed on the lever at *A*.

Solution I

Step 1: As shown on the free-body diagram, Fig. 5-14*b*, the short link *BD* is a *two-force member,* so the *resultant forces* at pins *D* and *B* must be equal, opposite, and collinear. Although the magnitude of the force is unknown, the line of action is known, since it passes through *B* and *D*.

The lever *ABC* is a three-force member, so the three nonparallel forces acting on it are concurrent at *O*, Fig. 5–14*c*. The distance *CO* must be 0.5 m, since the lines of action of **F** and the 400-N force are known. Hence, the (unknown) angle θ of $\mathbf{F}_A$ at *A* is

$$\theta = \tan^{-1}\left(\frac{0.7}{0.4}\right) = 60.3° \qquad\qquad Ans.$$

Step 2: The vector addition of the three forces is shown in Fig. 5–14*d*, where $\phi = 180° - 119.7° - 45° = 15.3°$. Applying the law of sines,

$$\frac{F_A}{\sin 45°} = \frac{400}{\sin 15.3°}$$

$$F_A = 1074.9 \text{ N} \qquad\qquad Ans.$$

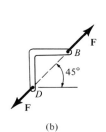

(b)

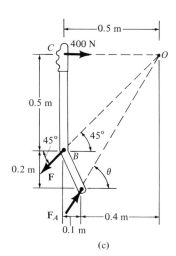

(c)

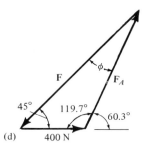

(d)

Fig. 5–14

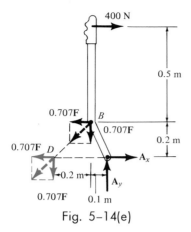

Fig. 5–14(e)

Solution II

Step 1: The free-body diagram of member ABC is shown in Fig. 5–14e. Note that $\mathbf{F}$ is directed from B to D, since DB is a two-force member. The three unknowns are A_x, A_y, and F.

Step 2: Summing moments about point A to obtain a direct solution for $\mathbf{F}$ yields

$$\zeta + \Sigma M_A = 0; \quad 400 \text{ N}(0.7 \text{ m}) - 0.707F(0.2 \text{ m}) - 0.707F(0.1 \text{ m}) = 0$$
$$400 \text{ N}(0.7 \text{ m}) - 0.707F(0.3 \text{ m}) = 0 \qquad (1)$$
$$F = 1320.1 \text{ N}$$

By the principle of transmissibility it is also possible to consider $\mathbf{F}$ to be located at D even though this point is not on the lever, Fig. 5–14e. In this case, the horizontal component of $\mathbf{F}$ creates *zero moment* about point A and the moment arm for the vertical component becomes 0.3 m. Hence, $\Sigma M_A = 0$ directly yields Eq. (1).

Summing forces to obtain A_x and A_y, using the result for $\mathbf{F}$, we have

$$\xrightarrow{+} \Sigma F_x = 0; \quad A_x - 0.707(1320.1 \text{ N}) + 400 \text{ N} = 0 \qquad A_x = 533.3 \text{ N}$$
$$+\uparrow \Sigma F_y = 0; \quad A_y - 0.707(1320.1 \text{ N}) = 0 \qquad A_y = 933.3 \text{ N}$$

Thus;

$$F_A = \sqrt{A_x^2 + A_y^2} = \sqrt{(533.3)^2 + (933.3)^2}$$
$$= 1074.9 \text{ N} \qquad \qquad Ans.$$
$$\theta = \tan^{-1}\left(\frac{A_y}{A_x}\right) = \tan^{-1}\left(\frac{933.3}{533.3}\right) = 60.3° \qquad Ans.$$

By comparison, solution I involves the least computation since the moment summation was not considered. Notice that implying that the force system be concurrent at O, Fig. 5–14c, necessitates that $\Sigma M_O = 0$. Furthermore, closure of the force triangle, Fig. 5–14d, implies that $\Sigma F_x = 0$ and $\Sigma F_y = 0$.

Problems

Neglect the weight of the members in the following problems unless specified.

5-13. Determine the reactive force components **P, T, R,** and **F** acting on cylinder *A* shown in Fig. 5–3*b*.

5-14. Determine the horizontal and vertical components of force developed by the pin at *A*, and the force in link *BD*, for the rod in Prob. 5–4.

5-15. Three smooth cylinders, each having a mass of $m = 10$ kg, rest on the 30° inclined plane *E*. Determine the reaction of plane *E* on each of the cylinders and the reaction of the 60° inclined plane *D* on cylinder *A*.

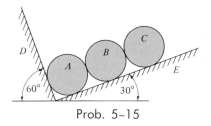

Prob. 5–15

5-15a. Solve Prob. 5–15 if each cylinder has a weight of $W = 20$ lb.

*5-16. Two smooth spheres, each having a mass of 200 g, are suspended from the same point by cords, each having a length of 100 mm. Determine the tension in each cord and the force acting between the spheres at their point of contact.

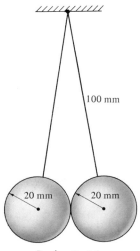

Prob. 5–16

5-17. Determine the horizontal and vertical components of reaction at the pin support *A* and the reaction at roller *B* for the loaded frame.

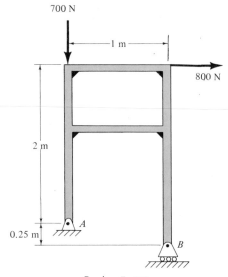

Prob. 5–17

5-18. Determine the magnitude and direction θ of the bone force $\mathbf{F}_B$ and the magnitude of the muscle force $\mathbf{T}$ necessary to hold the 10-kg ball in the equilibrium position shown. The ball has a center of mass at G.

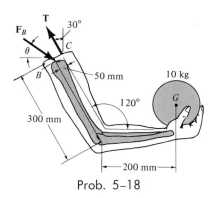

Prob. 5-18

5-19. Determine the tension in the cable and the horizontal and vertical components of reaction at the pin A. The pulley at D is frictionless.

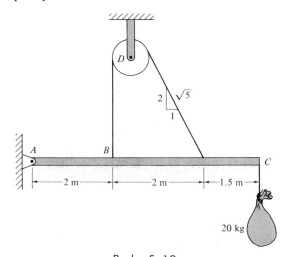

Prob. 5-19

*5-20. The forces acting on the jet aircraft, which is flying level at *constant velocity*, must be in equilibrium. If the aircraft has a mass of $m = 10$ Mg, and the jet provides a thrust which has a magnitude of $T = 5$ kN,

determine the lift provided by both the wings, $\mathbf{L}_w$, and tail, $\mathbf{L}_t$, and the air resistance or drag $\mathbf{F}_D$. The weight $\mathbf{W}$ of the aircraft acts through the center of gravity G as shown. Set $a = 1$ m, $b = 7$ m, and $c = 0.8$ m.

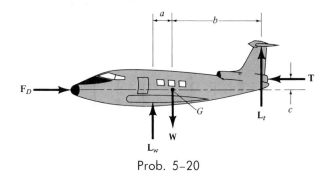

Prob. 5-20

*5-20a. Solve Prob. 5-20 if $W = 20(10^3)$ lb, $T = 1200$ lb, $a = 5$ ft, $b = 23$ ft, and $c = 2$ ft.

5-21. Determine the horizontal and vertical components of reaction at the pin support A and the force in link CE.

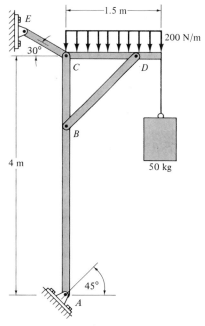

Prob. 5-21

5-22. While *slowly* walking, a man having a total mass of 80 kg places all his weight on *one foot*. Assuming that the normal force N_C of the ground acts on his foot at C, determine the resultant vertical compressive force F_B which the tibia T exerts on the astragalus B, and the vertical tension F_A in the achilles tendon A at the instant shown.

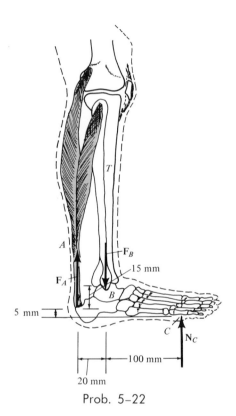

Prob. 5-22

extended to $l = 60$ m and $\theta = 45°$. Neglect the weight of the boom and the size of the pulley at E. Assume the crane does not overturn. *Note:* When $\theta = 60°$ BC is vertical, however, when $\theta = 45°$ this is not the case.

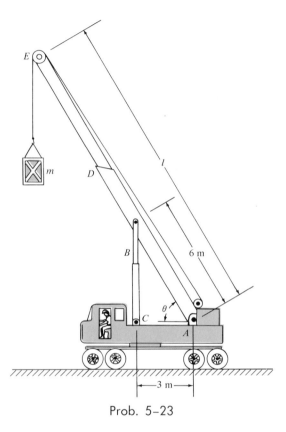

Prob. 5-23

5-23. The crane provides a long-reach capacity by using the telescopic boom segment DE. The entire boom is supported by a pin at A and by the telescopic hydraulic cylinder BC, which can be considered as a two-force member. The rated load capacity of the crane is measured by a maximum force developed in the hydraulic cylinder. If this maximum force is developed in the cylinder when the boom supports a mass $m = 4$ Mg and its length is $l = 50$ m and $\theta = 60°$, determine the greatest mass that can be supported when the boom length is

*5-24. Determine the horizontal and vertical components of reaction that the pin A and the roller B exert on the beam.

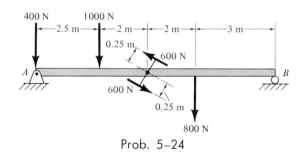

Prob. 5-24

5–25. Determine the horizontal and vertical components of reaction at A and the reaction at B required to support the truss. Set $F_1 = 600$ N, $F_2 = 2$ kN, $F_3 = 1$ kN, and $a = 3$ m.

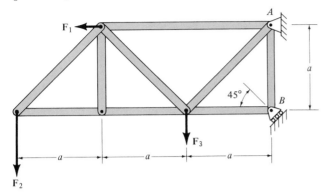

Prob. 5–25

5–25a. Solve Prob. 5–25 with $F_1 = 80$ lb, $F_2 = 500$ lb, $F_3 = 800$ lb, and $a = 10$ ft.

5–26. A lever valve is used to maintain a safe pressure in a boiler. If the maximum boiler pressure allowed is to be 2 MPa, determine the required mass m of block C that must be suspended from the stem so that the lever is on the verge of rotating upwards. For the solution, the boiler exerts a uniform pressure of 2 MPa on the *circular* cap at B as shown.

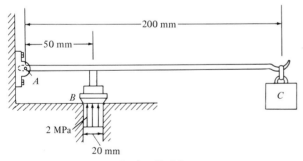

Prob. 5–26

5–27. The 25-kg lawn roller is to be lifted over the 100-mm-high step. Compare the magnitudes of force **P** required to (a) push it and (b) pull it over the step if in each case the force is directed at $\theta = 30°$ along the linkage AB as shown.

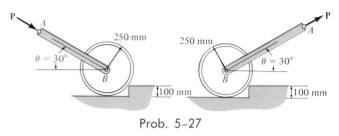

Prob. 5–27

*** 5–28.** Determine the minimum magnitude of force **P** and its associated angle θ required to pull the 25-kg roller over the 100-mm step in Prob. 5–27.

5–29. Calculate the tension in the cable at B required to hold the boom in the position shown. The cable passes over a frictionless pulley located at C. Neglect the size of this pulley.

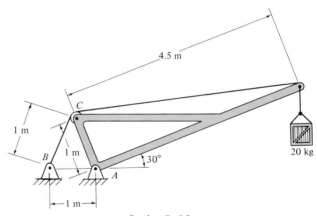

Prob. 5–29

5–30. The uniform ladder having a length $l = 4$ m, rests along the wall of a building at A and on the roof at B. If the ladder has a mass of $m = 10$ kg and the surfaces at A and B are smooth, determine the angle θ for equilibrium.

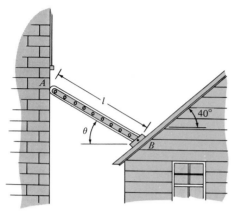

Prob. 5–30

5–30a. Solve Prob. 5–30 if the ladder has a weight of $W = 25$ lb and $l = 15$ ft.

5–31. A man having a total mass of 80 kg stands uniformly with both feet on a *smooth* ladder rung at A. Assuming the man's leg bone is only subjected to a compressive force F_B, determine this force and the tendon force $\mathbf{T}$. Also, show that for equilibrium, $\theta = 0°$.

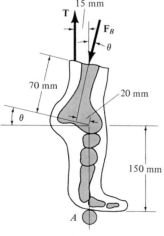

Prob. 5–31

* 5–32. When holding the 10-kg stone in equilibrium, the humerus H, assumed to be smooth, exerts normal forces $\mathbf{F}_C$ and $\mathbf{F}_A$ on the radius C and ulna A as shown. Determine these forces and the force $\mathbf{F}_B$ that the biceps B exert on the radius for equilibrium. The stone has a center of mass at G.

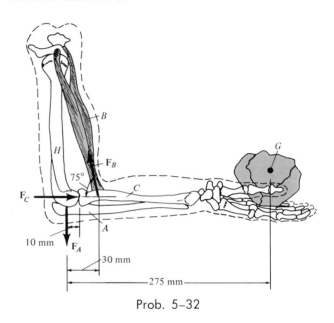

Prob. 5–32

5–33. Determine the tension in the supporting cables BC and DEF, and the reaction at the smooth incline A of the beam.

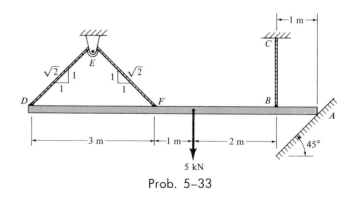

Prob. 5–33

184

5-34. A missile has a mass of 1.5 Mg, with center of gravity at G. It is hoisted into the firing position by means of a hydraulic cylinder located in arm AB. If the launching beam CD has a mass of 800 kg and a center of gravity at G', determine the force developed in AB when the beam is in the position shown.

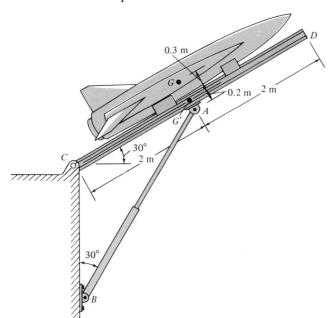

Prob. 5-34

5-35. Determine the horizontal and vertical components of force at the pin and the reaction at the roller support for the curved beam. Set $F_1 = 500$ N, $F_2 = 200$N, and $r = 2$ m.

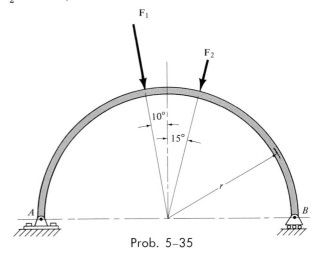

Prob. 5-35

5-35a. Solve Prob. 5-35 if $F_1 = 275$ lb, $F_2 = 145$ lb, and $r = 5$ ft.

*5-36. Determine the horizontal and vertical components of reaction at the pin A and the tension in cable BC for the boom. Two of these booms are used to equally support the uniform lifeboat, which has a mass of 400 kg.

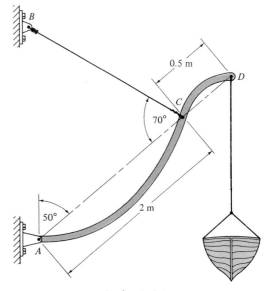

Prob. 5-36

5-37. A *Russell's traction* is used for immobilizing femoral fractures, C. If the lower leg has a mass of 7 kg, determine the mass m that must be suspended at D in order for the leg to be held in the position shown. Also, what is the tension force **F** in the femur and the distance $\bar{x}$ which locates the center of gravity G of the lower leg?

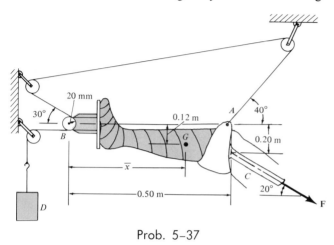

Prob. 5-37

5-38. The symmetrical shelf is subjected to a uniform load of 4 kPa. Support is provided by a bolt (or pin) located at each end A and A' and by the symmetrical brace arms, which bear against the smooth wall on both sides at B and B'. Determine the force resisted by each bolt at the wall and the force at B for equilibrium.

5-39. The *flying boom B* is used with a crane to position construction materials in coves and under overhangs. The horizontal "balance" of the boom is controlled by a 350-kg block D, which has a center of gravity at G and moves by internal sensing devices along the bottom flange F of the beam. Determine the position x of the block when the boom is used to lift the stone S, which has a mass of 75 kg. What is x when the stone is removed? The boom is uniform and has a mass of 100 kg.

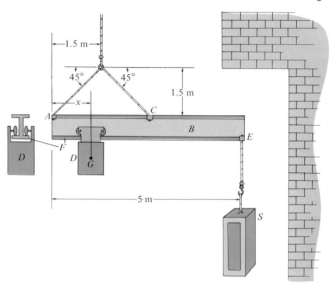

Prob. 5-39

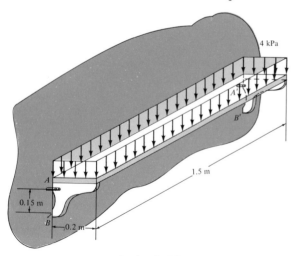

Prob. 5-38

186

*5-40. A cantilever beam, having an extended length of $a = 2$ m, is subjected to a vertical force $F = 6$ kN acting at its end. Assuming that the wall resists this load with linearly varying distributed loads over the length $b = 0.3$ m of the beam, determine the intensities w_1 and w_2 for equilibrium if $a = 2$ m. *Note:* For $b \ll a$, this distribution of load is replaced by a resultant force $\mathbf{F}_y$ and couple moment $\mathbf{M}$ as shown in Table 5–1, no. 9.

5-41. The crane AB and truck have a total mass of 50 Mg and a center of gravity at G. Determine the *greatest* mass m that can be suspended at C without tipping the truck.

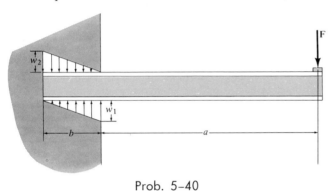

Prob. 5-40

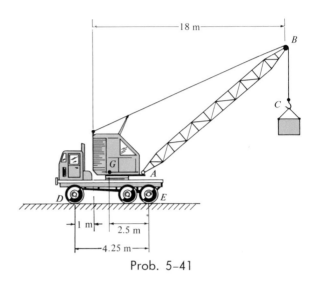

Prob. 5-41

*5-40a. Solve Prob. 5–40 if $F = 900$ lb, $a = 6$ ft, and $b = 1.5$ ft.

Spatial Force Systems Acting on a Rigid Body

5-4. Support Reactions

The first step in solving three-dimensional equilibrium problems, as in the case of two dimensions, is to draw a free-body diagram of the body (or group of bodies considered as a system). When drawing this diagram, it is necessary to include *all the reactions* that can occur at each point of support.

The reactive forces and couples acting at various types of supports and connections, when the members are viewed in three dimensions, are listed in Table 5–2. It is important both to recognize the symbols used to represent each of the supports and to clearly understand how the forces and couples are developed by each support. As in the two-dimensional case, a *force* is developed by a support that restricts the *translation* of the attached member, whereas a *couple* is developed when *rotation* of the attached member is prevented. For example, in Table 5–2, the ball-and-socket joint (4) prevents any translation of the connecting member; therefore, a force must act on the member at the point of connection. This

Table 5-2

3-D ANALYSIS

Types of Connection	Reaction	Number of Unknowns
(1) light cable		One unknown: magnitude of **F**; this force prevents translation of the attached member in the direction of the cable.
(2) smooth surface support		One unknown: magnitude of **F**; this force prevents translation of the member normal to the surface at the point of contact.
(3) roller on a smooth surface		One unknown: magnitude of **F**; this force prevents translation of the member normal to the surface at the point of contact.
(4) ball and socket		Three unknowns: magnitudes of three force components F_x, F_y, F_z; these forces prevent any translation of the member.
(5) single smooth bearing		Four unknowns: magnitudes of two force components F_x, F_z and the two couples M_x, M_z. The forces prevent translation of the member in the x-z plane, and the couple moments prevent rotation about any axis located in this plane.
(6) single smooth pin		Five unknowns: magnitudes of three force components F_x, F_y, F_z and two couples M_y, M_z. The forces prevent translation of the member in any general direction. The couple moments prevent rotation of the member about an axis located in the y-z plane.

Table 5–2 (Contd.)

Types of Connection	Reaction	Number of Unknowns
(7) 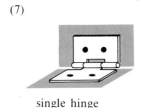 single hinge	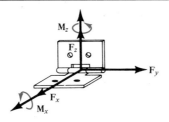	Five unknowns: magnitudes of three force components $\mathbf{F}_x$, $\mathbf{F}_y$, $\mathbf{F}_z$ and two couples $\mathbf{M}_x$, $\mathbf{M}_z$. The forces prevent translation of the member in any given direction. The couple moments prevent rotation about any axis located in the x-z plane.
(8) 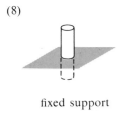 fixed support	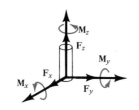	Six unknowns: magnitudes of three force components $\mathbf{F}_x$, $\mathbf{F}_y$, $\mathbf{F}_z$ and three couples $\mathbf{M}_x$, $\mathbf{M}_y$, $\mathbf{M}_z$. The forces prevent any translation of the member and the couples prevent any rotation.

force has three components, having unknown magnitudes, F_x, F_y, F_z. Provided these components are known, one can obtain the magnitude of force, $F = \sqrt{F_x^2 + F_y^2 + F_z^2}$, and its orientation, defined by the coordinate direction angles

$$\alpha = \cos^{-1}(F_x/F), \; \beta = \cos^{-1}(F_y/F), \; \gamma = \cos^{-1}(F_z/F)*$$

Since the connecting member is allowed to rotate freely about *any* axis, no moment is resisted at this support.

The *single smooth bearing* (5), Fig. 5–15a, is used to support the shaft subjected to $\mathbf{F}_x$ and $\mathbf{F}_z$. From the free-body diagram, Fig. 5–15b, it is seen that two force and two couple reactions are developed on the shaft by the support (Table 5–2 (5)). The force reactions $\mathbf{A}_x$ and $\mathbf{A}_z$ prevent translation of the shaft in the x and z directions, and the couples $\mathbf{M}_x$ and $\mathbf{M}_z$ prevent rotation about these axes. Furthermore, observe that the support cannot prevent translation or rotation of the shaft along the y axis. Consider now the same shaft supported by *two* smooth bearings A and B which are *properly aligned* on the shaft, Fig. 5–16a. In this case *only force reactions*

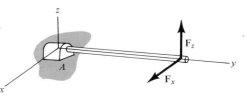

(a)

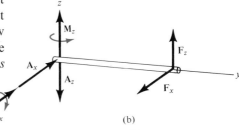

(b)

Fig. 5–15

*The three unknowns may also be represented as an unknown force magnitude F, and two unknown coordinate direction angles. The third direction angle can be obtained using the identity $\cos^2 \alpha + \cos^2 \beta + \cos^2 \gamma = 1$ (Eq. 2–10).

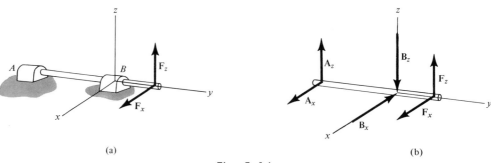

(a)

(b)

Fig. 5–16

are exerted on the shaft by the bearings, Fig. 5–16b. Moment reactions *cannot* occur, since the force reactions alone provide the necessary constraint. In other words, it is *not possible* for the applied forces F_x and F_y to cause the shaft to rotate about one of the bearings; this is *prevented* by constraining forces developed at the *other* bearing. For example, rotation of the shaft about the x axis caused by F_z, Fig. 5–16b, is prevented by the reaction A_z, whereas a rotation about the z axis caused by F_x is prevented by A_x. The same sort of situation applies for a body connected to two hinges. A *single hinge* (Table 5–2(7)) acting on the plate, Fig. 5–17a, may be subjected to "twisting" around the x and z axes due to an applied loading (not shown). To resist this, couples M_x and M_z are developed at the connection, Fig. 5–17a. If two properly aligned hinges A and B act on a plate, Fig. 5–17b, the hinge B, for example, cannot be twisted about the x or z axis, since there are constraining forces developed by hinge A. Therefore, no couples are developed at the hinges.

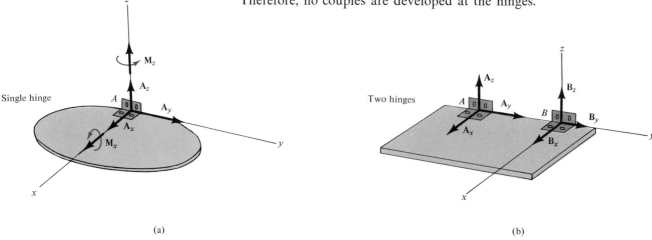

Single hinge

Two hinges

(a)

(b)

Fig. 5–17

5-5. Equations of Equilibrium

A necessary condition for the equilibrium of a rigid body subjected to a three-dimensional force system requires that both the *resultant* force and *resultant* couple acting on the body be equal to *zero*.

Vector Equations of Equilibrium.   The two conditions of equilibrium for a rigid body may be expressed mathematically in vector form as

$$\Sigma \mathbf{F} = \mathbf{0}$$
$$\Sigma \mathbf{M}_O = \mathbf{0}$$

(5–6)

where $\Sigma \mathbf{F}$ is the vector sum of all the external forces acting on the body and $\Sigma \mathbf{M}_O$ is the sum of the moments of all these forces about any point O located either on or off the body.

Scalar Equations of Equilibrium.   If each of the applied external forces and moments are expressed in Cartesian vector form and substituted into Eqs. 5–6, we have

$$\Sigma \mathbf{F} = \Sigma F_x \mathbf{i} + \Sigma F_y \mathbf{j} + \Sigma F_z \mathbf{k} = \mathbf{0}$$
$$\Sigma \mathbf{M}_O = \Sigma M_x \mathbf{i} + \Sigma M_y \mathbf{j} + \Sigma M_z \mathbf{k} = \mathbf{0}$$

Since the $\mathbf{i}, \mathbf{j},$ and $\mathbf{k}$ components are all independent from one another, the above equations are satisfied provided

$$\Sigma F_x = 0$$
$$\Sigma F_y = 0$$
$$\Sigma F_z = 0$$

(5–7a)

and

$$\Sigma M_x = 0$$
$$\Sigma M_y = 0$$
$$\Sigma M_z = 0$$

(5–7b)

These *six scalar equilibrium equations* may be used to solve for at most six unknowns shown on the free-body diagram. Equations 5–7a express the fact that the sum of the external force components acting in the $x, y,$ and z directions, respectively, must be zero and Eqs. 5–7b require the sum of the moment components directed along the $x, y,$ and z *axes,* respectively, to be zero.

The following two-step procedure should be used when solving spatial force equilibrium problems.

Step 1: Construct the free-body diagram for the body (or group of bodies considered as a single system). When drawing this diagram, it is important to include *all* the forces and couples that act *on* the body or group of bodies in question. These interactions are commonly caused by the externally applied loadings, contact forces exerted by adjacent bodies on the body, support reactions, and the weight of the body if it is significant compared to the magnitude of the other applied forces. Dimensions of the body, necessary for computing the moments of forces, are also included on the free-body diagram.

Step 2: Apply the equations of equilibrium. In many cases, problems can be solved by *direct application* of the six scalar equations $\Sigma F_x = 0$, $\Sigma F_y = 0$, $\Sigma F_z = 0$, $\Sigma M_x = 0$, $\Sigma M_y = 0$, $\Sigma M_z = 0$ (Eqs. 5–7); however, if the force components or moment arms seem difficult to determine, it is recommended that the solution be obtained by using vector equations: $\Sigma \mathbf{F} = \mathbf{0}, \Sigma \mathbf{M}_O = \mathbf{0}$ (Eqs. 5–6). In any case, it is *not necessary* that the set of axes chosen for force summation *coincide* with the set of axes chosen for moment summation. Instead, it is recommended that one *choose the direction of a moment axis such that it intersects the line of action of as many unknown forces as possible*. The moments of forces passing through points on this axis or forces which are parallel to the axis will then be zero. Furthermore, *any set of three nonorthogonal axes* may be chosen for either the force or moment summations. These axes must, however, *not be parallel* to one another, or linearly dependent equations will result. By the proper choice of axes, it may be possible to solve directly for an unknown quantity, or at least reduce the need for solving a large number of simultaneous equations for the unknowns.

5–6. Sufficient Conditions for Equilibrium

To ensure the equilibrium of a rigid body, it is not only necessary to satisfy the equations of equilibrium, but the body must also be *properly constrained* by its supports. As will be shown, proper constraining requires that (1) the lines of action of the support reactive forces do not intersect points on a common axis, and (2) the reactive forces must not all be parallel to one another. When the number of reactive forces needed to properly constrain the body in question is a *minimum*, the equations of equilibrium are *both* necessary and sufficient conditions for determining all the reactive forces. In particular, if *all* the unknown forces on a body can be determined by the equations of equilibrium, the problem is called *statically determinate*. However, if a problem exists in which these equations do not provide sufficient information for determining all the unknowns, the problem is referred to as *statically indeterminate.*

Redundant Constraints.   Statical indeterminacy can arise if a body has more constraints than are necessary (redundant) to maintain equilibrium. For example, the two-dimensional problem, Fig. 5-18a, and the three-dimensional problem, Fig. 5-18b, shown together with their free-body diagrams, are both statically indeterminate because of the redundancy in the number of support reactions. In the two-dimensional case, there are five unknowns, that is, M_A, A_x, A_y, B_y, and C_y, for which only three equilibrium equations can be written ($\Sigma F_x = 0$, $\Sigma F_y = 0$, and $\Sigma M_O = 0$, Eqs. 5-2). Two more equations are needed for a complete solution. Consequently, this problem is termed "indeterminate to the second degree" (five unknowns minus three equations). The three-dimensional problem has nine unknowns, for which only six equilibrium equations can be written (Eqs. 5-7). It is "indeterminate to the third degree." The additional equations needed to solve indeterminate problems of the type shown in Fig. 5-18 may be obtained from the conditions of *deformation* which occur between the loads and the internal movements of the body. These relations involve the physical properties of the body which are studied in subjects dealing with the mechanics of deformation, such as "strength of materials."

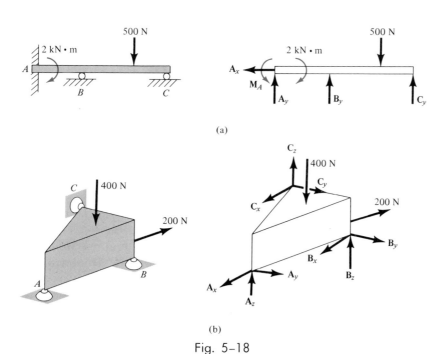

(a)

(b)

Fig. 5-18

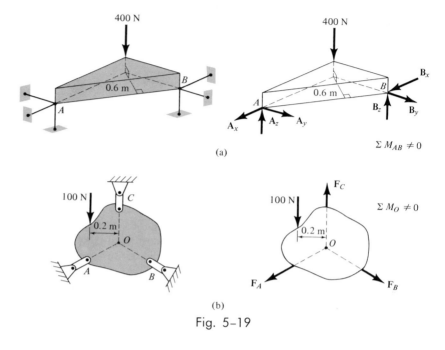

Fig. 5–19

Improper Constraints. In some cases, there may be as many unknown forces as there are equations of equilibrium; however, *instability* of the body can develop because of *improper constraining* action by the supports. In the case of three-dimensional problems, the body is improperly constrained if the support reactions *all intersect a common axis*. In two-dimensional problems, this axis is *perpendicular* to the plane of the forces and therefore appears as a point. Hence, when the reactive forces are *concurrent* at this point, the body is improperly constrained. Examples of both cases are given in Fig. 5–19. From the free-body diagrams it is seen that the summation of moments about axis AB, Fig. 5–19a, or point O, Fig. 5–19b, will *not* be equal to zero; thus rotation about axis AB or point O will take place.* Furthermore, in both cases it becomes *impossible* to solve *completely* for all the unknowns, since one can write a moment equation that does not involve any of the unknown support reactions. (This limits by one the number of available equilibrium equations.)

Another way in which partial constraining leads to instability occurs when the *reactive forces* are all *parallel*. Three- and two-dimensional examples of this are shown in Figs. 5–20a and 5–20b. In both cases, the summation of forces along the horizontal aa axis will not be equal to zero.

---

*For the three-dimensional problem, $\Sigma M_{AB} = (400\ \text{N})(0.6\ \text{m}) \neq 0$, and for the two-dimensional problem, $\Sigma M_O = (100\ \text{N})(0.2\ \text{m}) \neq 0$.

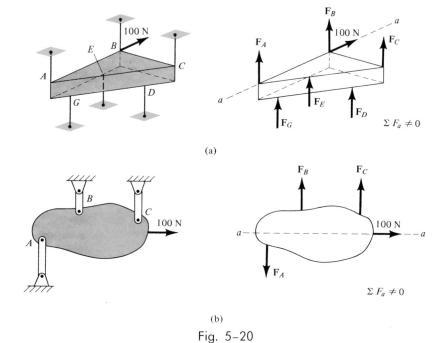

(a)

(b)

Fig. 5-20

Partial Constraints. In some cases, a body may have *fewer* reactive forces than equations of equilibrium that must be satisfied. The body then becomes only *partially constrained*. For example, consider the body shown in Fig. 5–21a with its corresponding free-body diagram in Fig. 5–21b. If O is a point not on the line of action of AB, the equations $\Sigma F_x = 0$ and $\Sigma M_O = 0$ will be satisfied by proper choice of the reactions $\mathbf{F}_A$ and $\mathbf{F}_B$. The equation $\Sigma F_y = 0$, however, will not be satisfied for the loading conditions and therefore equilibrium will not be maintained.

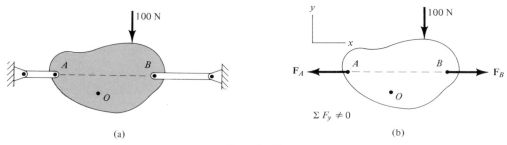

(a)

(b)

Fig. 5-21

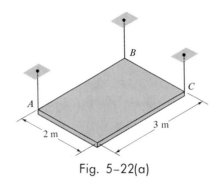

Fig. 5-22(a)

Example 5-12

The homogeneous plate shown in Fig. 5-22a has a mass of 100 kg. The plate is suspended in the horizontal plane by means of three parallel cords. Determine the tension in each of the cords.

Solution I

(***Vector Analysis***).   *Step 1:*   A free-body diagram of the plate is shown in Fig. 5-22b. The weight acts at the center of gravity of the plate (which coincides with the centroid). The magnitude of this force is $W = (100)(9.81) = 981$ N.

Step 2:   If the x, y, and z axes are oriented as shown in the figure, the vector-force equilibrium equation can be written as

$$\Sigma \mathbf{F} = \mathbf{0}; \qquad T_A \mathbf{k} + T_B \mathbf{k} + T_C \mathbf{k} - 981 \mathbf{k} = \mathbf{0}$$

or

$$\Sigma F_z = 0; \qquad T_A + T_B + T_C - 981 = 0 \qquad (1)$$

Summing moments of all the forces about the origin yields

$$\Sigma \mathbf{M}_A = \mathbf{0}; \qquad (\mathbf{r}_B \times T_B \mathbf{k}) + (\mathbf{r}_C \times T_C \mathbf{k}) + (\mathbf{r}_D \times \mathbf{W}) = \mathbf{0}$$

The position vectors can be determined from the geometry of the plate. Hence,

$$[3\mathbf{j} \times T_B \mathbf{k}] + [(2\mathbf{i} + 3\mathbf{j}) \times T_C \mathbf{k}] + [(1\mathbf{i} + 1.5\mathbf{j}) \times (-981 \mathbf{k})] = \mathbf{0}$$

or

$$(3T_B + 3T_C - 1471.5)\mathbf{i} + (-2T_C + 981)\mathbf{j} = \mathbf{0}$$

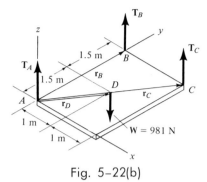

Fig. 5–22(b)

Equating to zero the corresponding **i** and **j** components gives the two scalar moment equations:

$$\Sigma M_x = 0; \qquad 3T_B + 3T_C - 1471.5 = 0 \qquad (2)$$
$$\Sigma M_y = 0; \qquad -2T_C + 981 = 0 \qquad (3)$$

Solving Eqs. (1) through (3), we obtain

$$T_A = 490.5 \text{ N} \qquad\qquad Ans.$$
$$T_B = 0 \qquad\qquad Ans.$$
$$T_C = 490.5 \text{ N} \qquad\qquad Ans.$$

Note that these reactions provide only the necessary *vertical support* for the loading. If a force acting in the *x-y* plane is applied to the plate, other supports would be required to maintain equilibrium. For this reason, the plate is only partially constrained.

Solution II

(*Scalar Analysis*). Since the three-dimensional geometry is rather simple, a *scalar analysis provides a direct solution to this problem*. From the free-body diagram, Fig. 5–22b, the force equilibrium equation $\Sigma F_z = 0$ yields Eq. (1). By definition, the moment of a force about an axis is equal to the product of the force magnitude and the perpendicular distance (moment arm) from the line of action of the force to the axis (Sec. 4–3). The direction of the moment is determined by the right-hand rule. Hence, summing the moments of the forces on the free-body diagram about the *x* and *y* axes yields Eqs. (2) and (3), respectively.

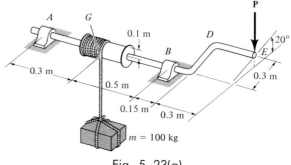

Fig. 5–23(a)

Example 5–13

The windlass shown in Fig. 5–23a is supported by two smooth bearings *A* and *B*, which are properly aligned on the shaft. Determine the magnitude of the vertical force **P** that must be applied to the handle to maintain equilibrium of the 100-kg crate. Also calculate the reactions at the bearings.

Solution
(*Scalar Analysis*).   *Step 1:*   The free-body diagram of the windlass is shown in Fig. 5–23b. Since the bearings at *A* and *B* are aligned correctly, *only* force reactions occur at these supports. Why are there no moment reactions?
Step 2:   For this problem, as in the last example, each of the five unknown force magnitudes shown on the free-body diagram can easily be determined by direct application of the *scalar* equations of equilibrium. Summing moments about the *x* axis, yields a direct solution for **P**. Why? For this scalar moment summation, it is necessary to compute the moment of each force as the product of the force magnitude and the *perpendicular distance* from the *x* axis to the line of action of the force. Using the right-hand rule and assuming positive moments act in the +**i** direction,

$$\Sigma (M_B)_x = 0; \qquad 981 \text{ N}(0.1 \text{ m}) - P(0.3 \cos 20° \text{ m}) = 0$$
$$P = 348.0 \text{ N} \qquad\qquad Ans.$$

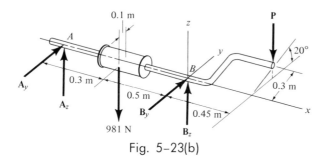

0.1 m

A

0.3 m

A_y

A_z

0.5 m

981 N

B_y

0.45 m

B_z

z

y

B

P

20°

0.3 m

x

Fig. 5–23(b)

Using this result and summing moments in a similar manner about the y and z axes yields

$$\Sigma(M_B)_y = 0; \quad -981 \text{ N}(0.5 \text{ m}) + A_z(0.8 \text{ m}) + (348.0 \text{ N})(0.45 \text{ m}) = 0$$
$$A_z = 417.4 \text{ N} \qquad\qquad Ans.$$

and

$$\Sigma(M_B)_z = 0; \qquad\qquad -A_y (0.8 \text{ m}) = 0$$
$$A_y = 0 \qquad\qquad Ans.$$

(Here a summation of moments was taken about axes which pass through the bearing support at B in order to eliminate the two unknown force components $\mathbf{B}_y$ and $\mathbf{B}_z$ from the moment equations.)

The reactions at B are obtained by a force summation, using the results computed above:

$$\Sigma F_y = 0; \qquad\qquad 0 + B_y = 0 \qquad B_y = 0 \qquad\qquad Ans.$$
$$\Sigma F_z = 0; \quad 417.4 - 981 + B_z - 348.0 = 0 \qquad B_z = 911.6 \text{ N} \qquad Ans.$$

As shown on the free-body diagram, the *supports* do not provide resistance against translation in the x direction. Hence, the windlass is only partially constrained.

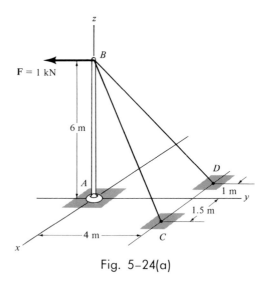

Fig. 5-24(a)

Example 5-14

Determine the tensions in the cables BC and BD, and the reaction at the support A for the mast shown in Fig. 5-24a.

Solution
(**Vector Analysis**).   *Step 1:*   The free-body diagram of the mast is shown in Fig. 5-24b. Since the support at A is a ball-and-socket joint, three unknown force components act at this point. The cable tensions $\mathbf{T}_{BC}$ and $\mathbf{T}_{BD}$ have a known direction and unknown magnitude. There are, therefore, five unknown force magnitudes shown on the free-body diagram.
Step 2:   Expressing each force on the free-body diagram in Cartesian vector form, we have

$$\mathbf{F} = \{-1000\mathbf{j}\}\ \text{N}$$

$$\mathbf{F}_A = A_x\mathbf{i} + A_y\mathbf{j} + A_z\mathbf{k}$$

$$\mathbf{T}_{BC} = T_{BC}\left(\frac{\mathbf{r}_{BC}}{r_{BC}}\right) = \frac{1.5T_{BC}}{\sqrt{54.25}}\mathbf{i} + \frac{4T_{BC}}{\sqrt{54.25}}\mathbf{j} - \frac{6T_{BC}}{\sqrt{54.25}}\mathbf{k}$$

$$= 0.204T_{BC}\mathbf{i} + 0.543T_{BC}\mathbf{j} - 0.815T_{BC}\mathbf{k}$$

$$\mathbf{T}_{BD} = T_{BD}\left(\frac{\mathbf{r}_{BD}}{r_{BD}}\right) = -\frac{1T_{BD}}{\sqrt{53}}\mathbf{i} + \frac{4T_{BD}}{\sqrt{53}}\mathbf{j} - \frac{6T_{BD}}{\sqrt{53}}\mathbf{k}$$

$$= -0.137T_{BD}\mathbf{i} + 0.549T_{BD}\mathbf{j} - 0.824T_{BD}\mathbf{k}$$

Applying the force equation of equilibrium gives

$$\Sigma \mathbf{F} = \mathbf{0}; \qquad \mathbf{F} + \mathbf{F}_A + \mathbf{T}_{BC} + \mathbf{T}_{BD} = \mathbf{0}$$
$$(A_x + 0.204 T_{BC} - 0.137 T_{BD})\mathbf{i} + (-1000 + A_y + 0.543 T_{BC} + 0.549 T_{BD})\mathbf{j}$$
$$+ (A_z - 0.815 T_{BC} - 0.824 T_{BD})\mathbf{k} = \mathbf{0}$$

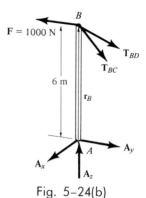

Fig. 5–24(b)

Hence,

$$\Sigma F_x = 0; \qquad A_x + 0.204 T_{BC} - 0.137 T_{BD} = 0 \qquad (1)$$
$$\Sigma F_y = 0; \qquad A_y + 0.543 T_{BC} + 0.549 T_{BD} - 1000 = 0 \qquad (2)$$
$$\Sigma F_z = 0; \qquad A_z - 0.815 T_{BC} - 0.824 T_{BD} = 0 \qquad (3)$$

Summing moments about the x, y, and z axes passing through point A to eliminate the three unknown force components passing through this point, we have

$$\Sigma \mathbf{M}_A = \mathbf{0}; \qquad (\mathbf{r}_B \times \mathbf{F}) + (\mathbf{r}_B \times \mathbf{T}_{BC}) + (\mathbf{r}_B \times \mathbf{T}_{BD}) = \mathbf{0}$$
$$\mathbf{r}_B \times (\mathbf{F} + \mathbf{T}_{BC} + \mathbf{T}_{BD}) = \mathbf{0}$$
$$6\mathbf{k} \times (-1000\mathbf{j} + 0.204 T_{BC}\mathbf{i} + 0.543 T_{BC}\mathbf{j} - 0.815 T_{BC}\mathbf{k}$$
$$-0.137 T_{BD}\mathbf{i} + 0.549 T_{BD}\mathbf{j} - 0.824 T_{BD}\mathbf{k}) = \mathbf{0}$$

Evaluating the cross product and combining terms yields

$$(-3.26 T_{BC} - 3.29 T_{BD} + 6000)\mathbf{i} + (1.22 T_{BC} - 0.822 T_{BD})\mathbf{j} = \mathbf{0}$$

so

$$\Sigma M_x = 0; \qquad -3.26 T_{BC} - 3.29 T_{BD} + 6000 = 0 \qquad (4)$$
$$\Sigma M_y = 0; \qquad 1.22 T_{BC} - 0.822 T_{BD} = 0 \qquad (5)$$

The moment equation about the z axis, $\Sigma M_z = 0$, is automatically satisfied for the given loading system since the lines of action of all the forces acting on the mast pass through the z axis.

Solving Eqs. (4) and (5) for T_{BC} and T_{BD} and substituting these values into Eqs. (1) through (3) yields

$$T_{BC} = 736.8 \text{ N} \qquad \qquad Ans.$$
$$T_{BD} = 1093.6 \text{ N} \qquad \qquad Ans.$$
$$A_x = 0.0 \text{ N} \qquad \qquad Ans.$$
$$A_y = 0.0 \text{ N} \qquad \qquad Ans.$$
$$A_z = 1501.6 \text{ N} \qquad \qquad Ans.$$

Note that the values of $A_x = A_y = 0$ could have been determined *by inspection*. Since the mast is a two-force member, the resultant force at A ($A_z = 1501.6$ N) must be directed along the mast and be equal but opposite to the resultant force at B, which is $\mathbf{F} + \mathbf{T}_{BC} + \mathbf{T}_{BD}$.

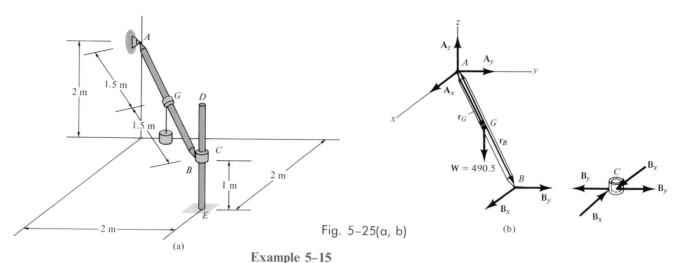

Fig. 5–25(a, b)

(a)

(b)

Example 5–15

Rod *AB*, shown in Fig. 5–25*a*, is used to support the 50-kg cylinder. Determine the components of reaction exerted on the rod by the ball-and-socket joints located at its ends *A* and *B*. The connection at *B* is attached to a collar *C*, which rests on the smooth rod *DE*. Neglect the mass of the rod and collar.

Solution I

(***Vector Analysis***). *Step 1:* The free-body diagrams of the rod and collar are shown in Fig. 5–25*b*. The components of force shown on the collar represent the forces developed by the ball-and-socket joint at *B* and rod *DE*. For equilibrium the *x* and *y* components must be of equal magnitude and opposite direction as shown. Since the *collar* is allowed to *translate* along rod *DE*, a vertical component of force, $\mathbf{B}_z$, cannot be developed at *B*. Notice that the principle of action, equal but opposite reaction, must be applied to $\mathbf{B}_x$ and $\mathbf{B}_y$ when constructing the two diagrams.

Step 2: Representing each of the forces on the free-body diagram of the rod in Cartesian vector form, we have

$$\mathbf{F}_A = A_x\mathbf{i} + A_y\mathbf{j} + A_z\mathbf{k}$$
$$\mathbf{F}_B = B_x\mathbf{i} + B_y\mathbf{j}$$
$$\mathbf{W} = \{-490.5\mathbf{k}\} \text{ N}$$

Applying the force equation of equilibrium,

$$\Sigma\mathbf{F} = 0; \qquad\qquad \mathbf{F}_A + \mathbf{F}_B + \mathbf{W} = 0$$
$$(A_x + B_x)\mathbf{i} + (A_y + B_y)\mathbf{j} + (A_z - 490.5)\mathbf{k} = 0$$

Thus,

$\Sigma F_x = 0;$	$A_x + B_x = 0$	(1)
$\Sigma F_y = 0;$	$A_y + B_y = 0$	(2)
$\Sigma F_z = 0;$	$A_z - 490.5 = 0$	(3)

Summing moments about the x, y, z axes passing through point A in order to eliminate the three components of force acting there, we have

$$\Sigma M_A = 0; \qquad\qquad (\mathbf{r}_G \times \mathbf{W}) + (\mathbf{r}_B \times \mathbf{F}_B) = 0$$

From Fig. 5-25b, the position vector $\mathbf{r}_B = \{2\mathbf{i} + 2\mathbf{j} - 1\mathbf{k}\}$ m. Since point G is located at the midpoint of the rod, then $\mathbf{r}_G = \frac{1}{2}\mathbf{r}_B = \{1\mathbf{i} + 1\mathbf{j} - 0.5\mathbf{k}\}$ m. Expressing the cross products in determinant form, we have

$$\begin{vmatrix} \mathbf{i} & \mathbf{j} & -\mathbf{k} \\ 1 & 1 & -0.5 \\ 0 & 0 & -490.5 \end{vmatrix} + \begin{vmatrix} \mathbf{i} & \mathbf{j} & \mathbf{k} \\ 2 & 2 & -1 \\ B_x & B_y & 0 \end{vmatrix} = 0$$

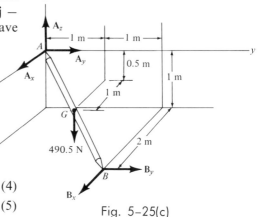

Fig. 5-25(c)

Expanding and rearranging terms yields

$$(B_y - 490.5)\mathbf{i} + (-B_x + 490.5)\mathbf{j} + (2B_y - 2B_x)\mathbf{k} = 0$$

so that

$$\Sigma M_x = 0; \qquad\qquad B_y - 490.5 = 0 \qquad (4)$$
$$\Sigma M_y = 0; \qquad\qquad -B_x + 490.5 = 0 \qquad (5)$$
$$\Sigma M_z = 0; \qquad\qquad 2B_y - 2B_x = 0 \qquad (6)$$

Solving Eqs. (1) through (6), we get

$$A_x = -490.5 \text{ N} \qquad\qquad Ans.$$
$$A_y = -490.5 \text{ N} \qquad\qquad Ans.$$
$$A_z = 490.5 \text{ N} \qquad\qquad Ans.$$
$$B_x = 490.5 \text{ N} \qquad\qquad Ans.$$
$$B_y = 490.5 \text{ N} \qquad\qquad Ans.$$

The negative signs indicate that $\mathbf{A}_x$, $\mathbf{A}_y$ act in a direction opposite to that shown on the free-body diagram, Fig. 5-25b. It should be noted that *six* equilibrium equations were written to solve for the *five unknowns*. Unique solutions were obtained however, since only two of Eqs. (4) through (6) are independent of one another. By inspection it is seen that the rod is only partially constrained, since the ball-and-socket supports do not provide resistance against rotation about its axis AB.

Solution II

(**Scalar Analysis**). Equations (1) through (6) can be established by *direct application of the six scalar equations of equilibrium*. As shown on the free-body diagram of the rod, Fig. 5-25c, a force summation in the x, y, and z directions yields Eqs. (1) to (3), respectively. When moments are summed about the x axis, only the moment of the weight, $(490.5 \text{ N})(1 \text{ m})(-\mathbf{i})$, and the moment of force $\mathbf{B}_y$, $(B_y)(1 \text{ m})(+\mathbf{i})$, need to be considered. $\mathbf{B}_x$ is parallel to the x axis and $\mathbf{A}_x$, $\mathbf{A}_y$, and $\mathbf{A}_z$ pass through the x axis. The result is given by Eq. (4). Verify that Eqs. (5) and (6) are obtained in a similar manner.

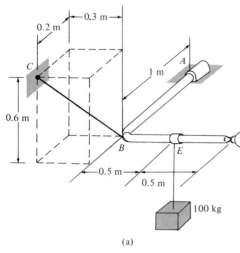

(a)

Example 5-16

The bent rod shown in Fig. 5–26a is supported at A by a smooth collar, at D by a ball-and-socket joint, and at B by means of cable BC. Using only *one equilibrium equation*, obtain a direct solution for the tension in cable BC. The smooth collar attachment at A is capable of exerting force components only in the z and y directions, since it is properly aligned on the shaft.

Solution

Step 1: The free-body diagram for the rod is shown in Fig. 5–26b. There are six unknowns: the three force components at the ball-and-socket joint, two at the smooth collar bearing, and the tension force in the cable.

Step 2: The cable tension $\mathbf{T}_B$ may be obtained *directly* by summing moments about an axis passing through points A and D. Why? The direction of the axis is defined by the unit vector $\mathbf{u}_A$, where

$$\mathbf{u}_A = \frac{\mathbf{r}_A}{r_A} = -0.707\mathbf{i} - 0.707\mathbf{j}$$

Hence, the sum of the moments about this axis is zero provided

$$\Sigma M_{AD} = \mathbf{u}_A \cdot \Sigma(\mathbf{r} \times \mathbf{F}) = 0$$

Here $\mathbf{r}$ represents a position vector drawn from *any point* on the axis AD to any point on the line of action of force $\mathbf{F}$. (See Eq. 4–28.) Thus, with reference to Fig. 5–26b, we can write

$$\mathbf{u}_A \cdot (\mathbf{r}_B \times \mathbf{T}_B + \mathbf{r}_E \times \mathbf{W}) = 0$$

$$(-0.707\mathbf{i} - 0.707\mathbf{j}) \cdot [(-1\mathbf{j}) \times (\tfrac{0.2}{0.7}T_B\mathbf{i} - \tfrac{0.3}{0.7}T_B\mathbf{j} + \tfrac{0.6}{0.7}T_B\mathbf{k})$$
$$+ (-0.5\mathbf{j}) \times (-981\mathbf{k})] = 0$$

$$(-0.707\mathbf{i} - 0.707\mathbf{j}) \cdot [(-0.857T_B + 490.5)\mathbf{i} + 0.286T_B\mathbf{k}] = 0$$

$$-0.707(-0.857T_B + 490.5) + 0 + 0 = 0$$

$$T_B = \frac{490.5}{0.857} = 572.3 \text{ N} \qquad\qquad Ans.$$

The advantage of using Cartesian vectors for this solution should be noted. Obviously, it would be very tedious to determine the perpendicular distances from the AD axis to the lines of action of $\mathbf{T}_B$ and $\mathbf{W}$ using scalar methods.

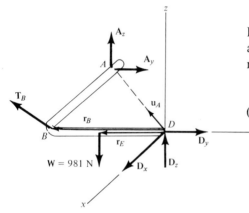

(b)

Fig. 5–26

Problems

5–42. The windlass supports the 50-kg mass. Determine the horizontal force **P** needed to hold the handle in the position shown, and the x, y, z components of reaction at the ball-and-socket joint *A* and the smooth bearing *B*. The bearing at *B* is in proper alignment and exerts only a force reaction on the windlass.

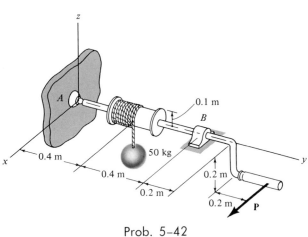

Prob. 5–42

5–43. Determine the x, y, z components of reaction at the ball supports *B* and *C* and the ball-and-socket *A* for the uniformly loaded plate.

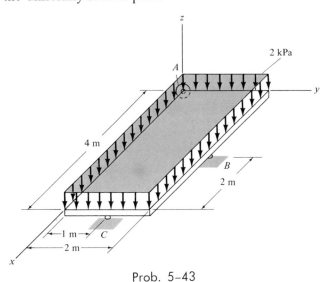

Prob. 5–43

*5–44. If the highway sign *S* is subjected to a uniform wind pressure of 700 Pa, determine the x, y, z components of reaction at the base of the post at *A*. The sign has a mass of 20 kg and a center of gravity located at its center.

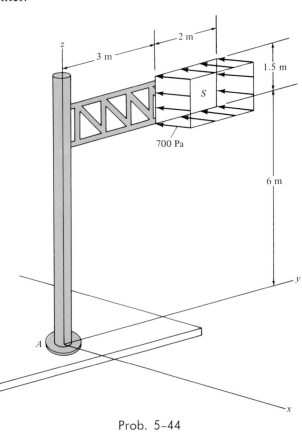

Prob. 5–44

5-45. The uniform rod AB has a mass of $m = 20$ kg and is supported at A by a ball-and-socket joint and leans against the smooth vertical wall at B. If a cable is attached at B to help support the rod, determine the x, y, z components of reaction at A and B acting on the rod. Set $a = 0.5$ m, $b = 2$ m, $c = 2.5$ m, and $d = 2$ m.

5-46. The wing of the jet aircraft and the engine have a total mass of 1.8 Mg and a center of gravity at G. If the vertical ground reaction on *each* tire T_1 and T_2 is 14 kN, determine the magnitude and direction of the resultant force and resultant moment which the fuselage exerts on the wing at the fixed point of attachment A.

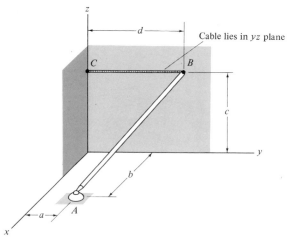

Prob. 5-45

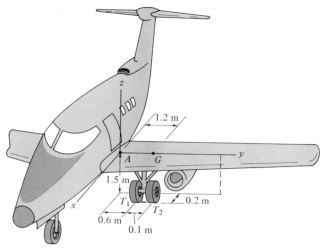

Prob. 5-46

5-45a. Solve Prob. 5-45 if the rod has a weight of $W = 30$ lb and $a = 2$ ft, $b = 5$ ft, $c = 6$ ft, and $d = 5$ ft.

5-47. Determine the x, y, z components of reaction at the two ball supports A and B and the ball-and-socket joint at C for the triangular plate.

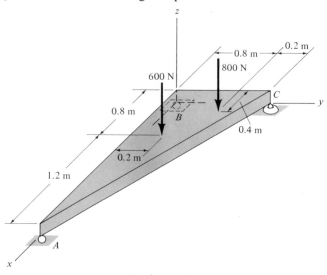

Prob. 5-47

*5-48. Determine the forces acting in cables AB and AC and the x, y, z components of reaction at the ball-and-socket joint D for the boom. The sphere at E has a mass of 150 kg.

5-50. A ball has a mass of $m = 2.5$ kg. If it rests on the surface of the smooth arc, determine the normal force per unit length along the ring of contact C which the support exerts on the ball. Set $r_1 = 75$ mm and $r_2 = 100$ mm.

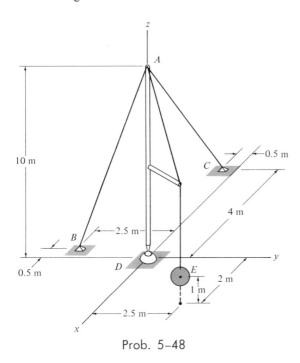

Prob. 5–48

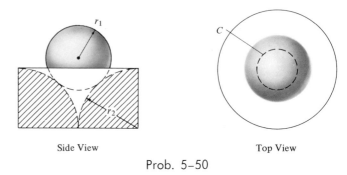

Side View   Top View

Prob. 5–50

5-50a. Solve Prob. 5–50 if the ball has a weight $W = 15$ lb and $r_1 = 0.5$ ft, $r_2 = 0.75$ ft.

5-51. The nonhomogeneous door of a large pressure vessel has a mass of 20 kg and a center of gravity at G. Determine the magnitude of the resultant force and resultant couple developed at the hinge A needed to support the door in the open position.

5-49. The cart supports the uniform crate having a mass of 40 kg. Determine the vertical reactions on the three casters at A, B, and C.

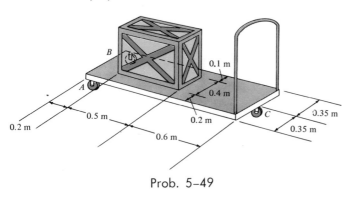

Prob. 5–49

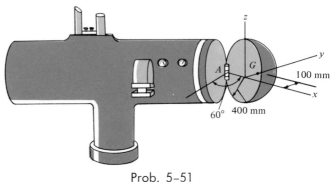

Prob. 5–51

*5-52. The homogeneous block has a mass of 50 kg and is supported at A by a ball-and-socket and at B by a smooth roller. Determine the tensions in cables CD and EF and the x, y, z components of reaction at A and B. The cable CD lies in a plane that is parallel to the yz plane.

5-54. The uniform 20-kg lid on a chest is propped open by the light rod CD. If the hinge at A prevents sliding of the lid along the x axis, whereas the hinge at B does not offer resistance in this direction, calculate the compressive force in CD and the x, y, z components of reaction at the hinges A and B. The hinges are in proper alignment and exert only force reactions on the lid.

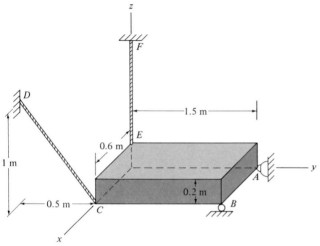

Prob. 5–52

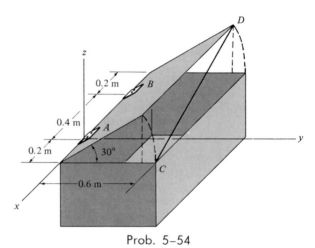

Prob. 5–54

5-53. The triangular plate is supported by a single hinge at A and a vertical cable BC. If the cable can sustain a maximum tension of 300 N, determine the maximum force $\mathbf{F}$ that may be applied to the plate. Also, compute the x, y, z components of reaction at A for this loading.

5-55. The semicircular plate supports a force having a magnitude of $F = 300$ N. Determine the tensions in cables BD and CD and the x, y, z components of reaction at the ball-and-socket joint at A. Set $r = 1.5$ m, $a = 3$ m, $b = 0.5$ m, and $c = 1$ m.

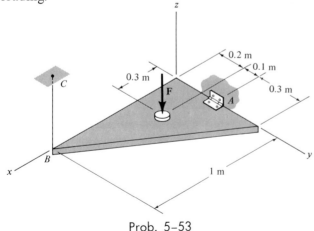

Prob. 5–53

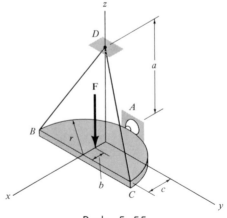

Prob. 5–55

5-55a. Solve Prob. 5-55 with $F = 75$ lb, $r = 5$ ft, $a = 9$ ft, $b = 1.5$ ft, and $c = 2.5$ ft.

*5-56. Determine the compressive force in strut AB and the x, y, z components of reaction at the hinge C of the 1-m-diameter hatch door. The door has a mass of 25 kg and a center of gravity at its midpoint G.

5-57. The bent rod is supported at A, B, and C by smooth bearings. Compute the x, y, z components of reaction at the bearings if the rod is subjected to a 300-N vertical force and a 50-N·m couple as shown. The bearings are in proper alignment and exert only force reactions on the rod.

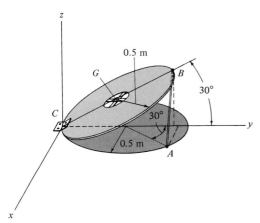

Prob. 5-56

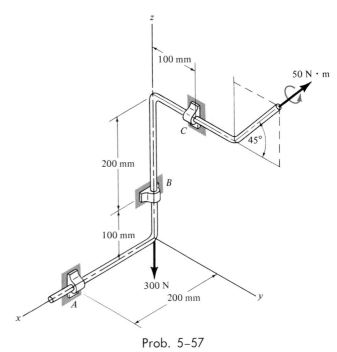

Prob. 5-57

5–58. The boom AC is supported at A by a ball-and-socket joint and by two cables BDC and CE. Cable BDC is continuous and passes over a frictionless pulley at D. Calculate the tension in the cables and the x, y, z components of reaction at A if a crate, having a mass of $m = 80$ kg, is suspended from the boom. Set $a = 2$ m and $b = 1.5$ m.

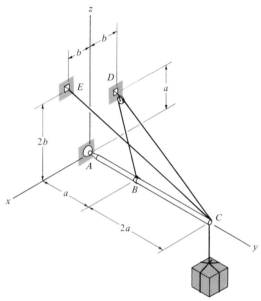

Prob. 5–58

5–59. The uniform door has a mass of 20 kg and is supported at A and B by hinges. The hinge at A supports only force in the horizontal plane, while B supports both horizontal and vertical components of force. To ensure that the door remains closed when no forces act on the handle, a mass, D, of 0.8 kg is attached to the door by means of the cord and pulley arrangement. Determine the magnitude of the horizontal force P that is applied perpendicular to the face of the door required to keep the door open in the position shown. What are the x, y, z components of reaction at hinges A and B? The hinges are properly aligned and exert no couple moments on the door.

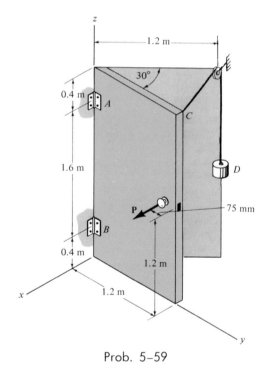

Prob. 5–59

Structural Analysis

6-1. Simple Trusses

A *truss* is a structure composed of slender members joined together at their end points. The members commonly used in construction consist of wooden struts, metal bars, angles, or channels. The joint connections are usually formed by bolting or welding the ends of the members to a common plate, called a *gusset plate,* as shown in Fig. 6-1, or by simply passing a large bolt or pin through each of the members.

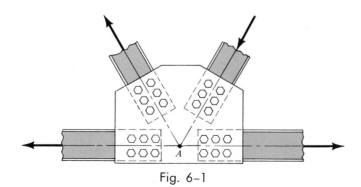

Fig. 6-1

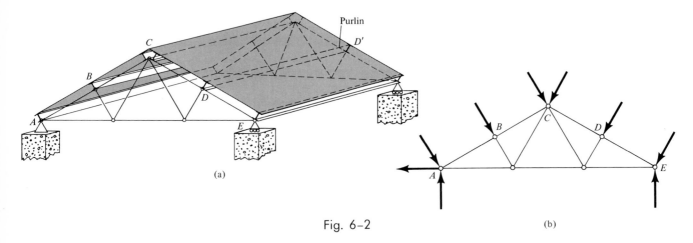

(a)

Fig. 6–2

(b)

Planar Trusses. *Planar* trusses lie in a single plane and are often used to support roofs and bridges. The truss *ABCDE*, shown in Fig. 6–2a, is an example of a typical roof-supporting truss. In this figure, the roof load is transmitted to the truss *at the joints* by means of a series of *purlins,* such as beam *DD'*. Since the imposed loading acts in the same plane as the truss, Fig. 6–2b, the analysis of the forces developed in the truss members is two-dimensional.

In the case of a bridge, such as shown in Fig. 6–3a, the load on the *deck* is first transmitted to *stringers,* then to *floor beams,* and finally to the *joints B, C,* and *D* of the two supporting side trusses. Like the roof truss, the bridge truss loading is coplanar, Fig. 6–3b.

When bridge or roof trusses extend over large distances, a rocker or roller is commonly used for supporting one end, joint *E* in Figs. 6–2a and 6–3a. This type of support allows freedom for expansion or contraction due to temperature or application of loads.

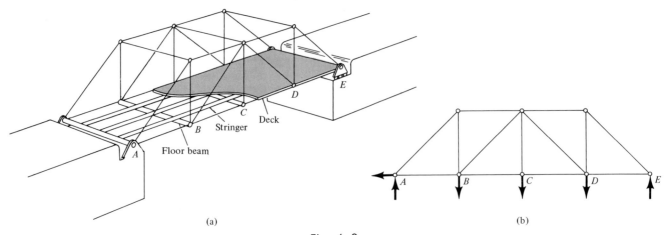

(a)

(b)

Fig. 6–3

Assumptions for Design. To design both the members and the connections of a truss, it is first necessary to determine the *forces* developed in each of the members when the truss is subjected to a given loading. In this regard, two important assumptions will be made:

1. *The members are joined together by smooth pins.* In cases where bolted or welded joint connections are used, this assumption is satisfactory provided the center lines of the joining members are concurrent at a point, as in the case of point *A* in Fig. 6–1.
2. *All the loadings are applied at the joints.* In most situations, such as for bridge and roof trusses, this assumption is true. Frequently in the force analysis, the weight of the members is neglected, since the force supported by the members is large in comparison with their weight. If the weight of the members is to be included in the analysis, it is satisfactory to apply half the weight of each member as a vertical force acting at each of its ends.

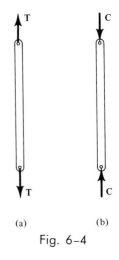

(a)   (b)

Fig. 6–4

In accordance with these two assumptions, *each truss member acts as a two-force member,* and therefore the forces acting at the ends of the member must be directed along the axis of the member. If the force tends to *elongate* the member, it is a *tensile force* (**T**), Fig. 6–4*a*; whereas, if the force tends to *shorten* the member, it is a *compressive force* (**C**), Fig. 6–4*b*. In the actual design of a truss it is important to state whether the nature of the force is tensile or compressive. Most often, the compression members must be made *thicker* than the tension members, because of the buckling or column effect that occurs in compression members.

Simple Truss. To prevent collapse, the framework of a truss must be rigid. Obviously, the four-bar frame *ABCD* in Fig. 6–5 will collapse unless a diagonal *AC* is used for support. The simplest framework which is rigid or stable is a *triangle*. Hence, a *simple truss* consists entirely of triangular elements in which each element is constructed by *starting* with a basic triangular element, such as *ABC* in Fig. 6–6, and adding two members (*AD* and *BD*) to form an additional triangular element. Hence it is seen that as each element of two members is connected, the number of joints is increased by one.

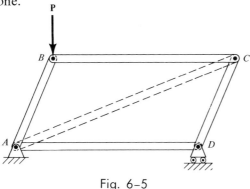

Fig. 6–5

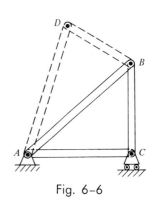

Fig. 6–6

The forces developed in simple truss members can be determined by means of the "method of joints" or the "method of sections." Once the force analysis of the truss has been completed, the size of the members and their connections can be determined using the theory of strength of materials.

6–2. The Method of Joints

If a truss is in equilibrium, each of its joints and members must be in equilibrium. In particular, the method of joints consists of satisfying the equilibrium conditions for all the forces exerted *on the pin* at each joint of the truss. Since the truss members are all straight two-force members lying in the same plane, the force system acting at each pin is *coplanar and concurrent.* Consequently, rotational or moment equilibrium is automatically satisfied at the joint, and it is therefore necessary that $\Sigma F_x = 0$ and $\Sigma F_y = 0$ to ensure translational or force equilibrium at the joint.

PROCEDURE FOR ANALYSIS

The analysis of the forces at a joint should be accompanied by an appropriate free-body diagram. When constructing this diagram, recall that the line of action of each member force acting on the joint is specified from the geometry of the truss, since the force in a member passes along the axis of the member.

In all cases, the method of joints should *begin* by applying $\Sigma F_x = 0$ and $\Sigma F_y = 0$ at a joint having at least one known force and at most two unknown forces. In this way, one obtains at most two algebraic equations which can be solved for the two unknowns. After having determined the forces acting on one joint, the forces on additional joints can be analyzed in turn. The correct "arrowhead" sense of direction of an unknown member force at a joint can, in many cases, be determined "by inspection." However, if the sense of the force direction is *assumed,* then after applying the equilibrium equations, the assumed directional sense can be verified from the numerical results. A *positive* answer indicates that the sense of direction is *correct;* whereas a *negative* answer indicates that the sense of direction shown on the free-body diagram must be *reversed.*

A simple example using the three-member truss in Fig. 6–7a will serve to numerically illustrate some of the preceding concepts. If the reactions at A and C had not been computed (although on the figure they are shown), the joint analysis would have to begin at joint B. Why? The free-body diagram of the pin at B is shown in Fig. 6–7b. Three forces act on the pin: the external force of 500 N and the two *unknown* forces developed by

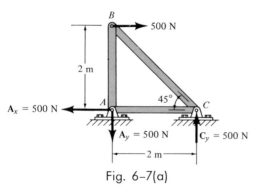

Fig. 6–7(a)

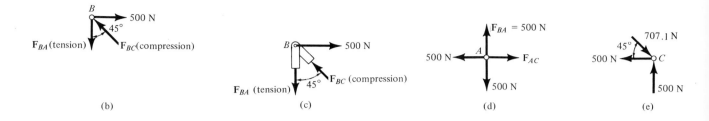

(b)　　　　　　　　　　　　　(c)　　　　　　　　　(d)　　　　　　　　　　(e)

members *BA* and *BC*. In this case, it is possible to determine the correct sense of direction of the unknowns by *inspection*. At joint *B*, $\mathbf{F}_{BC}$ is directed upward to the left as shown, since the *horizontal component* of this force must act to the left to balance the 500-N force which acts to the right. Indeed, when this is the case the solution for F_{BC} yields a *positive* result, i.e.,

$$\xrightarrow{+} \Sigma F_x = 0; \quad 500 - F_{BC} \sin 45° = 0 \quad\quad F_{BC} = 707.1 \text{ N} \quad (C)$$

Since the *vertical component* of $\mathbf{F}_{BC}$ acts upward, equilibrium requires that $\mathbf{F}_{BA}$ act downward.

$$+\uparrow\Sigma F_y = 0; \quad F_{BC} \cos 45° - F_{BA} = 0 \quad\quad F_{BA} = 500 \text{ N} \quad (T)$$

When the free-body diagrams of each of the *truss members* are drawn, Newton's third law—the principle of action, equal but opposite reaction—must always be applied. For example, $\mathbf{F}_{BC}$ acting on *pin B* in Fig. 6–7*f* represents the effect of member *BC on* the pin. Hence, it can be seen that a *"pushing on the pin"* (*B*) causes *compression* in the member (*BC*). In a similar manner, observe that *"pulling on the pin"* (*B*) causes *tension* in the member (*BA*). Perhaps an easier way of noting this effect is to isolate each joint with free-body diagrams showing small segments of the member connected to the pin, Figs. 6–7*c* and 6–7*g*. Clearly, the forces pushing or pulling on these small segments indicate the effect of the member being either in compression or tension.

Since the forces in members *BA* and *BC* have been determined, one can proceed to analyze the forces at joints *A* and *C*. The free-body diagram of joint *A* is shown in Fig. 6–7*d*. Here the effect of the pin *support* is represented by the reactions $A_x = 500$ N and $A_y = 500$ N. Since *BA* was found to be in tension, $\mathbf{F}_{BA}$ (500 N) acts *upward* at *A*. (Refer to Fig. 6–7*f* or 6–7*g*.) Clearly, $\Sigma F_y = 0$ is satisfied in Fig. 6–7*d*. Furthermore, $F_{AC} = 500$ N (T) in order to satisfy $\Sigma F_x = 0$. With the forces in all the members known, the results can be checked, in part, from the free-body diagram of the "last joint" *C*, Fig. 6–7*e*.

$$\xrightarrow{+} \Sigma F_x = 0; \quad\quad -500 + 707.1 \cos 45° \equiv 0$$
$$+\uparrow\Sigma F_y = 0; \quad\quad 500 - 707.1 \sin 45° \equiv 0$$

The following two examples further illustrate these concepts.

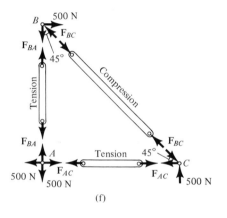

(f)

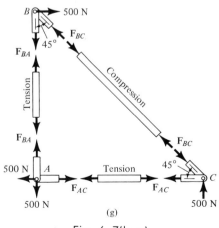

(g)

Fig. 6–7(b–g)

215

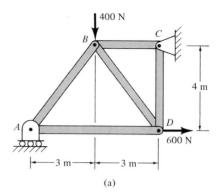

(a)

Example 6-1

A planar truss is subjected to the force system shown in Fig. 6–8a. Determine the force in each member of the truss. Indicate whether the members are in tension or compression. The truss is supported by a roller at A and a pin at C.

Solution

Since it is necessary to solve for the forces in all the members, the external reactions at joints A and C will be determined first. A free-body diagram of the entire truss is given in Fig. 6–8b. Applying the equations of equilibrium, we have

$$\xrightarrow{+}\Sigma F_x = 0; \qquad 600 - C_x = 0 \qquad C_x = 600 \text{ N}$$
$$\zeta + \Sigma M_C = 0; \quad -A_y(6) + 400(3) + 600(4) = 0 \qquad A_y = 600 \text{ N}$$
$$+\uparrow \Sigma F_y = 0; \qquad 600 - 400 - C_y = 0 \qquad C_y = 200 \text{ N}$$

Knowing these reactive forces, the analysis can start at joint A or C. The choice is arbitrary, since there are one known and two unknown member forces acting at each of these joints.

Joint A, Fig. 6–8c As shown on the free-body diagram, there are three forces that act at joint A. The slope of the members or inclination of the forces is determined from the geometry of the truss. Small sections of the attached members are shown on the free-body diagram to indicate the *compressive* effect of $\mathbf{F}_{AB}$ and the tensile effect of $\mathbf{F}_{AD}$. Applying the equations of equilibrium, we have

$$+\uparrow \Sigma F_y = 0; \qquad 600 - \tfrac{4}{5}F_{AB} = 0 \qquad F_{AB} = 750 \text{ N} \quad (C) \qquad \textit{Ans.}$$
$$\xrightarrow{+}\Sigma F_x = 0; \qquad F_{AD} - \tfrac{3}{5}(750) = 0 \qquad F_{AD} = 450 \text{ N} \quad (T) \qquad \textit{Ans.}$$

Joint D, Fig. 6–8d Since the force in AD is known, only two unknown member forces act at joint D. For the analysis, the free-body diagram of the *pin* at D is shown. (If desired, one may continue a practice of drawing

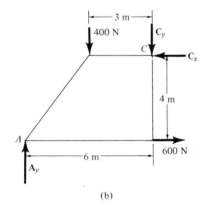

(b)

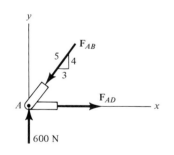

(c)

Fig. 6–8(a–c)

small sections of the members connected to the pins in order to visualize the effect of tensile and compressive forces.) Applying the equations of equilibrium, we have

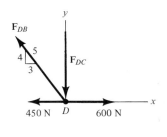

(d)

$$\xrightarrow{+}\Sigma F_x = 0; \quad -450 - \tfrac{3}{5}F_{DB} + 600 = 0 \qquad F_{DB} = 250 \text{ N} \quad \text{(T)} \qquad Ans.$$
$$+\uparrow\Sigma F_y = 0; \quad -F_{DC} + \tfrac{4}{5}(250) = 0 \qquad F_{DC} := 200 \text{ N} \quad \text{(C)} \qquad Ans.$$

Continuing in the same manner, we may next analyze the pin at joint C.

Joint C, Fig. 6-8e

$$\xrightarrow{+}\Sigma F_x = 0; \qquad F_{CB} - 600 = 0 \qquad F_{CB} = 600 \text{ N} \quad \text{(C)} \qquad Ans.$$
$$+\uparrow\Sigma F_y = 0; \qquad 200 - 200 \equiv 0 \quad \text{(check)}$$

All the forces in the truss members have now been determined. However, it is always prudent to analyze the "last joint" B in order to check, in part, the accuracy of the solution.

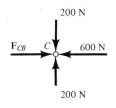

(e)

Joint B, Fig. 6-8f

$$\xrightarrow{+}\Sigma F_x = 0; \qquad 750(\tfrac{3}{5}) + 250(\tfrac{3}{5}) - 600 \equiv (\quad \text{(check)}$$
$$+\uparrow\Sigma F_y = 0; \qquad -400 + 750(\tfrac{4}{5}) - 250(\tfrac{4}{5}) \equiv \quad \text{(check)}$$

The results of this analysis are summarized in Fig. 6-8g, which shows the complete free-body diagram for each joint. Applying the "principle of action and reaction," the forces acting in the members are equal and opposite to those acting on the pins, as shown in the figure. The truss members are labeled as being either in tension or in compression.

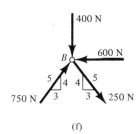

(f)

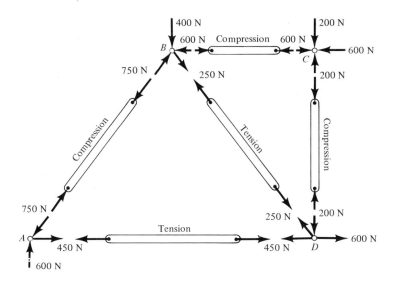

(g)

Fig. 6-8(d-g)

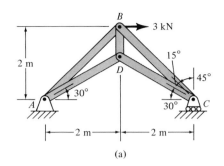

(a)

Example 6–2

Determine the forces acting in all the members of the truss shown in Fig. 6–9a.

Solution

It is first necessary to determine the external reactions. The free-body diagram of the entire truss is shown in Fig. 6–9b. Hence,

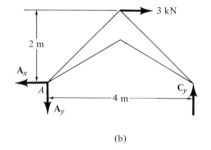

(b)

$$\xrightarrow{+} \Sigma F_x = 0; \qquad -A_x + 3 = 0 \qquad A_x = 3 \text{ kN}$$
$$\zeta + \Sigma M_A = 0; \qquad 3(2) - C_y(4) = 0 \qquad C_y = 1.5 \text{ kN}$$
$$+\uparrow \Sigma F_y = 0; \qquad 1.5 - A_y = 0 \qquad A_y = 1.5 \text{ kN}$$

The analysis may now begin at either joint C or A, since there is one known force and two unknown member forces acting at these joints.

Joint C, Fig. 6–9c

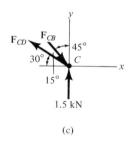

(c)

$$\xrightarrow{+} \Sigma F_x = 0; \qquad -F_{CD} \cos 30° + F_{CB} \sin 45° = 0$$
$$+\uparrow \Sigma F_y = 0; \qquad 1.5 + F_{CD} \sin 30° - F_{CB} \cos 45° = 0$$

These two equations must be solved *simultaneously* for each of the two unknowns. Note however that a *direct solution* for one of the unknown forces may be obtained by applying a force summation along an axis that is *perpendicular* to the direction of the other unknown force. For example, a force summation along the y' axis, which is perpendicular to the direction of $\mathbf{F}_{CD}$, Fig. 6–9d, yields a direct solution for F_{CB}.

$$+ \nearrow \Sigma F_{y'} = 0; \; 1.5 \cos 30° - F_{CB} \sin 15° = 0 \quad F_{CB} = 5.02 \text{ kN} \quad \text{(C)} \quad \textit{Ans.}$$

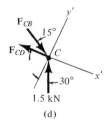

(d)

Fig. 6–9(a–d)

In a similar fashion, summing forces along the y'' axis, Fig. 6–9e, yields a direct solution for F_{CD}.

$+\nearrow \Sigma F_{y''} = 0$;

$$1.5 \cos 45° - F_{CD} \sin 15° = 0 \qquad F_{CD} = 4.10 \text{ kN} \quad \text{(T)} \qquad Ans.$$

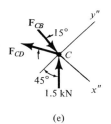

(e)

Joint D, Fig. 6–9f

$\overset{+}{\rightarrow}\Sigma F_x = 0$; $\qquad -F_{DA} \cos 30° + 4.10 \cos 30° = 0$

$$F_{DA} = 4.10 \text{ kN} \quad \text{(T)} \qquad\qquad Ans.$$

$+\uparrow\Sigma F_y = 0$; $\quad F_{DB} - 2(4.10 \sin 30°) = 0 \qquad F_{DB} = 4.10 \text{ kN} \quad \text{(T)} \qquad Ans.$

Joint B, Fig. 6–9g

$\overset{+}{\rightarrow}\Sigma F_x = 0$; $\qquad F_{BA} \cos 45° - 5.02 \cos 45° + 3 = 0$

$$F_{BA} = 0.777 \text{ kN} \quad \text{(C)} \qquad\qquad Ans.$$

$+\uparrow\Sigma F_y = 0$; $\quad 0.777 \sin 45° - 4.10 + 5.02 \sin 45° \equiv 0 \quad \text{(check)}$

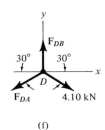

(f)

The results are checked in part for accuracy by analyzing the "last joint," A.

Joint A, Fig. 6–9h

$\overset{+}{\rightarrow}\Sigma F_x = 0$; $\quad -3 - 0.777 \sin 45° + 4.10 \cos 30° \equiv 0 \quad \text{(check)}$

$+\uparrow\Sigma F_y = 0$; $\quad -0.777 \cos 45° + 4.10 \sin 30° - 1.5 \equiv 0 \quad \text{(check)}$

A summary of the force analysis is given in Fig. 6–9i.

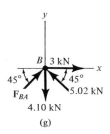

(g)

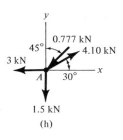

(h)

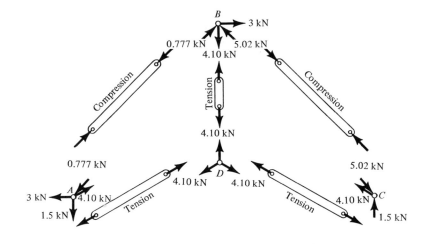

(i)

Fig. 6–9(e–i)

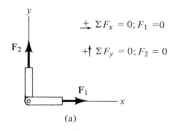

$$\underset{+}{\rightarrow} \Sigma F_x = 0; F_1 = 0$$

$$+\uparrow \Sigma F_y = 0; F_2 = 0$$

F_1

(a)

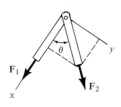

$$\nwarrow^{+} \Sigma F_y = 0; F_2 \sin\theta = 0; F_2 = 0, \text{ since } \sin\theta \neq 0$$

$$\nearrow^{+} \Sigma F_x = 0; F_1 + 0 = 0; F_1 = 0$$

(b)

Fig. 6-10

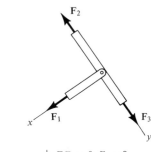

$$\nearrow^{+} \Sigma F_x = 0, F_1 = 0$$

$$\nwarrow^{+} \Sigma F_y = 0, F_2 = F_3$$

(a)

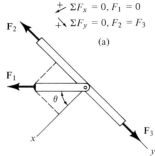

$$\nearrow^{+} \Sigma F_x = 0; F_1 \cos\theta = 0; F_1 = 0 \text{ since } \cos\theta \neq 0$$

$$\nwarrow^{+} \Sigma F_y = 0; F_2 = F_3$$

(b)

Fig. 6-11

Zero-Force Members. Truss analysis, by using the method of joints, is greatly simplified if one is able to first determine those members which support *no loading*. These *zero-force members* are used to increase the stability of the truss during construction, and to provide support if the applied loading is changed.

The zero-force members of a truss can generally be determined by inspection of the joints. For example, if two members are connected at a right angle to each other at a joint *that has no external load,* as shown in Fig. 6-10a, the force in each member must be zero in order to maintain equilibrium. Furthermore, this is true regardless of the angle, say θ, between the members ($\theta \neq 0°$; $\theta \neq 180°$), Fig. 6-10b. Zero-force members also occur at joints having geometries as shown in Figs. 6-11a and 6-11b. In both cases, *no external load acts on the joint,* so that a force summation in the x direction, which is *perpendicular* to the two collinear members, requires that $F_1 = 0$. Particular attention should be directed to these conditions of joint geometry and loading, since the truss analysis can be considerably simplified by *first* spotting the zero-force members.

Example 6-3

Using the method of joints, show that members *FP, PG, HJ, JG, GK,* and *LE* of the *Fink truss,* shown in Fig. 6-12a, are subjected to zero force when the truss is loaded as shown.

Solution

If a free-body diagram of joint *F* is considered so that the y axis is directed along member *FP*, Fig. 6-12b, then the lines of action of forces $\mathbf{F}_{FE}$ and $\mathbf{F}_{FG}$ are both along the x axis. Therefore, the equation $\Sigma F_y = 0$ is satisfied provided $F_{FP} = 0$; i.e., *FP* is a zero-force member.

A free-body diagram of joint *P* is shown in Fig. 6-12c. By orienting the x axis along members *PE* and *PK*, then equilibrium in the y direction is satisfied provided $\Sigma F_y = F_{PG} \cos\theta = 0$. Hence, $F_{PG} = 0$, since θ, the angle that member *PG* makes with the y axis, is not 90°. In a similar manner, by considering, in turn, free-body diagrams of joints *H, J, G,* and *L,* complete this example and show that the force in members *HJ, JG, GK,* and *LE* is zero.

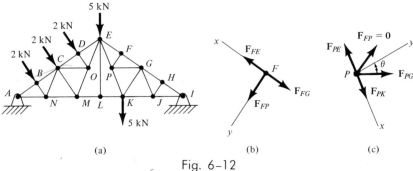

(a)

(b)

(c)

Fig. 6-12

Problems

6-1. Compute the force in each member of the truss and indicate whether the members are in tension or compression. All the members are 3 m long.

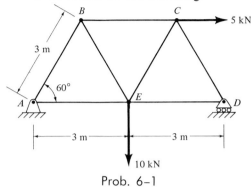

Prob. 6-1

6-2. Determine the force in each member of the truss, and indicate whether the members are in tension or compression.

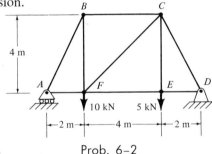

Prob. 6-2

6-3. The *Howe bridge truss* is subjected to the loading shown. Determine the force in members *IB, BH,* and *HD,* and indicate whether the members are in tension or compression. Specify all the zero-force members.

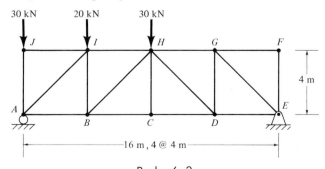

Prob. 6-3

*6-4. For the given loading, determine the zero-force members in the *Howe roof truss.* Explain your answers using the appropriate joint free-body diagrams.

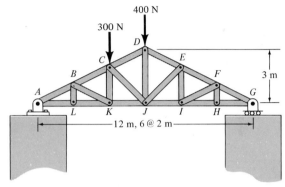

Prob. 6-4

6-5. Determine the force in each member of the truss and indicate whether the members are in tension or compression. Set $P = 800$ N, $a = 4$ m, $b = 3$ m, and $c = 1$ m.

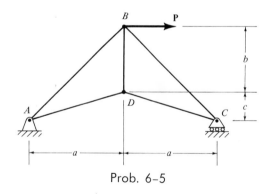

Prob. 6-5

6-5a. Solve Prob. 6-5 with $P = 1500$ lb, $a = 20$ ft, $b = 15$ ft, and $c = 5$ ft.

221

6-6. Two side trusses support the deck, which is subjected to a uniform load of 1.6 kPa. Determine the force developed in each member of truss *ABCD*. For the analysis assume that the central 2 m of load, *EFGH*, is carried by the floor beam *BB'*, and the end floor beams *AA'* and *CC'* equally carry the rest of the load. *Hint:* First reduce the loading to a system of coplanar parallel forces acting at joints *A*, *B*, and *C* of the truss.

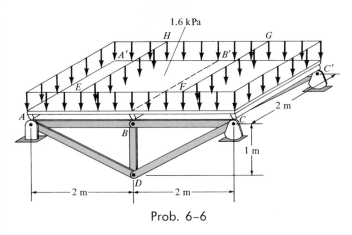

Prob. 6–6

6-7. Determine the forces developed in members *EC*, *FC*, and *FE*, of the bridge truss, and indicate whether the members are in tension or compression.

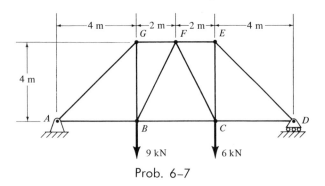

Prob. 6–7

6-8. Determine the force in each member of the crane truss and indicate whether the members are in tension or compression. The pulley at *D* is frictionless.

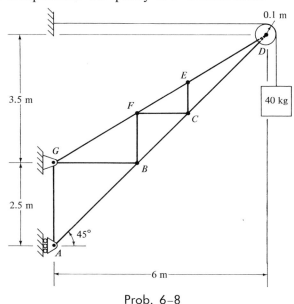

Prob. 6–8

6-9. The *Howe roof truss* supports the vertical loading shown. Determine the force in members *BL*, *LC*, and *KC*, and indicate whether the members are in tension or compression.

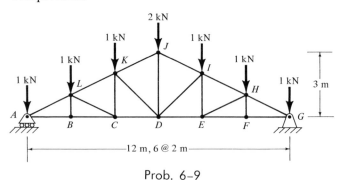

Prob. 6–9

222

6-10. The *scissors truss* is used to support the roof load. Determine the force in members FD and BF, and indicate whether the members are in tension or compression. Set $F_1 = 800$ N, $F_2 = 600$ N, and $a = 3$ m.

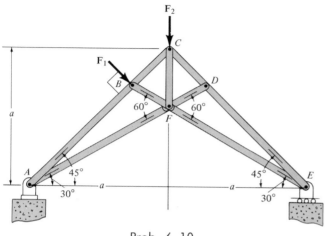

Prob. 6-10

6-10a. Solve Prob. 6-10 if $F_1 = 200$ lb, $F_2 = 175$ lb, and $a = 10$ ft.

6-11. Determine the force in each member of the truss and indicate whether the members are in tension or compression.

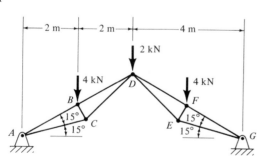

Prob. 6-11

* 6-12. Determine the force in each member of the truss and indicate whether the members are in tension or compression. The pulley at C is frictionless.

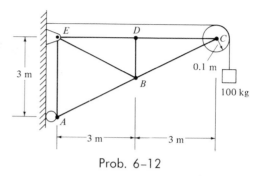

Prob. 6-12

6-13. Determine the forces in members QF and LE of the truss and indicate whether the members are in tension or compression.

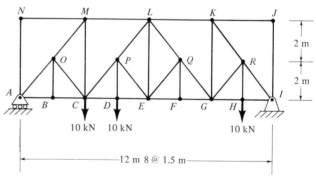

Prob. 6-13

223

6–3. The Method of Sections

A second method for finding the forces in the members of a truss, the *method of sections,* consists of passing an *imaginary section* through the truss, thus cutting it into two parts. Provided the entire truss is in equilibrium, then each of the two parts must also be in equilibrium; and, as a result, the three equations of equilibrium may be applied to either one of these two parts to determine the member forces at the "cut section."

PROCEDURE FOR ANALYSIS

When applying the method of sections to determine the force in a member, a decision should *first* be made as to *how* to "cut" or section the truss. Since only *three independent equilibrium equations* are available ($\Sigma F_x = 0$, $\Sigma F_y = 0$, $\Sigma M_O = 0$), one should try to select a section that passes through not more than *three* members in which the forces are unknown. A free-body diagram of one of the parts of the sectioned truss is then drawn. The line of action of each member force is specified from the *geometry of the truss,* since the force in a member passes along the axis of the member. If there are three unknowns on the free-body diagram, one can generally apply the equations of equilibrium in such a manner as to yield a *direct solution* for each of the unknowns, rather than having to solve simultaneous equations. In this regard, moments should be summed about a point that lies at the intersection of the lines of action of two unknown forces, so that the third unknown force is determined *directly* from the moment equilibrium equation. If two of the unknown forces are *parallel,* forces may be summed *perpendicular* to the direction of these unknowns to determine, *directly,* the third unknown force.

The truss shown in Fig. 6–13a will serve to numerically illustrate some of these concepts. Here it is necessary to determine the force in members

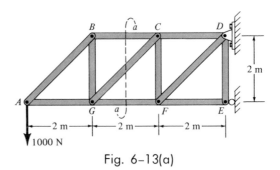

Fig. 6–13(a)

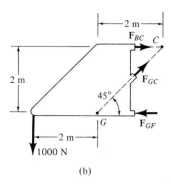

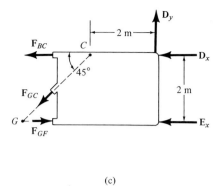

(b)   (c)

Fig. 6–13(b, c)

GF, *GC*, and *BC*. Using section *aa*, the free-body diagrams of the two parts of the truss are shown in Figs. 6–13*b* and 6–13*c*. In effect, the section has "exposed" the *internal forces* in the members as *external forces* on each of the free-body diagrams. When constructing these two diagrams, note that the member forces acting on one side of the truss must be equal but opposite to those acting on the other side—Newton's third law. As shown, the members assumed to be in *tension* (*BC* and *GC*) are subjected to a "pull," whereas the member in *compression* (*GF*) is subjected to a "push."

The three unknown member forces may be obtained by applying the three equilibrium equations using the free-body diagram in Fig. 6–13*b*. If the free-body diagram in Fig. 6–13*c* is considered, the three external reactions $\mathbf{D}_x$, $\mathbf{D}_y$, and $\mathbf{E}_x$ will have to be determined *first*. Why? (This, of course, is done in the usual manner, by considering a free-body diagram of the *entire truss*.) To obtain a direct solution for F_{BC}, moments must be summed about point *G*, Fig. 6–13*b*. Why?

$$\zeta + \Sigma M_G = 0; \qquad F_{BC}(2 \text{ m}) - 1000 \text{ N}(2 \text{ m}) = 0$$
$$F_{BC} = 1000 \text{ N} \quad (T)$$

In a similar manner, to obtain F_{GF} directly, sum moments about *C*.

$$\zeta + \Sigma M_C = 0; \qquad -(1000 \text{ N})(4 \text{ m}) + F_{GF}(2 \text{ m}) = 0$$
$$F_{GF} = 2000 \text{ N} \quad (C)$$

Finally, to determine F_{GC} directly, sum forces in the *y* direction. Why?

$$+\uparrow \Sigma F_y = 0; \qquad -1000 \text{ N} + F_{GC} \sin 45° = 0$$
$$F_{GC} = 1414 \text{ N} \quad (T)$$

If the forces in only a *few members* of a truss are to be found, as in this example, the method of sections generally provides the most *direct* means of obtaining these forces. In the general case, however, the most efficient solution may be obtained using *both* the method of sections *and* the method of joints as indicated by the following example.

Example 6–4

Determine the force in each of the members *GE*, *GC*, *BC*, and *EC* of the bridge truss shown in Fig. 6–14*a*. Indicate whether the members are in tension or compression.

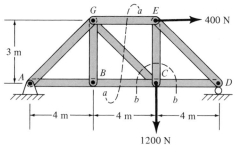

Fig. 6–14(a)

Solution

In order to use the method of sections to solve this problem, it is *first* necessary to determine the external reactions at *A* or *D*. Why? A free-body diagram of the entire truss is shown in Fig. 6–14*b*. Applying the equations of equilibrium, we have

$$\xrightarrow{+} \Sigma F_x = 0; \qquad 400 \text{ N} - A_x = 0 \qquad A_x = 400 \text{ N}$$

$$\zeta + \Sigma M_A = 0;$$

$$-1200 \text{ N}(8 \text{ m}) - 400 \text{ N}(3 \text{ m}) + D_y(12 \text{ m}) = 0 \qquad D_y = 900 \text{ N}$$

$$+\uparrow \Sigma F_y = 0; \quad A_y - 1200 \text{ N} + 900 \text{ N} = 0 \qquad A_y = 300 \text{ N}$$

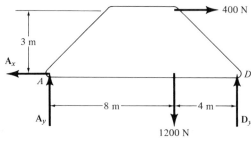

Fig. 6–14(b)

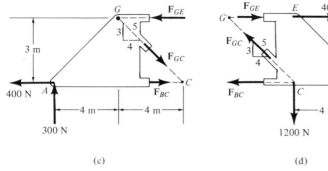

(c)

Fig. 6–14(c–d)

Knowing these reactions, the truss can now be sectioned along the dashed line *aa* in Fig. 6–14*a*, which yields the free-body diagrams in Figs. 6–14*c* and 6–14*d*. For the analysis the free-body diagram in Fig. 6–14*c* will be used, since it involves the least number of forces.

Summing moments about point *G* eliminates the two unknown forces $\mathbf{F}_{GE}$ and $\mathbf{F}_{GC}$ and yields a direct solution for F_{BC}.

$$\zeta + \Sigma M_G = 0; \quad -300 \text{ N}(4 \text{ m}) - 400 \text{ N}(3 \text{ m}) + F_{BC}(3 \text{ m}) = 0$$
$$F_{BC} = 800 \text{ N} \quad (\text{T}) \qquad\qquad Ans.$$

In the same manner, by summing moments about point *C* we obtain a direct solution for F_{GE}.

$$\zeta + \Sigma M_C = 0; \quad -300 \text{ N}(8 \text{ m}) + F_{GE}(3 \text{ m}) = 0$$
$$F_{GE} = 800 \text{ N} \quad (\text{C}) \qquad\qquad Ans.$$

Since $\mathbf{F}_{BC}$ and $\mathbf{F}_{GE}$ have no vertical components of force, summing forces in the *y* direction directly yields F_{GC}, i.e.,

$$+\uparrow \Sigma F_y = 0; \quad 300 \text{ N} - \tfrac{3}{5}F_{GC} = 0 \quad F_{GC} = 500 \text{ N} \quad (\text{T}) \qquad\qquad Ans.$$

The same results can be obtained by applying the equations of equilibrium to the free-body diagram shown in Fig. 6–14*d*. With this diagram how would you directly obtain each unknown force using a *single equation* of equilibrium for each calculation?

The easiest way to obtain the force in *EC* is to consider the section *bb*, which isolates joint *C*, Fig. 6–14*a*. The free-body diagram of this joint or section is shown in Fig. 6–14*e*. Applying the equation of vertical force equilibrium, yields

$$+\uparrow \Sigma F_y = 0; \quad -1200 \text{ N} + (\tfrac{3}{5})500 \text{ N} + F_{CE} = 0$$
$$F_{CE} = 900 \text{ N} \quad (\text{T}) \qquad\qquad Ans.$$

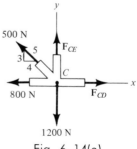

Fig. 6–14(e)

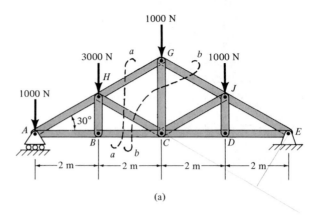

(a)

Example 6-5

Determine the force in member GC of the truss shown in Fig. 6–15a. Indicate whether the member is in tension or compression.

Solution

A free-body diagram of the entire truss is shown in Fig. 6–15b. It is first necessary to solve for the external reactions. Why?

$\zeta + \Sigma M_A = 0$;

$$-3000 \text{ N}(2 \text{ m}) - 1000 \text{ N}(4 \text{ m}) - 1000 \text{ N}(6 \text{ m}) + E_y(8 \text{ m}) = 0$$

$$E_y = 2000 \text{ N}$$

$\overset{+}{\rightarrow} \Sigma F_x = 0; \qquad\qquad E_x = 0$

Using the computed value of E_y,

$+\uparrow \Sigma F_y = 0$;

$$A_y - 1000 \text{ N} - 3000 \text{ N} - 1000 \text{ N} - 1000 \text{ N} + 2000 \text{ N} = 0$$

$$A_y = 4000 \text{ N}$$

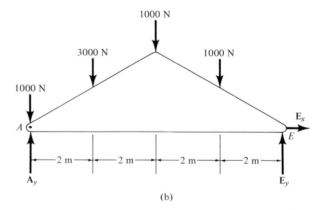

Fig. 6–15(a, b)

(b)

By the method of sections, any imaginary vertical section that cuts through *GC*, Fig. 6–15*a*, will also have to cut through three other members whose forces are unknown. For example, section *bb* cuts through *GJ*, *GC*, *HC*, and *BC*. If a free-body diagram of the left side of this section is considered, Fig. 6–15*d*, it is possible to determine F_{GJ} by summing moments about *C* to eliminate the other three unknowns; however, F_{GC} cannot be determined from the remaining two equilibrium equations. It is therefore necessary to determine F_{BC} or F_{HC} *before* using section *bb*. For example, it is possible to determine F_{HC} by considering the adjacent section *aa*, Fig. 6–15*a*. A free-body diagram of the portion of the truss to the left of this section is given in Fig. 6–15*c*. Summing moments about point *A*, one can obtain F_{HC} directly. A simple way to do this is to resolve $\mathbf{F}_{HC}$ into its rectangular components and, by the principle of transmissibility, to extend this force to point *C*, as shown. The moments of $\mathbf{F}_{HC} \cos 30°$, $\mathbf{F}_{BC}$, and $\mathbf{F}_{HG}$ are all zero about point *A*. Hence,

$$\zeta + \Sigma M_A = 0; \quad -3000 \text{ N}(2 \text{ m}) + F_{HC} \sin 30° (4 \text{ m}) = 0$$
$$F_{HC} = 3000 \text{ N} \quad (C)$$

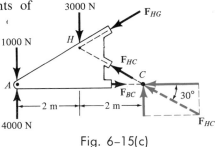

Fig. 6–15(c)

The force in *GC* can now be obtained from Fig. 6–15*d* by summing moments about *E*. Again extending $\mathbf{F}_{HC}$ to point *C* and resolving it into rectangular components as shown, we have

$$\zeta + \Sigma M_E = 0; \quad 1000 \text{ N}(8 \text{ m}) - 4000 \text{ N}(8 \text{ m}) + 3000 \text{ N}(6 \text{ m})$$
$$+ 1000 \text{ N}(4 \text{ m}) + F_{GC}(4 \text{ m}) - F_{HC} \sin 30° (4 \text{ m}) = 0$$

Substituting $F_{HC} = 3000$ N and solving yields

$$F_{GC} = 2000 \text{ N} \quad (T) \qquad\qquad Ans.$$

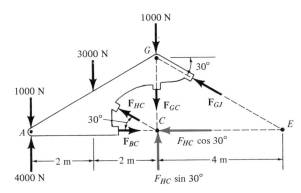

Fig. 6–15(d)

Problems

6-14. Determine the forces in members *EB* and *ED* of the *Warren truss* and indicate whether the members are in tension or compression.

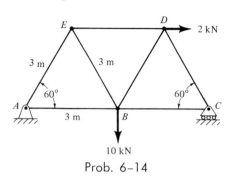

10 kN

Prob. 6-14

6-15. Determine the force in members *CF* and *GC* of the truss and indicate whether the members are in tension or compression. Set $F_1 = 5$ kN, $F_2 = 10$ kN, and $a = 2$ m.

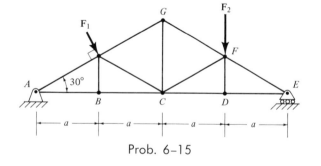

Prob. 6-15

6-15a. Solve Prob. 6-15 with $F_1 = 500$ lb, $F_2 = 1200$ lb, and $a = 6$ ft.

*** 6-16.** Determine the forces in members *GF* and *BC* of the truss and indicate whether the members are in tension or compression.

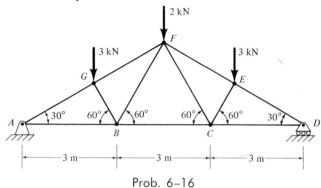

Prob. 6-16

6-17. The *Pratt bridge truss* is subjected to the loading shown. Determine the force in members *LD* and *BC*, and indicate whether these members are in tension or compression.

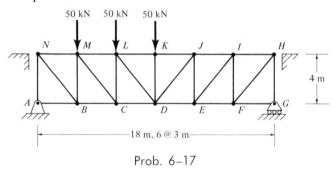

Prob. 6-17

6-18. Determine the force in members *HG*, *CG*, and *CF* of the truss and indicate whether the members are in tension or compression.

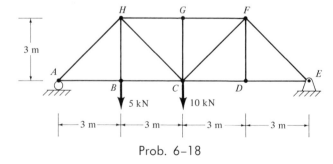

Prob. 6-18

6-19. Determine the forces in members *BE* and *EF* of the truss and indicate whether the members are in tension or compression. After sectioning the truss, solve for each force *directly*, using a single equation of equilibrium to determine each force.

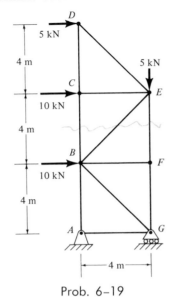

Prob. 6-19

*6-20. Determine the force in members *BC* and *MC* of the *K truss* and indicate whether the members are in tension or compression. Set $F_1 = 2$ kN, $F_2 = 5$ kN, $F_3 = 7$ kN, $a = 3$ m, and $b = 2$ m.

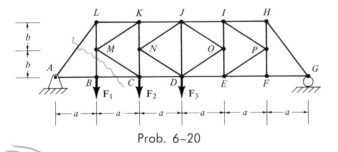

Prob. 6-20

*6-20a. Solve Prob. 6-20 with $F_1 = 1200$ lb, $F_2 = 1500$ lb, $F_3 = 1800$ lb, $a = 15$ ft, and $b = 10$ ft.

6-21. Compute the force in members *GF* and *GD* of the truss and indicate whether the members are in tension or compression.

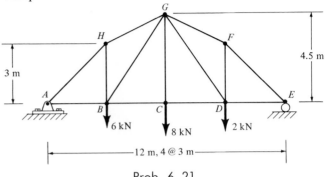

Prob. 6-21

6-22. Determine the force in members *OE*, *LE*, and *LK* of the *Baltimore truss* and indicate whether the members are in tension or compression.

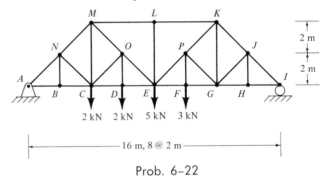

Prob. 6-22

6-23. Determine the forces in members *CJ* and *CB* of the aircraft frame and indicate whether the members are in tension or compression. All triangles, except *ALK*, are equilateral. The frame is connected to the fuselage by a short link at *A* and by a pin at *J*.

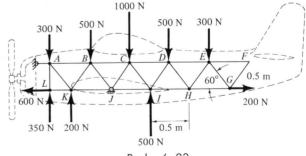

Prob. 6-23

231

*6–24. Determine the force developed in members BC and CH of the roof truss and indicate whether the members are in tension or compression.

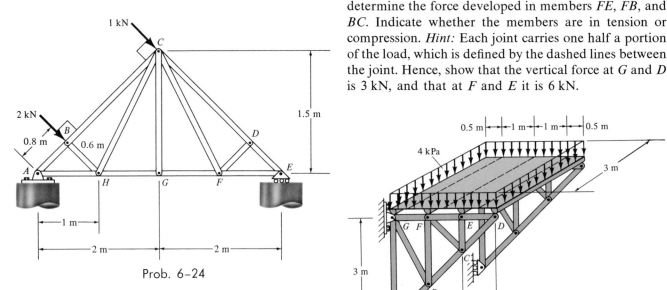

Prob. 6–24

6–26. Two trusses support the edges of a plate that is subjected to a uniform load of 4 kPa. If the plate transmits only vertical loads to a truss at joints D, E, F, and G, determine the force developed in members FE, FB, and BC. Indicate whether the members are in tension or compression. *Hint:* Each joint carries one half a portion of the load, which is defined by the dashed lines between the joint. Hence, show that the vertical force at G and D is 3 kN, and that at F and E it is 6 kN.

Prob. 6–26

6–25. Determine the force in members CF, GF, and CD of the *Warren roof truss* and indicate whether the members are in tension or compression. Set $F_1 = 2$ kN, $F_2 = 10$ kN, $F_3 = 5$ kN, $a = 4$ m, $b = 3$ m, and $c = 1$ m.

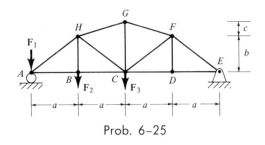

Prob. 6–25

6–25a. Solve Prob. 6–25 with $F_1 = 600$ lb, $F_2 = 1800$ lb, $F_3 = 1500$ lb, $a = 10$ ft, $b = 7.5$ ft, and $c = 2.5$ ft.

6-27. The truss is used to support an outdoor theater screen. If a wind pressure, acting on the face of the screen, creates the loading shown on the joints, determine the force in members *BJ* and *CH*. Indicate whether the members are in tension or compression.

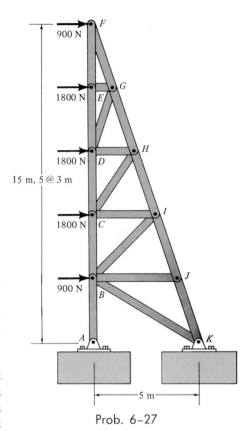

Prob. 6-27

*6-4. Space Trusses

A *space truss* consists of members joined together at their ends to form a stable three-dimensional structure. The simplest element of a space truss is a *tetrahedron*, formed by connecting six members together, as shown in Fig. 6-16. Any additional members added to this basic element would be redundant in supporting the force **P**. A *simple space truss* can be built from this basic tetrahedral element by adding three additional members and a joint forming a system of multiconnected tetrahedrons.

Assumptions for Design. The members of a space truss may be treated as two-force members provided the external loading is applied at the joints and the joints consist of ball-and-socket connections. These assumptions are justified if the welded or riveted connections of the joined members intersect at a common point and the weight of the members can be neglected. In cases where the weight of a member is to be included in the analysis, it is generally satisfactory to apply it as a vertical force, half of its magnitude applied at each end of the member.

PROCEDURE FOR ANALYSIS

Either the method of sections or the method of joints can be used to determine the forces developed in the members of a simple space truss.

If only a *few* member forces are to be determined, the method of sections may be used. When an imaginary section is passed through a truss, and the truss is separated into two parts, the force system acting on one of the parts must satisfy the *six* scalar equilibrium equations: $\Sigma F_x = 0, \Sigma F_y = 0, \Sigma F_z = 0, \Sigma M_x = 0, \Sigma M_y = 0, \Sigma M_z = 0$ (Eqs. 5-7). By proper choice of the section and axes for summing forces and moments, many of the unknown member forces in a space truss can be computed *directly,* using a single equilibrium equation.

Generally, the forces in *all* the members of the truss must be deter-

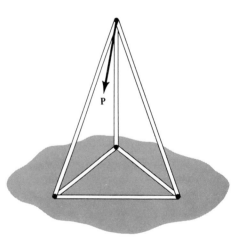

Fig. 6-16

mined, in which case the method of joints is most suitable for the analysis. When using the method of joints, it is necessary to solve the three scalar equilibrium equations $\Sigma F_x = 0$, $\Sigma F_y = 0$, $\Sigma F_z = 0$ at each joint. The solution of many simultaneous equations can be avoided if the force analysis begins at a joint having at least one known force and at most three unknown forces. If the three-dimensional geometry of the force system at the joint is complicated, it is recommended that Cartesian vectors be used in the solution. The following example numerically illustrates this procedure.

Example 6-6

Determine the forces acting in each of the members of the space frame shown in Fig. 6–17a. Indicate whether the members are in tension or compression.

Solution

Since there is one known force and three unknown forces acting at joint A, the force analysis of the truss will begin by applying the method of joints to joint A.

Joint A, Fig. 6–17b   Expressing each of the forces that act on the free-body diagram of joint A in vector notation, yields

$$\mathbf{P} = \{-4\mathbf{j}\} \text{ kN}$$
$$\mathbf{F}_{AB} = F_{AB}\mathbf{j}$$
$$\mathbf{F}_{AC} = -F_{AC}\mathbf{k}$$
$$\mathbf{F}_{AE} = F_{AE}\left(\frac{\mathbf{r}_{AE}}{r_{AE}}\right) = F_{AE}\frac{(2\mathbf{i} + 2\mathbf{j} - 2\mathbf{k})}{2\sqrt{3}}$$

For equilibrium,

$$\Sigma \mathbf{F} = 0; \qquad \mathbf{P} + \mathbf{F}_{AB} + \mathbf{F}_{AC} + \mathbf{F}_{AE} = 0$$

$$-4\mathbf{j} + F_{AB}\mathbf{j} - F_{AC}\mathbf{k} + \frac{1}{\sqrt{3}}F_{AE}\mathbf{i} + \frac{1}{\sqrt{3}}F_{AE}\mathbf{j} - \frac{1}{\sqrt{3}}F_{AE}\mathbf{k} = 0$$

$$\frac{1}{\sqrt{3}}F_{AE}\mathbf{i} + \left(-4 + F_{AB} + \frac{1}{\sqrt{3}}F_{AE}\right)\mathbf{j} + \left(-F_{AC} - \frac{1}{\sqrt{3}}F_{AE}\right)\mathbf{k} = 0$$

Equating the **i**, **j**, and **k** components to zero gives

$$\Sigma F_x = 0; \qquad\qquad \frac{1}{\sqrt{3}}F_{AE} = 0$$

$$\Sigma F_y = 0; \qquad\qquad -4 + F_{AB} + \frac{1}{\sqrt{3}}F_{AE} = 0$$

$$\Sigma F_z = 0; \qquad\qquad -F_{AC} - \frac{1}{\sqrt{3}}F_{AE} = 0$$

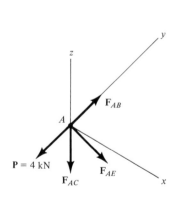

(a)

(b)

Fig. 6–17(a–b)

Solving these equations, we have

$$F_{AC} = F_{AE} = 0 \qquad \qquad \textit{Ans.}$$

$$F_{AB} = 4 \text{ kN} \quad (\text{T}) \qquad \qquad \textit{Ans.}$$

Since F_{AB} is known, joint B may be analyzed next.

Joint B, Fig. 6-17c

$$\Sigma \mathbf{F} = \mathbf{0}; \qquad 2\mathbf{k} + \mathbf{F}_{AB} + \mathbf{R}_B + \mathbf{F}_{BD} + \mathbf{F}_{BE} = \mathbf{0}$$

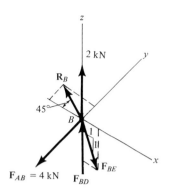

(c)

$$2\mathbf{k} - 4\mathbf{j} - R_B \cos 45° \, \mathbf{i} + R_B \sin 45° \, \mathbf{j} + F_{BD}\mathbf{k}$$
$$+ \frac{1}{\sqrt{2}} F_{BE}\mathbf{i} - \frac{1}{\sqrt{2}} F_{BE}\mathbf{k} = \mathbf{0}$$

$$\left(-R_B \cos 45° + \frac{1}{\sqrt{2}} F_{BE} \right)\mathbf{i} + \left(-4 + R_B \sin 45° \right)\mathbf{j}$$
$$+ \left(2 + F_{BD} - \frac{1}{\sqrt{2}} F_{BE} \right)\mathbf{k} = \mathbf{0}$$

Thus,

$$\Sigma F_x = 0; \qquad -R_B \cos 45° + \frac{1}{\sqrt{2}} F_{BE} = 0$$

$$\Sigma F_y = 0; \qquad -4 + R_B \sin 45° = 0$$

$$\Sigma F_z = 0; \qquad 2 + F_{BD} - \frac{1}{\sqrt{2}} F_{BE} = 0$$

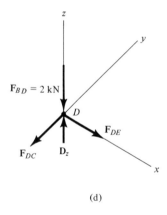

(d)

Solving,

$$R_B = F_{BE} = 5.66 \text{ kN} \quad (\text{T}) \qquad \qquad \textit{Ans.}$$

$$F_{BD} = 2 \text{ kN} \quad (\text{C}) \qquad \qquad \textit{Ans.}$$

The *scalar* equations of equilibrium may be applied directly to the force system shown on the free-body diagrams of joints D and C, since the force components are easily determined. Hence,

Joint D, Fig. 6-17d

$$\Sigma F_x = 0; \qquad F_{DE} = 0 \qquad \qquad \textit{Ans.}$$
$$\Sigma F_y = 0; \qquad F_{DC} = 0 \qquad \qquad \textit{Ans.}$$
$$\Sigma F_z = 0; \qquad D_z - 2 = 0 \qquad D_z = 2 \text{ kN}$$

Joint C, Fig. 6-17e

$$\Sigma F_x = 0; \qquad F_{CE} \sin 45° = 0 \qquad F_{CE} = 0 \qquad \textit{Ans.}$$
$$\Sigma F_y = 0; \qquad C_y = 0$$

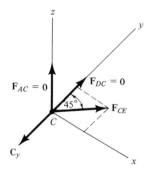

(e)

Fig. 6-17(c-e)

235

Problems

*6–28. Determine the force developed in each of the members of the space truss and indicate whether the members are in tension or compression.

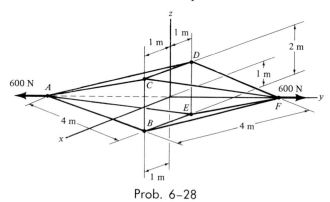

Prob. 6–28

6–29. The space truss is used to support vertical forces at joints B, C, and D. Determine the force developed in each of the members and indicate whether the members are in tension or compression.

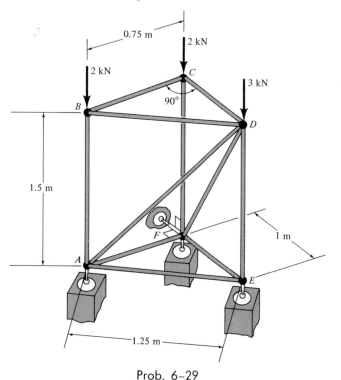

Prob. 6–29

6–30. Determine the force developed in each of the members of the space truss and indicate whether the members are in tension or compression. The crate has a mass of $m = 50$ kg and $a = 2$ m.

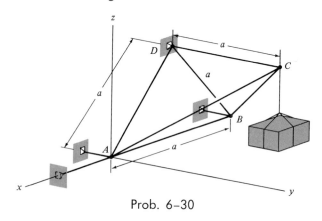

Prob. 6–30

6–30a. Solve Prob. 6–30 if the crate has a weight of $W = 150$ lb and $a = 8$ ft.

6–31. Two space trusses are used to equally support the uniform 50-kg sign. Determine the force developed in members AB, AC, and BC of truss $ABCD$, and indicate whether these members are in tension or compression. Horizontal short links support the truss at joints B, C, and D.

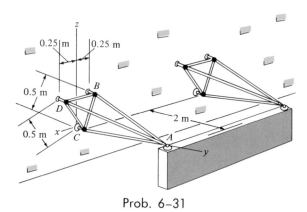

Prob. 6–31

*6-32. Determine the force in each of the members of the space truss when joint A is subjected to the force $\mathbf{P}$. Indicate whether the members are in tension or compression. The support reactions at B, C, and D are to be determined.

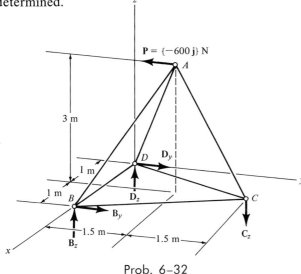

Prob. 6-32

6-5. Frames and Machines

Frames and machines are two common types of structures which are often composed of pin-connected members. *Frames* are generally stationary and are used to support loads, whereas *machines* contain moving parts and are designed to transmit and alter the effect of forces. In general, both types of structures contain *multiforce members,* i.e., members that are subjected to more than two forces. Unlike straight two-force members that can only elongate or contract under load, Fig. 6–18a, multiforce members can *bend,* Fig. 6–18b. Because of this, the method of joints or the method of sections, as used for truss analysis, cannot, in general, be applied to determine the forces in frames and machines. A partial analysis of these structures may be made, however, by removing the pin connections at the joints, and considering each member as a *separate* rigid body. The forces acting at the joints may then be determined by applying the three equations of equilibrium to *each member.* Provided that the structure is properly constrained and contains no more supports or members than are necessary to prevent collapse, the force analysis will be statically determinate. Hence, the equations of equilibrium for each member are both necessary and sufficient conditions for equilibrium. Once the forces at the joints are obtained, it is possible to *design* the size of the members, connections, and supports by using the theory of "strength of materials."

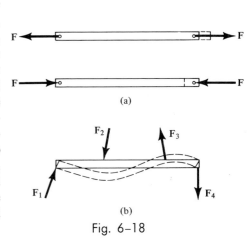

Fig. 6–18

The following three-step procedure should be used to determine the *joint reactions* for frames and machines (structures) composed of multiforce members.

Step 1: In some cases, depending upon what is required in the problem, draw a free-body diagram of the entire structure, apply the three equations of equilibrium, and solve for as many of the unknown *external reactions* as possible. It may not be possible to solve for all the external reactions on the free-body diagram, since the whole structure is generally composed of many rigid bodies, consisting of struts, beams, pulleys, etc.

Step 2: Draw a free-body diagram for each member of the structure. Forces common to two members act with equal magnitudes, but opposite directions on the respective free-body diagrams of the members. Recall that all *two-force members,* regardless of their shape, have only *one unknown,* i.e., the magnitude of force that acts along a line joining the end points of the member (Sec. 5–3).* The unknown forces acting at the joints of multiforce members should be represented by their rectangular components. In many cases it is possible to tell by inspection the proper "arrowhead" sense of direction of the unknown forces; however, if this seems difficult, the directional sense can be assumed.

Step 3: Apply the three equations of equilibrium to the force system shown on *each* of the free-body diagrams and solve for the unknown forces. Many times, the solution for the unknowns will be straightforward if moments are summed about a point that lies at the intersection of the lines of action of as many unknown forces as possible. If after obtaining the solution an unknown force magnitude is found to be a positive quantity, its sense of direction, as shown on the free-body diagram, is correct; if it is negative, the directional sense of the force is reversed.

The examples that follow illustrate this three-step procedure. All of these examples should be *thoroughly understood* before proceeding to solve the problems.

Example 6–7

Determine the tension in the cables and also the force **P** required to support the 100-kg block, utilizing the frictionless pulley system shown in (a) Fig. 6–19*a*; (b) Fig. 6–20*a*.

Solution

The solution of both of these problems can begin at *Step 2.*

*Try to form a habit of *immediately spotting* the two-force members in the structure, since this considerably simplifies the force analysis of the structure. Examples include members *BC* in Prob. 6–34, *AB* in Prob. 6–35, and *ACD* in Prob. 6–46.

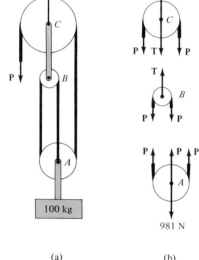

(a)   (b)

Fig. 6–19

Part (a).

Step 2: A free-body diagram of each pulley and a portion of the contacting cable is given in Fig. 6–19b. Since the cable is *continuous* and the pulleys are frictionless, the cable has a *constant tension* of P N acting throughout its length. (See Example 5–8.) The link connection between pulleys B and C is a two-force member and it therefore has an unknown tension of T N acting in it. Notice that the *principle of action and reaction* must be carefully observed for forces $\mathbf{P}$ and $\mathbf{T}$ when drawing the *separate* free-body diagrams, as shown in Fig. 6–19b.

Step 3: Applying the vertical force equilibrium equation to each pulley, we have for pulley A,

$$+\uparrow\Sigma F_y = 0; \qquad 3P - 981 = 0 \qquad P = 327 \text{ N} \qquad\qquad Ans.$$

for pulley B,

$$+\uparrow\Sigma F_y = 0; \qquad T - 2P = T - 654 = 0 \qquad T = 654 \text{ N} \qquad Ans.$$

and for pulley C,

$$+\uparrow\Sigma F_y = 0; \; R - 2P - T = R - 2(327) - 654 = 0 \; R = 1308 \text{ N} \quad Ans.$$

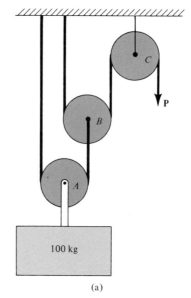

(a)

Part (b).

Step 2: A free-body diagram of each pulley and a portion of its contacting cable is shown in Fig. 6–20b. Observe how the principle of action and reaction is applied to forces $\mathbf{T}_1$ and $\mathbf{P}$ on the separate free-body diagrams.

Step 3: Applying the vertical force equilibrium equation to each pulley, we have, for pulley A,

$$+\uparrow\Sigma F_y = 0; \qquad 2T_1 - 981 = 0 \qquad T_1 = 490.5 \text{ N} \qquad\qquad Ans.$$

for pulley B,

$$+\uparrow\Sigma F_y = 0; \; 2P - T_1 = 2P - 490.5 = 0 \qquad P = 245.2 \text{ N} \qquad Ans.$$

and for pulley C,

$$+\uparrow\Sigma F_y = 0; \; T_2 - 2P = T_2 - 490.5 = 0 \qquad T_2 = 490.5 \text{ N} \qquad Ans.$$

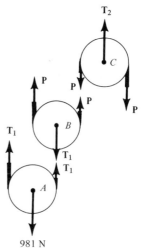

(b)

Fig. 6–20

239

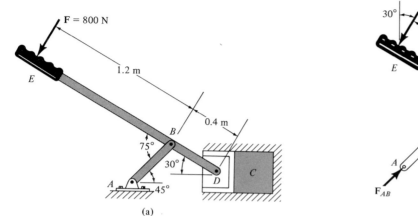

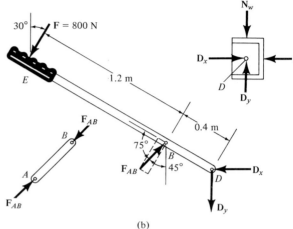

(a)

(b)

Fig. 6–21

Example 6–8

The piston-and-link mechanism shown in Fig. 6–21*a* is used as a toggle press for purposes of compressing material contained within the cylinder *C*. If a force of 800 N is applied perpendicular to the handle of the lever, determine the compressive force exerted by the piston and the horizontal and vertical components of force at pin *D*. Assume that the surface of contact between the piston and the cylinder wall is smooth.

Solution

The solution of this problem will begin at *Step 2*.

Step 2: A free-body diagram of each of the three members is shown in Fig. 6–21*b*. Member *AB* is a two-force member, and hence only the magnitude of force acting at its two end points is unknown. Since there is a pin connection at *D*, there are two unknown components D_x and D_y acting on both the piston and the lever. As shown, the principle of action and reaction of forces must be applied at joints *B* and *D* when drawing the separate free-body diagrams. In particular, four components of force act on the piston: D_x and D_y represent the effect of the pin (or lever *EBD*), N_W is the *resultant force* of the wall, and **P** is the resultant compressive force of the material within the cylinder.

Step 3: Applying the three equations of equilibrium to the lever *EBD*:

$$\zeta + \Sigma M_D = 0; \quad -800 \text{ N}(1.6 \text{ m}) + F_{AB} \sin 75° \, (0.4 \text{ m}) = 0$$
$$F_{AB} = 3312.9 \text{ N}$$

Using this result, we have

$$\overset{+}{\rightarrow} \Sigma F_x = 0; \quad -800 \sin 30° \text{ N} + (3312.9) \sin 45° \text{ N} - D_x = 0$$
$$D_x = 1942.2 \text{ N} \qquad \qquad Ans.$$
$$+\uparrow \Sigma F_y = 0; \quad -800 \cos 30° \text{ N} + (3312.9) \cos 45° \text{ N} - D_y = 0$$
$$D_y = 1649.4 \text{ N} \qquad \qquad Ans.$$

For the piston,

$$\xrightarrow{+}\Sigma F_x = 0; \qquad -P + 1942.2 \text{ N} = 0$$
$$P = 1942.2 \text{ N} \qquad \qquad Ans.$$

The analysis indicates that by using this "machine," a small input force (0.800 kN) applied to the handle yields a much larger output force (1.94 kN) at the piston.

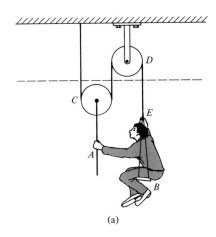

(a)

Example 6–9

A man having a mass of 50 kg supports himself by means of the cable and pulley system shown in Fig. 6–22a. If the seat has a mass of 10 kg, determine the equilibrium force that he must exert on the cable at A and the force applied to the seat. Neglect the weight of the cables and pulleys.

Solution I

The analysis of this problem will begin at *Step 2*.

Step 2: The free-body diagrams of the man, seat, and pulley C are shown in Fig. 6–22b. The *two* cables are subjected to tensions $\mathbf{T}_A$ and $\mathbf{T}_E$, respectively. The man is subjected to three forces: his weight $W = 50(9.81) = 490.5$ N; the tension of cable AC, $\mathbf{T}_A$; and the reaction of the seat, $\mathbf{N}_s$.

Step 3: Applying the equation of vertical force equilibrium to the man, we have

$$+\uparrow\Sigma F_y = 0; \qquad T_A + N_s - 490.5 \text{ N} = 0 \qquad (1)$$

For the seat,

$$+\uparrow\Sigma F_y = 0; \qquad T_E - N_s - 98.1 \text{ N} = 0 \qquad (2)$$

For pulley C,

$$+\uparrow\Sigma F_y = 0; \qquad 2T_E - T_A = 0 \qquad (3)$$

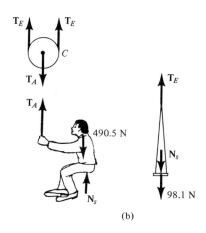

(b)

The magnitude of force $\mathbf{T}_E$ can be determined by adding Eqs. (1) and (2) to eliminate N_s and then using Eq. (3). Using the result, the other unknowns are easily obtained.

$$T_A = 392.4 \text{ N} \qquad \qquad Ans.$$
$$T_E = 196.2 \text{ N}$$
$$N_s = 98.1 \text{ N} \qquad \qquad Ans.$$

Solution II

By using the dashed section shown in Fig. 6–22a, a free body of the man, pulley, and seat can be considered as a *single system*, Fig. 6–22c. Here $\mathbf{N}_s$ and $\mathbf{T}_A$ are *internal* forces and hence are not included in the "combined" free-body diagram. Applying $\Sigma F_y = 0$ yields a *direct* solution for T_E.

$$+\uparrow\Sigma F_y = 0; \quad 3T_E - 98.1 \text{ N} - 490.5 \text{ N} = 0 \qquad T_E = 196.2 \text{ N}$$

The other unknowns can be obtained from Eqs. (2) and (3).

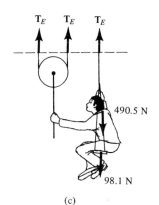

(c)

Fig. 6–22

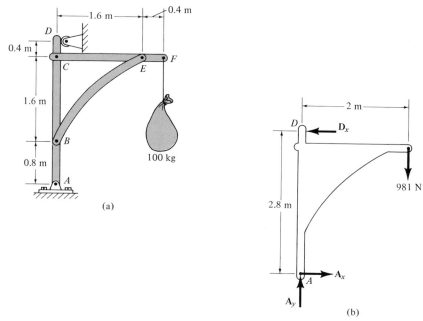

Fig. 6–23(a–b)

Example 6–10

Determine the horizontal and vertical components of force which the pins at B and C exert on member $ABCD$ of the loaded frame shown in Fig. 6–23a.

Solution

Step 1: A free-body diagram of the entire frame is shown in Fig. 6–23b. Applying the three equations of equilibrium to obtain the external reactions yields

$$\zeta + \Sigma M_A = 0; \quad -981 \text{ N}(2 \text{ m}) + D_x(2.8 \text{ m}) = 0 \quad D_x = 700.7 \text{ N}$$
$$\xrightarrow{+} \Sigma F_x = 0; \quad A_x - 700.7 \text{ N} = 0 \quad A_x = 700.7 \text{ N}$$
$$+\uparrow \Sigma F_y = 0; \quad A_y - 981 \text{ N} = 0 \quad A_y = 981 \text{ N}$$

Step 2: Using the above values for the reactions, the free-body diagram of each frame member is shown in Fig. 6–23c. Notice that member BE is a two-force member. There are two components of force at C, since two connecting members are pinned together there. As shown, the principle of action and reaction of forces must be applied at joints B, C, and E when the separate free-body diagrams are drawn.

Step 3: If the three equations of equilibrium are applied to member CEF, we have

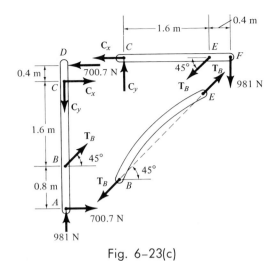

Fig. 6–23(c)

$\zeta +\Sigma M_C = 0;$ $-981 \text{ N}(2 \text{ m}) - (T_B \sin 45°)(1.6 \text{ m}) = 0$

$$T_B = -1734.4 \text{ N} \qquad Ans.$$

$\xrightarrow{+}\Sigma F_x = 0;$ $-C_x - (-1734.4 \cos 45° \text{ N}) = 0$

$$C_x = 1226.2 \text{ N} \qquad Ans.$$

$+\uparrow\Sigma F_y = 0;$ $C_y - (-1734.4 \sin 45° \text{ N}) - 981 \text{ N} = 0$

$$C_y = -245.5 \text{ N} \qquad Ans.$$

Since the magnitudes of forces $\mathbf{T}_B$ and $\mathbf{C}_y$ were calculated as negative quantities, they were assumed to be acting in the wrong sense of direction on the free-body diagrams, Fig. 6–23c. The correct direction of these forces might have been determined "by inspection" *before* applying the equations of equilibrium to member *CEF*. As shown in Fig. 6–23c, moment equilibrium about point *E* on member *CEF* indicates that $\mathbf{C}_y$ must actually act *downward* to counteract the moment created by the 981-N force about point *E*. Similarly, summing moments about point *C*, it is seen that the vertical component of force $\mathbf{T}_B$ must actually act *upward*.

A free-body diagram of member *ABCD*, showing the magnitude and *correct direction* of all the computed forces acting on it, is given in Fig. 6–23d. To *check* the previous calculations the three equations of equilibrium may be applied to this member:

$\xrightarrow{+}\Sigma F_x = 0;$ $-700.7 \text{ N} + 1226.2 \text{ N} - 1734.4 \cos 45° \text{ N} + 700.7 \text{ N} \equiv 0$

$+\uparrow\Sigma F_y = 0;$ $245.5 \text{ N} - 1734.4 \sin 45° \text{ N} + 981 \text{ N} \equiv 0$

$\zeta +\Sigma M_B = 0;$ $700.7 \text{ N}(2 \text{ m}) - 1226.2 \text{ N}(1.6 \text{ m}) + 700.7 \text{ N}(0.8 \text{ m}) \equiv 0$

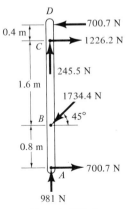

Fig. 6–23(d)

Example 6–11

The hydraulic, truck-mounted crane shown in Fig. 6–24a is used to lift a beam that has a mass of 100 kg. Determine the horizontal and vertical components of force acting at the joints of each of the members of the crane if the beam is held in the position shown.

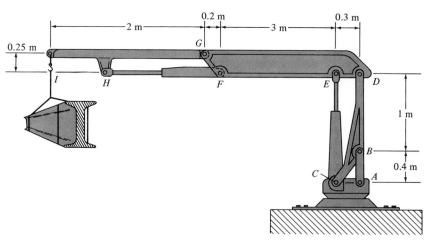

Fig. 6–24(a)

Solution

The analysis of this problem will begin at *Step 2*.

Step 2: By inspection, members *BA*, *EC*, and *FH* are all two-force members, since they are pin-connected and subjected to no intermediate load. The free-body diagrams are shown in Fig. 6–24b.

Step 3: Each of the unknown forces can be determined *directly* by using previously computed results and applying the equations of equilibrium, in turn, to members *GHI*, *DEFG*, and *DBC*. The computations follow:

Member *GHI*:

$$\zeta + \Sigma M_G = 0; \quad -F_{FH}(0.25 \text{ m}) + 981 \text{ N}(2 \text{ m}) = 0 \quad F_{FH} = 7848 \text{ N} \quad Ans.$$

$$\xrightarrow{+} \Sigma F_x = 0; \quad 7848 \text{ N} - G_x = 0 \quad G_x = 7848 \text{ N} \quad Ans.$$

$$+\uparrow \Sigma F_y = 0; \quad G_y - 981 \text{ N} = 0 \quad G_y = 981 \text{ N} \quad Ans.$$

Member *DEFG*:

$$\zeta + \Sigma M_D = 0; \quad -F_{EC}(0.3 \text{ m}) + 981 \text{ N}(3.5 \text{ m}) + 7848 \text{ N}(0.25 \text{ m}) = 0$$

$$F_{EC} = 17\,985 \text{ N} = 18.0 \text{ kN} \qquad \text{Ans.}$$

$$\xrightarrow{+} \Sigma F_x = 0; \quad D_x - 7848 \text{ N} + 7848 \text{ N} = 0 \qquad D_x = 0 \qquad \text{Ans.}$$

$$+\uparrow \Sigma F_y = 0; \quad -D_y + 17\,985 \text{ N} - 981 \text{ N} = 0$$

$$D_y = 17\,004 \text{ N} = 17.0 \text{ kN} \qquad \text{Ans.}$$

Member *DBC*:

$$\zeta + \Sigma M_C = 0; \quad -F_{BA}(0.3 \text{ m}) + 17\,004 \text{ N}(0.3 \text{ m}) + (0)(1.4 \text{ m}) = 0$$

$$F_{BA} = 17.0 \text{ kN} \qquad \text{Ans.}$$

$$\xrightarrow{+} \Sigma F_x = 0; \quad 0 - C_x = 0 \qquad C_x = 0 \qquad \text{Ans.}$$

$$+\uparrow \Sigma F_y = 0; \quad 17.0 \text{ kN} - 17.0 \text{ kN} + C_y = 0 \qquad C_y = 0 \qquad \text{Ans.}$$

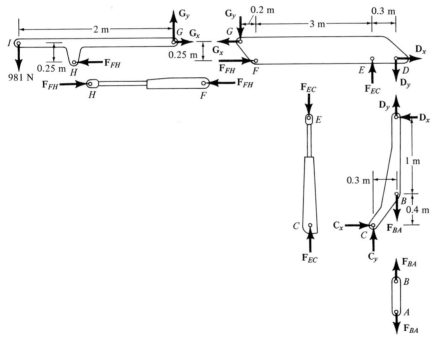

Fig. 6-24(b)

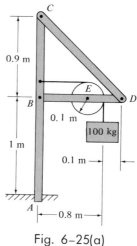

Fig. 6–25(a)

Example 6–12

Determine the horizontal and vertical force components acting at the pin connections B and C of the loaded frame shown in Fig. 6–25a.

Solution

Step 1:   The free-body diagram of the entire frame is shown in Fig. 6–25b. Applying the three equations of equilibrium yields

$\zeta + \Sigma M_A = 0;$ $M_A - 981 \text{ N}(0.8 \text{ m}) = 0$ $M_A = 784.8 \text{ N} \cdot \text{m}$

$\overset{+}{\rightarrow} \Sigma F_x = 0;$ $A_x = 0$

$+\uparrow \Sigma F_y = 0;$ $A_y - 981 \text{ N} = 0$ $A_y = 981 \text{ N}$

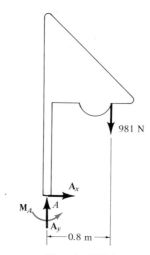

Fig. 6–25(b)

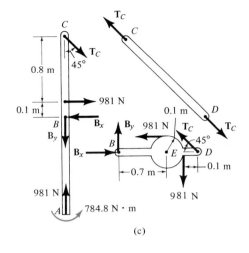

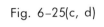

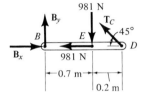

Fig. 6-25(c, d)

(c)

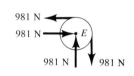

(d)

Step 2: Using these values for the reactions, the free-body diagram for each of the members is shown in Fig. 6–25c. Notice that member *CD* is a two-force member. Member *BED*, the pulley, and a portion of the contacting cable are considered to be acting together as a single body. Since the cable is *removed* from member *ABC*, a force of 981 N must pull horizontally on this member.

Step 3: Applying the three equations of equilibrium to member *BED* yields

$$\zeta +\Sigma M_B = 0; \quad -981 \text{ N}(0.8 \text{ m}) + 981 \text{ N}(0.1 \text{ m}) + T_C \sin 45°\ (0.9 \text{ m}) = 0$$
$$T_C = 1079.2 \text{ N} \qquad\qquad\qquad Ans.$$

$$\xrightarrow{+}\Sigma F_x = 0;$$
$$B_x - 981 \text{ N} - 1079.2 \cos 45°\text{ N} = 0 \qquad B_x = 1744.0 \text{ N} \quad Ans.$$

$$+\uparrow\Sigma F_y = 0;$$
$$1079.2 \sin 45°\text{ N} - 981 \text{ N} + B_y = 0 \qquad B_y = 218.0 \text{ N} \qquad Ans.$$

It is suggested that the above calculations be checked by applying the three equations of equilibrium to member *ABC*.

An alternative approach may be used to solve for the forces acting on member *BED*. If the pulley is *removed*, it must be held in equilibrium by 981-N force components exerted *on* it by the pin at point *E*, Fig. 6–25d. By inspection, both force and moment equilibrium of the pulley are satisfied. By Newton's third law, equal, but opposite, 981-N force components are shown on the free-body diagram of member *BED*, Fig. 6–25d. Using this diagram, the reactions B_x, B_y, and T_C may be obtained.

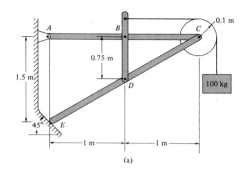

(a)

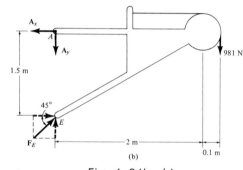

(b)

Fig. 6–26(a, b)

Example 6–13

The frame shown in Fig. 6–26a is supported at A by a pin and at E by a smooth inclined plane. Determine the horizontal and vertical components of force that each of the pin connections exerts on the frame. The pulley at C is frictionless.

Solution

Step 1: A free-body diagram of the entire frame is shown in Fig. 6–26b. Writing the three equations of equilibrium, we have

$$\zeta + \Sigma M_A = 0; \quad -981 \text{ N}(2.1 \text{ m}) + F_E \cos 45° (1.5 \text{ m}) = 0$$
$$F_E = 1942.6 \text{ N}$$

(Note that the component $F_E \sin 45°$ is not included in the moment equation since the line of action of this force passes through point A.) Using the computed value of F_E,

$$\xrightarrow{+} \Sigma F_x = 0; \quad 1942.6 \cos 45° \text{ N} - A_x = 0 \quad A_x = 1373.4 \text{ N} \qquad Ans.$$
$$+\uparrow \Sigma F_y = 0; \quad -A_y + 1942.6 \sin 45° \text{ N} - 981 \text{ N} = 0$$
$$A_y = 392.4 \text{ N} \qquad Ans.$$

Step 2: A free-body diagram of each member, including the pins at D and C, and the pulley is shown in Fig. 6–26c. The force $\mathbf{F}_E$ is represented on the free-body diagram of member EDC in terms of its two components having magnitudes of $(1942.6)(0.707) = 1373.4 \text{ N}$.

Each of the free-body diagrams should be carefully studied. In particular, note that *three force interactions* occur *on* the pin located at C: the force components $\mathbf{C}_x$ and $\mathbf{C}_y$ representing the force exerted by member ABC, the force components $\mathbf{C}'_x$ and $\mathbf{C}'_y$ representing the force exerted by member EDC, and finally the two 981-N force components exerted by the pulley. Writing the equilibrium equations $\Sigma F_x = 0$ and $\Sigma F_y = 0$ for the pin at C will enable one to obtain two equations that relate the unknown force magnitudes C_x, C'_x, and C_y, C'_y (Eqs. (4) and (5) in *Step 3*). When only *two force interactions* occur at a pin connection (such as pin D in Fig. 6–26c), to satisfy $\Sigma F_x = 0$ and $\Sigma F_y = 0$ the force components acting must be equal in magnitude but opposite in direction. For this reason the free-body diagrams for *pins* that connect *only two members were not considered* in the force analysis of the previous examples. In all cases, the length of the pin is considered *short*, so that any moment created by the force interactions on the pin can be neglected.

Step 3: Applying the three equations of equilibrium, for member ABC,

$$\xrightarrow{+} \Sigma F_x = 0; \quad -1373.4 \text{ N} + B_x + C_x = 0 \qquad (1)$$
$$+\uparrow \Sigma F_y = 0; \quad -392.4 \text{ N} + B_y - C_y = 0 \qquad (2)$$
$$\zeta + \Sigma M_B = 0; \quad 392.4 \text{ N}(1 \text{ m}) - C_y(1 \text{ m}) = 0 \qquad (3)$$

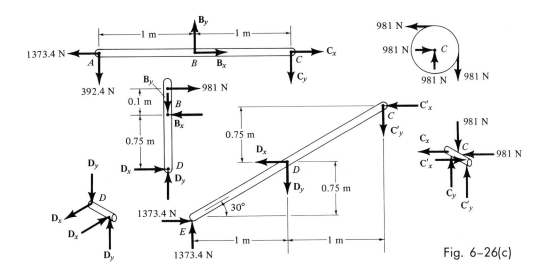

Fig. 6-26(c)

for pin C,

$$\xrightarrow{+}\Sigma F_x = 0; \qquad -C_x + C'_x - 981\,\text{N} = 0 \qquad\qquad (4)$$

$$+\uparrow\Sigma F_y = 0; \qquad C_y + C'_y - 981\,\text{N} = 0 \qquad\qquad (5)$$

and for member EDC,

$$\xrightarrow{+}\Sigma F_x = 0; \qquad 1373.4\,\text{N} - D_x - C'_x = 0 \qquad\qquad (6)$$

$$+\uparrow\Sigma F_y = 0; \qquad 1373.4\,\text{N} - D_y - C'_y = 0 \qquad\qquad (7)$$

$$\lefthalfcup+\Sigma M_D = 0; \quad -1373.4\,\text{N}(1\,\text{m}) + 1373.4\,\text{N}(0.75\,\text{m}) - C'_y(1\,\text{m})$$
$$+ C'_x(0.75\,\text{m}) = 0 \quad (8)$$

Eight independent equations have been written, which may be solved for the eight unknowns: B_x, B_y, C_x, C_y, C'_x, C'_y, D_x, and D_y. The solution is straightforward since each equation may be solved directly for an unknown. This is done by first solving Eq. (3) for C_y, then proceeding to obtain the remaining unknowns using Eqs. (2), (5), (8), (4), (6), (7), and (1), in that order. The final results are:

$$B_x = 1111.8\,\text{N} \qquad\qquad Ans.$$
$$B_y = 784.8\,\text{N} \qquad\qquad Ans.$$
$$C_x = 261.6\,\text{N} \qquad\qquad Ans.$$
$$C_y = 392.4\,\text{N} \qquad\qquad Ans.$$
$$C'_x = 1242.6\,\text{N} \qquad\qquad Ans.$$
$$C'_y = 588.6\,\text{N} \qquad\qquad Ans.$$
$$D_x = 130.8\,\text{N} \qquad\qquad Ans.$$
$$D_y = 784.8\,\text{N} \qquad\qquad Ans.$$

It is suggested that these calculations be checked by applying the equations of equilibrium to member BD.

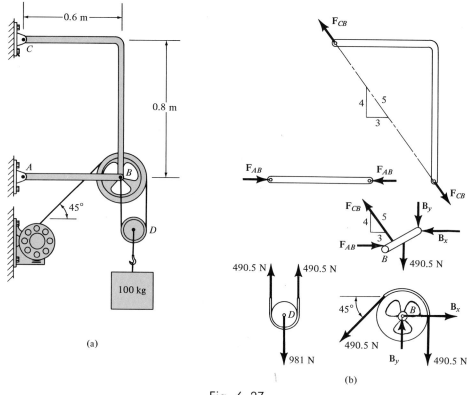

Fig. 6–27

Example 6–14

The 100-kg block is held in equilibrium by means of the pulley and continuous cable system shown in Fig. 6–27a. If the cable is attached to the pin at B, compute the forces which this pin exerts on each of its connecting members.

Solution

The analysis of this problem will begin at *Step 2*.

Step 2: A free-body diagram of each member of the frame is shown in Fig. 6–27b. By inspection, members AB and CB are two-force members. Furthermore, the cable must be subjected to a force of 490.5 N in order to hold pulley D in equilibrium. A free-body diagram of the pin at B is needed, since *four interactions* occur at this pin. These are caused by the attached cable (490.5 N), member AB ($\mathbf{F}_{AB}$), member CB ($\mathbf{F}_{CB}$), and the pulley B ($\mathbf{B}_x$ and $\mathbf{B}_y$).

Step 3: Applying the equations of force equilibrium to pulley B, we have

$\xrightarrow{+}\Sigma F_x = 0; \quad B_x - 490.5 \cos 45° \text{ N} = 0 \qquad B_x = 346.8 \text{ N}$ \hfill *Ans.*

$+\uparrow\Sigma F_y = 0; \qquad B_y - 490.5 \sin 45° \text{ N} - 490.5 \text{ N} = 0$

$$B_y = 837.3 \text{ N} \qquad \textit{Ans.}$$

Using these results, equilibrium of the pin requires that

$+\uparrow\Sigma F_y = 0; \quad \frac{4}{5}F_{CB} - 837.3 \text{ N} - 490.5 \text{ N} = 0 \qquad F_{CB} = 1659.7 \text{ N} \quad \textit{Ans.}$

$\xrightarrow{+}\Sigma F_x = 0; \quad F_{AB} - \frac{3}{5}(1659.7 \text{ N}) - 346.8 \text{ N} = 0 \qquad F_{AB} = 1342.6 \text{ N} \ \textit{Ans.}$

It may be noted that the two-force member CB is subjected to bending as caused by the force $\mathbf{F}_{CB}$. From the standpoint of design, it would be better to make this member *straight* (from C to B) so that the force $\mathbf{F}_{CB}$ would only create tension in the member.

Before solving the following problems, it is suggested that a brief review be made of all the previous examples. This may be done by trying to locate the two-force members, drawing the free-body diagrams, and conceiving ways of applying the equations of equilibrium to obtain the solution. Then after this is done see Step 1 of Example 7–2 for a further illustration of the above method of analysis.

Problems

Neglect the weight of the members unless stated otherwise.

6-33. Determine the horizontal and vertical components of force that the pins A, B, and C exert on the frame.

6-34. Determine the horizontal and vertical components of force that the pins A and C exert on the frame.

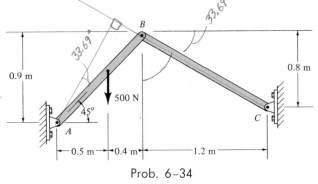

Prob. 6-33   Prob. 6-34

6-35. If $M = 1.5$ kN $\cdot$ m and $a = 0.3$ m, determine the horizontal and vertical components of force that the pins at A and C exert on the mechanism.

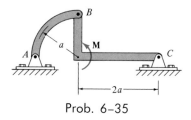

Prob. 6-35

6-35a. Solve Prob. 6-35 if $M = 750$ lb $\cdot$ ft and $a = 2$ ft.

* **6-36.** Determine the tension in member BC of the frame and the horizontal and vertical components of force that the pin at D exerts on member ABD.

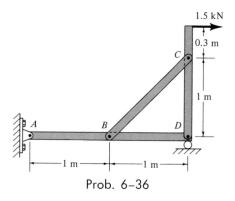

Prob. 6-36

6-37. Determine the horizontal and vertical components of force that the pins at A, B, and C exert on the two members of the arch.

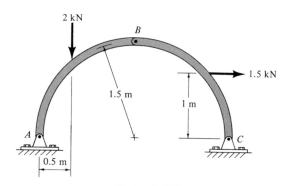

Prob. 6-37

6-38. The jack shown supports a 350-kg automobile engine. Determine the compression in the hydraulic cylinder C and the magnitude of force that pin B exerts on member BDE.

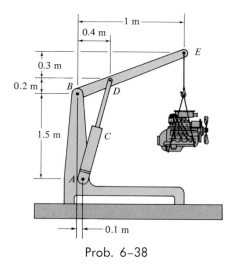

Prob. 6-38

6-39. Determine the horizontal and vertical components of force that the pins at A, B, and C exert on their connecting members.

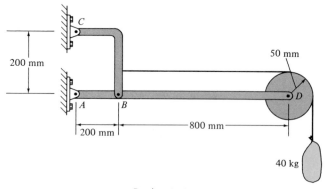

Prob. 6-39

252

*6-40. Determine the compressive force developed in member BC and the magnitude of force that the pin at A exerts on member AC of the frame. The force has a magnitude of $F = 700$ N and $a = 400$ mm, $b = 300$ mm.

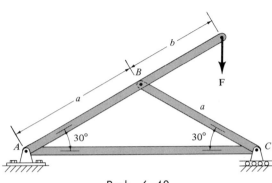

Prob. 6–40

*6-40a. Solve Prob. 6–40 if $F = 200$ lb, and $a = 3$ ft, $b = 2.5$ ft.

6-41. A force of 150 N is applied to the handles of the pliers. Determine the force developed on the smooth bolt B and the reaction that the pin A exerts on its attached members.

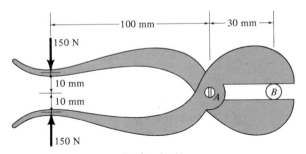

Prob. 6–41

6-42. Determine the clamping force exerted on the smooth pipe at B if a force of 100 N is applied to the handles of the pliers. The pliers are pinned together at A.

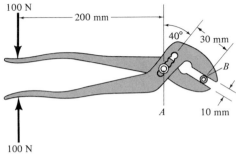

Prob. 6–42

6-43. Determine the force P required to maintain equilibrium of each of the pulley systems.

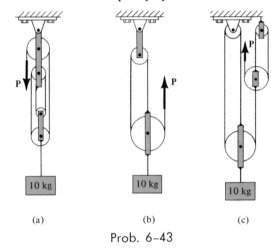

(a)   (b)   (c)

Prob. 6–43

*6-44. Compute the reactions at supports A, B, C, and D.

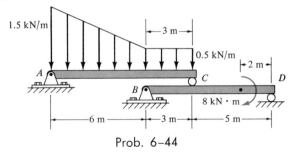

Prob. 6–44

253

6-45. Compute the tension **T** in the cord, and determine the angle θ that the pulley-supporting link *AB* makes with the vertical. Neglect the mass of the pulleys and the link. The block has a mass of $m = 30$ kg and the cord is attached to the pin at *B*. The pulleys have radii of $r_1 = 50$ mm and $r_2 = 25$ mm.

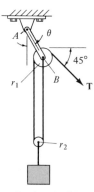

Prob. 6-45

6-45a. Solve Prob. 6-45 if the block weighs $W = 150$ lb, and $r_1 = 2$ in., $r_2 = 1$ in.

6-46. The car has a mass of 1.8 Mg and a center of gravity at *G*. If it is supported by a hydraulic jack *AD*, determine the compression in the cylinder *C* and the horizontal and vertical components of force at pin *B*.

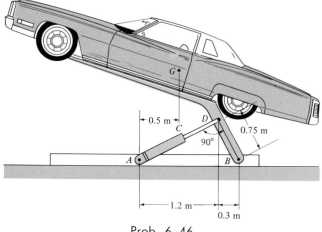

Prob. 6-46

6-47. Determine the horizontal and vertical components of force that the pins at *A*, *B*, and *C* exert on the frame.

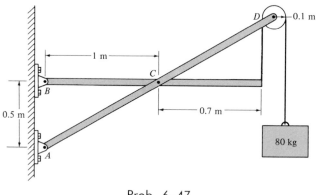

Prob. 6-47

*6-48.** The mechanism is subjected to a horizontal force of 800 N. Determine the magnitude of the vertical force **P** that must be applied to the smooth block at *B*, required for equilibrium. Members *ACE* and *BCD* are each 600 mm long and are pinned together at their center *C*.

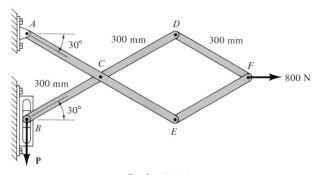

Prob. 6-48

254

6-49. Determine the resultant force that the pins at A and B exert on the frame members. The pulley at C has a radius of 100 mm and the pulley at D has a radius of 50 mm. The two members, the pulley and cord are attached to the pin at C.

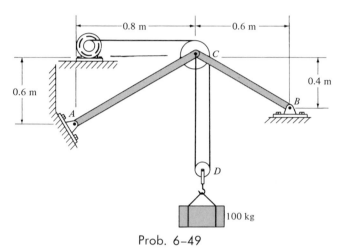

Prob. 6–49

6-50. The clamp is used to grip two smooth boards by initially turning screw A. The screw at B is then used to increase this gripping force. If the force exerted by each screw is directed along its axis, determine the force that the screw at B exerts on C and the horizontal and vertical components of force developed at pin D if the compressive force on the boards due to the action of the clamp is $F = 1.2$ kN. Set $a = 100$ mm and $b = 125$ mm.

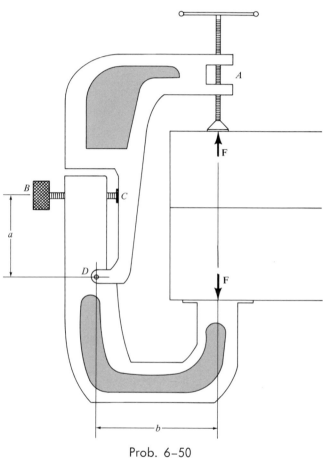

Prob. 6–50

6-50a. Solve Prob. 6–50 if $F = 300$ lb, $a = 3$ in., and $b = 4$ in.

6-51. Determine the horizontal and vertical components of force that the pins at *A*, *B*, and *D* exert on the *A frame*.

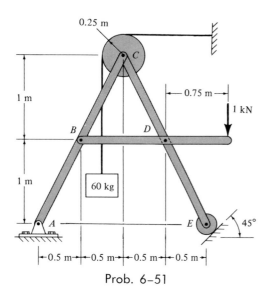

Prob. 6-51

* 6-52. The tractor shovel shown carries a 500-kg load of soil, having a center of gravity at *G*. Compute the forces developed in the hydraulic cylinders *IJ* and *BC* due to this loading.

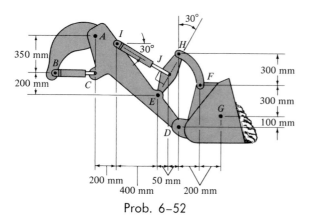

Prob. 6-52

6-53. For the tractor mechanism shown, determine the force created in the hydraulic cylinders *EF* and *AD* in order to hold the shovel in equilibrium. The shovel load has a mass of 1.25 Mg and a center of gravity at *G*. All joints are pin-connected.

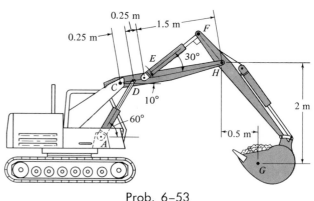

Prob. 6-53

6-54. The scissor lift consists of *two* sets of cross members and *two* hydraulic cylinders, symmetrically located on *each side* of the platform. If the platform has a uniform mass of 40 kg, with a center of gravity at G_1, and a 90-kg load is centrally located on the platform, with center of gravity G_2, determine the force in each of the hydraulic cylinders, *DE*, for equilibrium. Rollers are located at *B* and *D*.

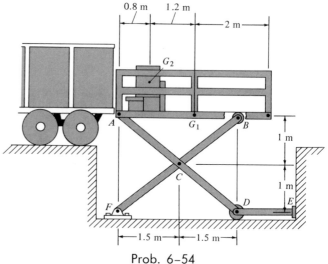

Prob. 6-54

6-55. The smooth disk D has a mass of $m_D = 15$ kg. Determine the horizontal and vertical components of reaction that the pin at B exerts on the curved beam and the reaction at the roller C. Set $a = 100$ mm and $r = 500$ mm.

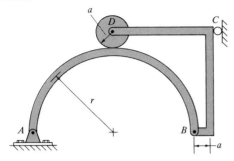

Prob. 6-55

6-55a. Solve Prob. 6-55 if the disk has a weight of $W_D = 40$ lb, and $a = 1$ ft, $r = 5$ ft.

*6-56. Determine the force that the jaws J of the metal cutters exert on the smooth cable C if 200-N forces are applied to the handles. The jaws are pinned at E and A, and D and B.

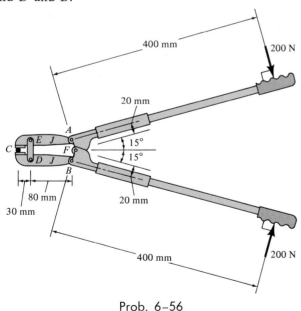

Prob. 6-56

6-57. The *poidometer* feeds a measured amount of granular material in a uniform stream onto the conveyor belt. The weight of material on the belt creates a force **F** that actuates a scale beam *JEDI*, which in turn lowers or lifts a smooth gate G and thereby controls the flow from the feed hopper H. The gate G has a mass of 10 kg and is confined by smooth guides at K and L and a pin at J. Neglecting friction and the weights of the other members, determine the mass m that must be placed on the scale at I to balance a belt force $F = 80$ N.

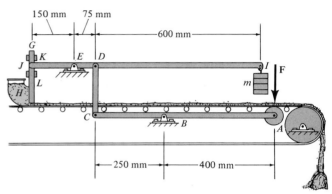

Prob. 6-57

6-58. If a force of $P = 500$ N is applied perpendicular to the handle of the mechanism, determine the magnitude of force **F** for equilibrium. The members are pin-connected at A, B, C, and D.

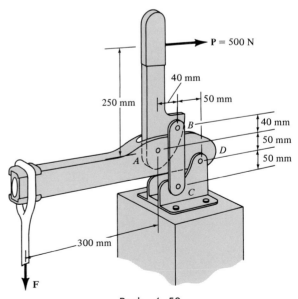

Prob. 6-58

6-59. The winch is used to lift a scaffold by applying a force **P** perpendicular to the handle. The handle is pinned at A and rotates the drum D by means of the driving pawl BF. Determine the minimum magnitude of force **P** needed to lift a scaffold that requires a cable tension of 6 kN. If the force **P** and pawl BF are released, what force does the loose locking pawl CE develop on the drum to support the load? *Hint:* The pawls BF and CE are two-force members.

*6-60. Determine the tension in chain CD and the force that the pin at B exerts on member CB of the frame. The cylinder has a mass of $m = 80$ kg. Set $r = 0.1$ m and $a = 0.75$ m.

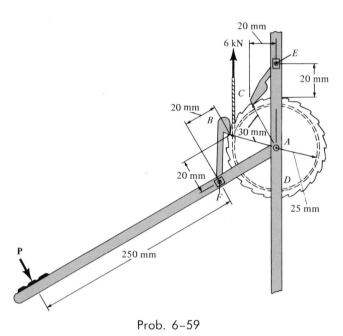

Prob. 6–59

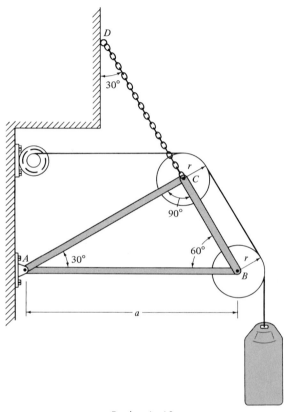

Prob. 6–60

*6-60a. Solve Prob. 6–60 if the cylinder weighs $W = 200$ lb, and $r = 0.5$ ft, $a = 4$ ft.

6-61. The 10-kg rectangular furnace door has a mass center at *G*. Two links, such as *CDE*, are located at both sides of the door and are rotated clockwise in order to close the door to the dashed position. Determine the resultant force **F** that must be applied perpendicular to both handles in order to close the door.

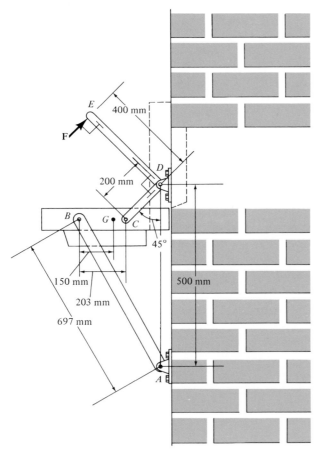

Prob. 6-61

Internal Forces

7-1. Internal Forces Developed in Structural Members

In this section the methods for computing the *internal forces* acting in various types of structural members will be discussed. This type of analysis is *necessary* for design, since the size of the member's cross-sectional area depends upon both the type and the amount of load developed at the cross section.

Internal Loadings. When the *external loading* on a multiforce member is applied perpendicular to the member's longitudinal axis, the *internal loading* in the member generally consists of a resultant force and couple. To show how these resultants can be developed, consider the "simply supported" beam in Fig. 7–1a, which is subjected to the forces $\mathbf{F}_1$ and $\mathbf{F}_2$. The support reactions $\mathbf{A}_x$, $\mathbf{A}_y$, and $\mathbf{B}_y$ are determined by applying the equations of equilibrium, using the free-body diagram of the entire beam,

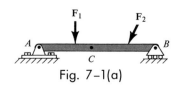

Fig. 7–1(a)

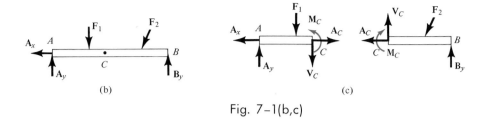

(b)　　　　　　　　　　　　　　　　(c)

Fig. 7–1(b,c)

Fig. 7–1b. To determine the *internal* resultant loading at point *C* along the axis of the beam (or at any other location), it is necessary to pass an imaginary section through the beam, cutting it into two segments at that point, Fig. 7–1c. Doing this "exposes" the internal forces as external on the free-body diagram of each segment. Since both of these segments (*AC* and *CB*) were prevented from translating *and* rotating *before* the beam was sectioned, equilibrium of each section is maintained if rectangular force components A_C and V_C, and a resultant couple M_C, are developed at the cut section. In accordance with Newton's third law of motion, these loadings must be equal in magnitude and opposite in direction on each of the segments. The magnitude of each unknown can be determined by applying the three equations of equilibrium to either segment *AC* or *CB*. A *direct solution* for A_C is obtained by applying $\Sigma F_x = 0$; V_C is obtained directly from $\Sigma F_y = 0$; and M_C is determined by summing moments about point C, $\Sigma M_C = 0$, in order to eliminate the moments of the unknowns A_C and V_C.

In structural mechanics, the force components A_C, acting normal to the beam at the cut section, and V_C, acting tangent to the section, are termed the *axial force* and *shear force*, respectively. The couple M_C is referred to as the *bending moment*. In general, the magnitudes for each of these loadings will be different at various points along the axis of the beam. In all cases, however, the *method of sections* can be used to determine their values. For application, it is very important that the load acting on the beam, or structural member, *be kept at its exact position before the beam is sectioned*. For example, in the case of the "cantilever beam" *AB* shown in Fig. 7–2a, the internal forces and couple acting at *C* are obtained from either section *AC* or *CB* of Fig. 7–2b. *After* the section is made, either portion of the distributed load can *then* be simplified before applying the

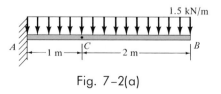

Fig. 7–2(a)

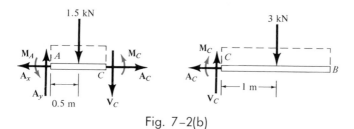

Fig. 7-2(b)

equations of equilibrium to one of the segments. Quite obviously, if the entire distributed load was simplified *before* the section was made, erroneous free-body diagrams for *AC* and *CB* would result.* A *direct solution* for A_C, V_C, and M_C is possible if segment *CB* is chosen for analysis. If segment *AC* is chosen, reactions $\mathbf{A}_x$, $\mathbf{A}_y$, and $\mathbf{M}_A$ will have to be obtained *first,* since only three equations of equilibrium are available for each segment.

PROCEDURE FOR ANALYSIS

The following three-step procedure should be used when applying the method of sections to determine the internal axial force, shear force, and bending moment at a specific point in a member.

Step 1: If the internal forces and moments are to be determined at a point located *between* two points of support, determine the external reactions acting at the supports of the member. If the member is part of a frame, the pin reactions can be determined by the methods of Sec. 6–5.

Step 2: *Before* simplifying any distributed loadings that act on the member, pass an imaginary section through the member, perpendicular to its axis at the point where the internal loading is to be determined; then draw a free-body diagram of one of the "cut" segments on either side of the section. Indicate the unknowns **A, V,** and **M** at the section.

Step 3: Apply the three equations of equilibrium to obtain **A, V,** and **M.** In most cases, moments should be summed about the section point in order to eliminate the unknowns **A** and **V** and thereby obtain a direct solution for **M.** If the solution of the equilibrium equations yields a quantity having a negative magnitude, the assumed directional sense of the quantity is opposite to that shown on the free-body diagram.

The following example problems numerically illustrate this procedure.

---

*The force resultant for *AB* is (1.5 kN/m)(3 m) = 4.5 kN. Since this force acts 1.5 m from *A*, then, if it was placed on the beam, no external load would act on segment *AC* after sectioning *AB*. Certainly, this does not represent the true loading situation.

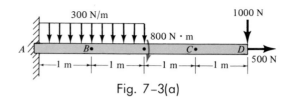

Fig. 7-3(a)

Example 7-1

Determine the axial force, shear force, and bending moment acting at points B and C of the cantilever beam loaded as shown in Fig. 7-3a. The 500-N axial force is applied along the central axis of the beam.

Solution

For purposes of illustration, *Step 1* will be applied, although it is not necessary for the solution of this problem. Why?

Step 1: A free-body diagram of the entire beam is shown in Fig. 7-3b. The distributed load has been replaced by its resultant force in accordance with the principles outlined in Sec. 4-10. Applying the three equations of equilibrium yields

$$\xrightarrow{+}\Sigma F_x = 0; \qquad -A_x + 500 \text{ N} = 0 \qquad A_x = 500 \text{ N}$$
$$+\uparrow\Sigma F_y = 0; \quad A_y - 600 \text{ N} - 1000 \text{ N} = 0 \qquad A_y = 1600 \text{ N}$$
$$\zeta+\Sigma M_A = 0; \quad -600 \text{ N}(1 \text{ m}) - 800 \text{ N}\cdot\text{m} - 1000 \text{ N}(4 \text{ m}) + M_A = 0$$
$$M_A = 5400 \text{ N}\cdot\text{m}$$

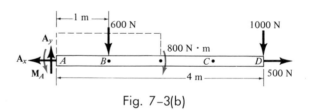

Fig. 7-3(b)

Step 2: Passing an imaginary cutting section, perpendicular to the axis of the beam, through point C, yields the free-body diagrams of the beam segments AC and CD shown in Fig. 7-3c.

Step 3: The magnitudes of the internal forces A_C and V_C and the internal moment M_C can be obtained by applying the equations of equilibrium to either segment AC or CD. For segment CD we have

$$\xrightarrow{+}\Sigma F_x = 0; \qquad 500 \text{ N} - A_C = 0 \qquad A_C = 500 \text{ N} \qquad \text{Ans.}$$
$$+\uparrow\Sigma F_y = 0; \qquad V_C - 1000 \text{ N} = 0 \qquad V_C = 1000 \text{ N} \qquad \text{Ans.}$$
$$\zeta+\Sigma M_C = 0; \quad -M_C - 1000 \text{ N}(1 \text{ m}) = 0 \qquad M_C = -1000 \text{ N}\cdot\text{m} \qquad \text{Ans.}$$

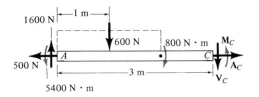
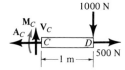

Fig. 7–3(c)

or for segment AC we have

$\xrightarrow{+} \Sigma F_x = 0;$ $A_C - 500 \text{ N} = 0$ $A_C = 500 \text{ N}$ *Ans.*

$+\uparrow \Sigma F_y = 0;$ $1600 \text{ N} - 600 \text{ N} - V_C = 0$ $V_C = 1000 \text{ N}$ *Ans.*

$\zeta + \Sigma M_C = 0;$

$M_C - 800 \text{ N} \cdot \text{m} + 600 \text{ N}(2 \text{ m}) - 1600 \text{ N}(3 \text{ m}) + 5400 \text{ N} \cdot \text{m} = 0$

$M_C = -1000 \text{ N} \cdot \text{m}$ *Ans.*

As shown, segment CD provides the *simplest* solution for the internal forces since there are *fewer loads* on the free-body diagram of this segment. Also, the results were obtained *without* having to first obtain the external reactions at A (*Step 1*). Because the magnitude of $\mathbf{M}_C$ has been computed as a *negative* quantity, its sense of direction is *opposite* to that assumed on the free-body diagrams in Fig. 7–3c. Therefore, the correct direction of $\mathbf{M}_C$ is clockwise on segment AC and counterclockwise on segment CD.

Passing a section through point B, yields the free-body diagram of the beam segment BD shown in Fig. 7–3d. *After sectioning the beam,* the portion of the distributed load acting on the segment is replaced by its resultant force. Applying the three equations of equilibrium, we have

$\xrightarrow{+} \Sigma F_x = 0;$ $-A_B + 500 \text{ N} = 0$ $A_B = 500 \text{ N}$ *Ans.*

$+\uparrow \Sigma F_y = 0;$ $V_B - 300 \text{ N} - 1000 \text{ N} = 0$ $V_B = 1300 \text{ N}$ *Ans.*

$\zeta + \Sigma M_B = 0;$ $-M_B - 300 \text{ N}(0.5 \text{ m}) - 800 \text{ N} \cdot \text{m} - 1000 \text{ N}(3 \text{ m}) = 0$

$M_B = -3950 \text{ N} \cdot \text{m}$ *Ans.*

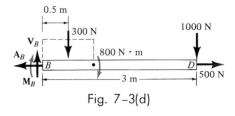

Fig. 7–3(d)

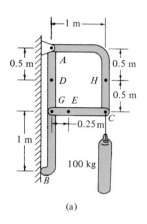

(a)

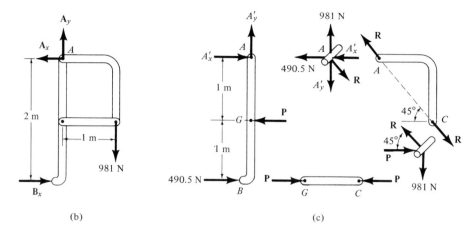

(b)   (c)

Fig. 7–4(a–c)

Example 7–2

Determine the axial force, shear force, and bending moment acting at points E, H, and D of the frame shown in Fig. 7–4a. Neglect the weights of the members. The cord on the 100-kg cylinder is attached to the pin at C.

Solution

Step 1: The free-body diagram of the entire frame is shown in Fig. 7–4b. Applying the three equations of equilibrium yields

$$+\uparrow\Sigma F_y = 0; \qquad A_y - 981 \text{ N} = 0 \qquad A_y = 981 \text{ N}$$
$$\big(+\Sigma M_A = 0; \; -981 \text{ N}(1 \text{ m}) + B_x(2 \text{ m}) = 0 \qquad B_x = 490.5 \text{ N}$$
$$\xrightarrow{+}\Sigma F_x = 0; \qquad 490.5 \text{ N} - A_x = 0 \qquad A_x = 490.5 \text{ N}$$

A free-body diagram of each member of the frame, including the pins at A and C, is shown in Fig. 7–4c. Since the weights of the members are neglected, GC and CA are both two-force members, and thus the forces acting at their respective end points must be equal, opposite, and collinear. The 981- and 490.5-N forces shown on the free-body diagram of the pin at A represent the effect of the *pin support* at A acting *on* the pin. (These are the same forces $\mathbf{A}_x$ and $\mathbf{A}_y$ shown on the free-body diagram in Fig. 7–4b. There they represent the effect of the pin support on the entire frame.) Note that a separate free-body diagram of each pin has been included here since there are *three* interactions at each pin. For example, at pin A the forces are caused by members AC, AGB, and the support.

Applying the force equations of equilibrium to the pin at C gives

$$+\uparrow\Sigma F_y = 0; \quad R \sin 45° - 981 \text{ N} = 0 \qquad R = 1387.6 \text{ N}$$
$$\xrightarrow{+}\Sigma F_x = 0; \quad P - 1387.6 \cos 45° \text{ N} = 0 \qquad P = 981 \text{ N}$$

Knowing the magnitude of **P**, we may calculate A'_x and A'_y by applying the equations of equilibrium to member AGB:

266

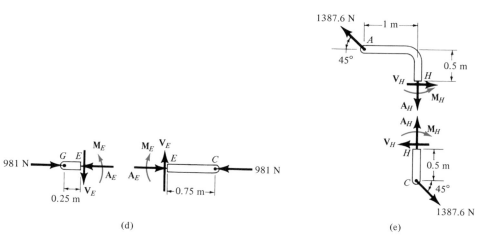

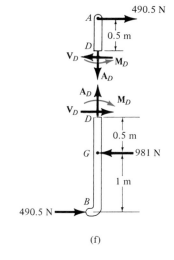

Fig. 7–4(d–f)

$+\uparrow\Sigma F_y = 0;$ $\qquad\qquad A_y' = 0$

$\rightarrow\Sigma F_x = 0;$ $\quad 490.5\ \text{N} - 981\ \text{N} + A_x' = 0$ $\qquad A_x' = 490.5\ \text{N}$

$\large\zeta +\Sigma M_G = 0;$ $\quad 490.5\ \text{N}(1\ \text{m}) - 490.5\ \text{N}(1\ \text{m}) \equiv 0$ $\quad$ (check)

The above calculations may be *checked* by applying the magnitudes of the computed forces on the pin at A and showing that these forces satisfy the force equations of equilibrium for the pin.

Steps 2, 3: Member GC is sectioned at E and the free-body diagrams of the two segments are shown in Fig. 7–4d. Applying the three equations of equilibrium to either segment reveals that

$\rightarrow\Sigma F_x = 0;$ $\qquad\qquad A_E = 981\ \text{N}$ $\qquad\qquad$ *Ans.*

$+\uparrow\Sigma F_y = 0;$ $\qquad\qquad V_E = 0$ $\qquad\qquad$ *Ans.*

$\large\zeta +\Sigma M_E = 0;$ $\qquad\qquad M_E = 0$ $\qquad\qquad$ *Ans.*

Steps 2, 3: The free-body diagrams shown in Fig. 7–4e are obtained when member AC is sectioned at H. Applying the three equations of equilibrium to section CH yields

$\rightarrow\Sigma F_x = 0;$ $\quad 1387.6 \cos 45°\ \text{N} - V_H = 0$ $\quad V_H = 981\ \text{N}$ $\qquad$ *Ans.*

$+\uparrow\Sigma F_y = 0;$ $-1387.6 \sin 45°\ \text{N} + A_H = 0$ $\qquad A_H = 981\ \text{N}$ $\qquad$ *Ans.*

$\large\zeta +\Sigma M_H = 0;$ $\ 1387.6 \cos 45°\ \text{N}(0.5\ \text{m}) - M_H$ $\qquad M_H = 490.5\ \text{N} \cdot \text{m}$ $\ $ *Ans.*

Steps 2, 3: The free-body diagrams of sections AD and DGB are shown in Fig. 7–4f. Applying the equations of equilibrium to segment AD gives

$+\uparrow\Sigma F_y = 0;$ $\qquad\qquad A_D = 0$ $\qquad\qquad$ *Ans.*

$\rightarrow\Sigma F_x = 0;$ $\qquad 490.5\ \text{N} - V_D = 0$ $\qquad V_D = 490.5\ \text{N}$ $\qquad$ *Ans.*

$\large\zeta +\Sigma M_D = 0;$ $-490.5\ \text{N}(0.5\ \text{m}) + M_D = 0$ $\qquad M_D = 245.2\ \text{N} \cdot \text{m}$ $\quad$ *Ans.*

Problems

7-1. Determine the internal axial force, shear force, and moment at points C and D.

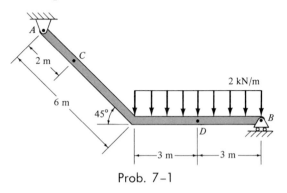

Prob. 7-1

7-2. Determine the internal axial force, shear force, and moment at points D and E of the frame.

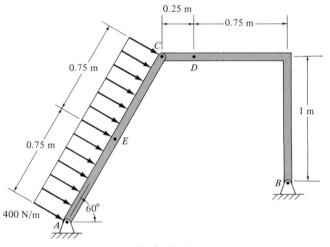

Prob. 7-2

7-3. Determine the internal axial force, shear force, and moment at point D in the frame.

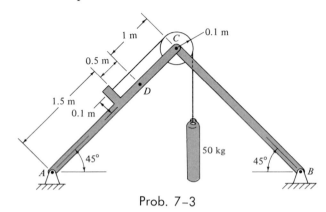

Prob. 7-3

*7-4. Determine the internal axial force, shear force, and moment at points E and F of the frame.

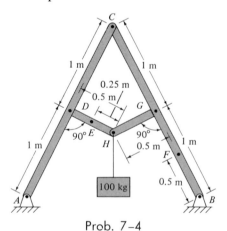

Prob. 7-4

7-5. Determine the axial force, shear force, and moment at point A if the clamp exerts a compressive force of $F = 900$ N on board B. Force **F** acts along the central axis of the screw. Set $a = 100$ mm and $b = 125$ mm.

7-7. Determine the internal axial force, shear force, and moment at points D and E of the frame.

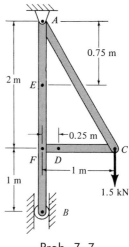

Prob. 7-7

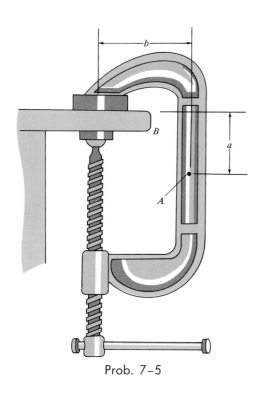

Prob. 7-5

7-5a. Solve Prob. 7-5 if $F = 230$ lb, and $a = 2.5$ in., $b = 3$ in.

7-6. Determine the internal axial force, shear force, and moment at points D and E of the frame.

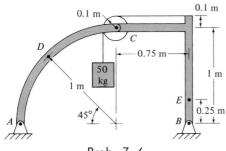

Prob. 7-6

* **7-8.** Determine the internal axial force, shear force, and moment at points F and E.

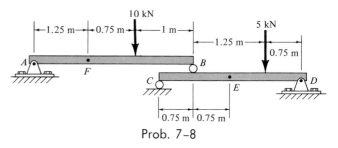

Prob. 7-8

7-9. Determine the internal axial force, shear force, and moment at points C, D, and E of the frame. Specifically, points D and E are located just to the left and just to the right of the concentrated force of 10 kN.

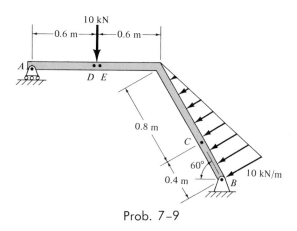

Prob. 7-9

7-10. Determine the internal axial force, shear force, and moment at point D of the beam. Set $w = 500$ N/m and $a = 0.5$ m.

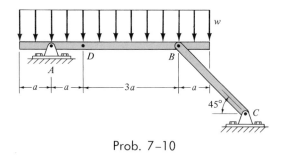

Prob. 7-10

7-10a. Solve Prob. 7-10 if $w = 50$ lb/ft and $a = 2$ ft.

7-11. The jib crane carries a load of 12 kN on the trolley T, which rolls along the bottom flange of the beam CD. Determine the position x of the trolley that will create the greatest bending moment in the column at E. As a result of end constraints, 0.3 m $\leqslant x \leqslant 3.9$ m.

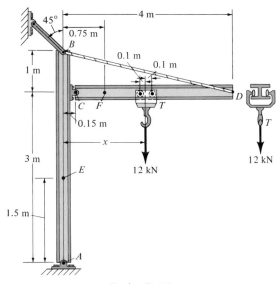

Prob. 7-11

*7-12. Determine the position x of the trolley in Prob. 7-11 that will create the greatest shear force at F.

*7–2. Shear and Bending-Moment Diagrams for a Beam

Beams are structural members which are designed to support loadings applied perpendicular to their axis. In general, beams are long, straight bars, having a constant cross-sectional area. The actual design of such members requires a detailed knowledge of the *variation* of the internal shear force V and bending moment M acting at *each point* along the axis of the beam. After the force and bending-moment analysis is complete, one can use the theory of strength of materials to determine the beam's required cross-sectional area.

The *variations* of V and M as a function of the position x of an arbitrary point along the beam's axis can be obtained by using the method of sections discussed in Sec. 7–1. Here however, it is necessary to locate the imaginary section at a distance x from the end of the beam rather than at a specified point. If the results are plotted, the graphical variations of V and M as a function of x are termed the *shear diagram* and *bending-moment diagram*, respectively. The internal axial force will not be considered in the following discussion for two reasons. In most cases, the loads applied to a beam act perpendicular to the beam's axis and hence produce only internal shear and bending moment. For design purposes, the beam's resistance to shear, and particularly to bending, is more important than its ability to resist axial force.

Sign Convention. Before presenting a method for determining the shear and bending moment as functions of x and later plotting these functions (shear and bending-moment diagrams), it is first necessary to establish a *sign convention* so as to define "positive" and "negative" shear force and bending moment acting in the beam. [This is analogous to assigning coordinate directions x positive to the right and y positive upward, for plotting a function $y = f(x)$.] The sign convention to be adopted here is illustrated in Fig. 7–5a. On the *left-hand face* (L.H.F.) of a beam segment, the internal shear force $\mathbf{V}$ acts downward and the internal moment $\mathbf{M}$ acts counterclockwise. In accordance with Newton's third law, an equal and opposite shear force and bending moment must act on the *right-hand face* (R.H.F.) of a segment. Perhaps an easy way to remember this sign convention is to isolate a small beam segment and note that *positive shear tends to rotate the segment clockwise* (Fig. 7–5b) and a *positive bending moment tends to bend the segment, if it were elastic, concave upward* (Fig. 7–5c).

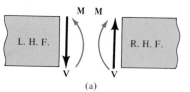

(a)

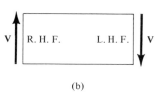

(b)

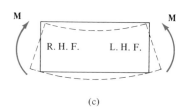

(c)

Fig. 7–5

The following three-step procedure should be applied when constructing the shear and bending-moment diagrams for a beam.

Step 1:   Determine all the external forces acting on the beam and resolve these forces into components acting perpendicular and parallel to the beam's axis.

Step 2:   Section the beam perpendicular to its axis at an arbitrary distance x measured from the beam's left end; and from a free-body diagram of the *left segment,* determine the unknowns V and M at the cut section as functions of x. When drawing the free-body diagram, **V** and **M** should be shown acting in their *positive direction,* in accordance with the sign convention given in Fig. 7–5. V is obtained from $\Sigma F_y = 0$ and M is obtained by summing moments about point S located at the cut section, $\Sigma M_S = 0$.

In general, the internal shear and bending-moment functions will either be discontinuous, or their slope will be discontinuous at points where a distributed load changes or where concentrated forces or couples are applied. Because of this, shear and bending-moment functions must be determined for *each segment* of the beam located between any two discontinuities of loading. For example, sections located at x_1, x_2, and x_3 will have to be used to fully describe V and M throughout the entire length of the beam in Fig. 7–6. These functions are valid *only* within regions $0 \leqslant x_1 < a$, $a < x_2 < b$, and $b < x_3 \leqslant l$.

Step 3:   Plot the shear diagram (V versus x) and the moment diagram (M versus x). If computed values of the functions describing V and M are *positive,* the magnitudes are plotted above the x axis, whereas negative values are plotted below the axis. Generally, it is convenient to plot the shear and bending-moment diagrams directly below the free-body diagram of the beam.

The following examples numerically illustrate this procedure.

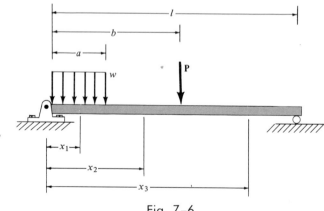

Fig. 7–6

Example 7–3

Draw the shear and bending-moment diagrams for the beam shown in Fig. 7–7a.

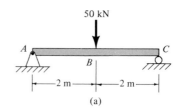

(a)

Solution

Step 1: Using the equations of equilibrium, one can show that the support reactions are $A_y = C_y = 25$ kN, acting upward, and $A_x = 0$, Fig. 7–7d.

Step 2: If the beam is sectioned an arbitrary distance x from point A, extending within the region AB, the free-body diagram of the left segment is shown in Fig. 7–7b. The unknowns **V** and **M** are indicated acting in the *positive direction* on the right-hand face of the element according to the established sign convention. Why? Summing forces in the vertical direction, we have

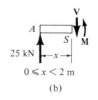

(b)

$$+\uparrow \Sigma F_y = 0; \qquad 25 - V = 0$$
$$V = 25 \text{ kN} \qquad (1)$$

Summing moments about point S yields

$$\zeta + \Sigma M_S = 0; \qquad M - 25(x) = 0$$
$$M = 25x \text{ kN} \cdot \text{m} \qquad (2)$$

A free-body diagram for a left segment of the beam extending a distance x within the region BC is shown in Fig. 7–7c. As always, **V** and **M** are shown acting in the positive sense. Summing forces, we have

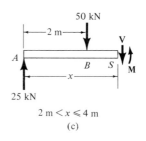

(c)

$$+\uparrow \Sigma F_y = 0; \qquad 25 - 50 - V = 0$$
$$V = -25 \text{ kN} \qquad (3)$$

Summing moments about point S, we obtain

$$\zeta + \Sigma M_S = 0; \qquad M + 50(x - 2) - 25(x) = 0$$
$$M = (100 - 25x) \text{ kN} \cdot \text{m} \qquad (4)$$

Comparing the results, do you see why it was necessary to section the beam within each of the two regions AB and BC?

Step 3: When the above shear-force and bending-moment Eqs. (1) through (4) are plotted within the regions in which they are valid, the shear and bending-moment diagrams shown in Fig. 7–7d are obtained. The shear diagram indicates that the internal shear force is always 25 kN (positive) along beam segment AB. Just to the right of point B, the shear force changes sign and remains at a constant value of -25 kN for segment BC. The moment diagram starts at zero, increases linearly to point C at $x = 2$ m, where $M_{max} = 25$ kN(2 m) $= 50$ kN $\cdot$ m, and thereafter decreases back to zero.

It is seen in Fig. 7–7d that the graph of the shear and moment diagrams is discontinuous at points of concentrated force, i.e., points A, B, and C. For this reason, as stated earlier, it is necessary to express both the shear and bending-moment functions separately for regions between concen-

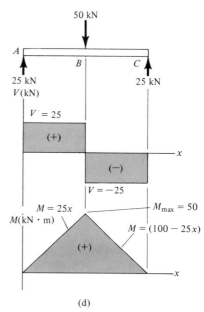

(d)

Fig. 7–7

273

trated loads. It should be realized, however, that all loading discontinuities are mathematical, arising from the *idealization of a concentrated force and couple*. Physically, all loads are applied over a finite area, and if this load variation could be accounted for, all shear and bending-moment diagrams would actually be continuous over the beam's entire length.

Example 7-4

Draw the shear and bending-moment diagrams for the beam shown in Fig. 7–8*a*.

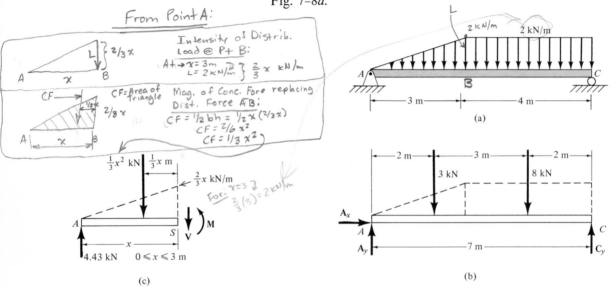

From Point A:
Intensity of Distrib.
Load @ Pt B:
$A + \rightarrow x = 3m$
$L = 2 kN/m$ } $\frac{2}{3} x$ kN/m

CF = Area of Triangle
Mag. of Conc. Fore replacing
Dist. Force AB:
$CF = \frac{1}{2} bh = \frac{1}{2} x (\frac{2}{3}x)$
$CF = \frac{2}{6} x^2$
$CF = \frac{1}{3} x^2$

(a)

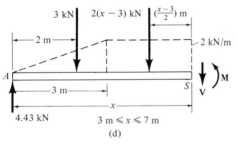

$\frac{1}{3}x^2$ kN $\frac{1}{3}x$ m

$\frac{2}{3}x$ kN/m

For: $x = 3$
$\frac{2}{3}(3) = 2 kN/m$

4.43 kN $0 \leqslant x \leqslant 3$ m

(c)

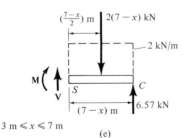

3 kN $2(x-3)$ kN $\left(\frac{x-3}{2}\right)$ m

2 m

2 kN/m

A

3 m

x

4.43 kN 3 m $\leqslant x \leqslant 7$ m

(d)

$\left(\frac{7-x}{2}\right)$ m $2(7-x)$ kN

2 kN/m

M

S   C

$(7-x)$ m   6.57 kN

3 m $\leqslant x \leqslant 7$ m

(e)

Fig. 7–8(a–e)

(b)

Solution

Step 1: A free-body diagram of the entire beam is shown in Fig. 7–8*b*. To compute the external reactions, the distributed load has been temporarily reduced to two concentrated forces. Applying the three equations of equilibrium, we have

$$\xrightarrow{+} \Sigma F_x = 0; \qquad\qquad A_x = 0$$

$$\curvearrowright + \Sigma M_A = 0; \; -3 \text{ kN}(2 \text{ m}) - 8 \text{ kN}(5 \text{ m}) + C_y(7 \text{ m}) = 0$$

$$C_y = 6.57 \text{ kN}$$

$$+\uparrow \Sigma F_y = 0; \quad A_y - 3 \text{ kN} - 8 \text{ kN} + 6.57 \text{ kN} = 0 \qquad A_y = 4.43 \text{ kN}$$

Step 2: A free-body diagram for the left segment of the beam extending within region *AB* is shown in Fig. 7–8*c*. The distributed loading acting on this segment, which has an intensity of $\frac{2}{3}x$ at its end, has been replaced by a concentrated force *after* the segment is isolated as a free-body diagram. Since the distributed load now has a length x, the *magnitude* of the concentrated force is equal to $\frac{1}{2}(x)(\frac{2}{3}x) = \frac{1}{3}x^2$. This force *acts through the centroid* of the distributed loading area, a distance $x/3$ from the right end. Applying the two equations of equilibrium yields

$$+\uparrow\Sigma F_y = 0; \qquad 4.43 - \frac{x^2}{3} - V = 0$$

$$V = \left(4.43 - \frac{x^2}{3}\right) \text{kN} \qquad (1)$$

$$\zeta +\Sigma M_S = 0; \qquad M + \frac{x^2}{3}\left(\frac{x}{3}\right) - 4.43x = 0$$

$$M = \left(4.43x - \frac{x^3}{9}\right)\text{kN}\cdot\text{m} \qquad (2)$$

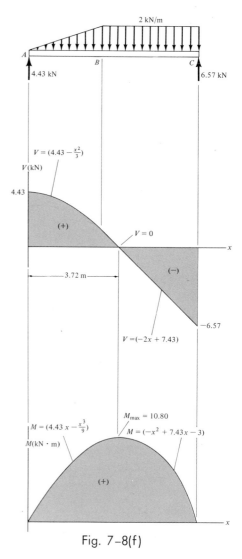

Fig. 7–8(f)

Since the distributed loading *changes slope* at point B, Fig. 7–8f, another section of the beam must be chosen within region BC. The free-body diagram for the *left segment* that extends within this region is shown in Fig. 7–8d. Notice how the distributed loading has been simplified on this diagram. Applying the equilibrium equations gives

$$+\uparrow\Sigma F_y = 0; \qquad 4.43 - 3 - 2(x - 3) - V = 0$$

$$V = (-2x + 7.43) \text{kN} \qquad (3)$$

$$\zeta +\Sigma M_S = 0; \quad M + 2(x - 3)\left(\frac{x-3}{2}\right) + 3(x - 2) - 4.43x = 0$$

$$M = (-x^2 + 7.43x - 3)\text{kN}\cdot\text{m} \qquad (4)$$

A second method for obtaining these *same* shear and bending-moment equations for segment BC consists of isolating the *right segment* of the beam. Since the origin of the x axis has been established at point A, the length of this segment is $(7 - x)$ m. The free-body diagram is shown in Fig. 7–8e. According to the established sign convention, note that the internal shear and bending moment acting at the cut section are indicated in the positive direction. For equilibrium, we again obtain

$$+\uparrow\Sigma F_y = 0; \qquad V - 2(7 - x) + 6.57 = 0$$

$$V = (-2x + 7.43) \text{kN} \qquad (3)$$

$$\zeta +\Sigma M_S = 0; \quad -M - 2(7 - x)\left(\frac{7-x}{2}\right) + 6.57(7 - x) = 0$$

$$M = (-x^2 + 7.43x - 3)\text{kN}\cdot\text{m} \qquad (4)$$

Step 3: The shear and bending-moment diagrams shown in Fig. 7–8f are obtained by plotting Eqs. (1) through (4) within the regions in which they are valid.

The point of *zero shear* can be found using Eq. (3):

$$V = -2x + 7.43 = 0$$

$$x = 3.72 \text{ m}$$

The value of x happens to represent the point on the beam where the *maximum moment* occurs (see Sec. 7–3). Using Eq. (4), we have

$$M_{\text{max}} = [-(3.72)^2 + 7.43(3.72) - 3] \text{kN}\cdot\text{m}$$

$$= 10.80 \text{ kN}\cdot\text{m}$$

Problems

7-13. Draw the shear and bending-moment diagrams for the beam.

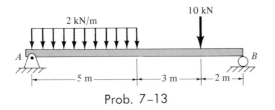

Prob. 7-13

7-14. Draw the shear and bending-moment diagrams for the cantilever beam.

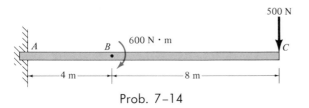

Prob. 7-14

7-15. Draw the shear and bending-moment diagrams for the beam. Set $F = 4\,\text{kN}$, and $a = 2\,\text{m}$, $b = 2.5\,\text{m}$.

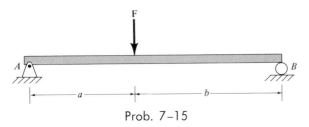

Prob. 7-15

7-15a. Solve Prob. 7-15 if $F = 2000\,\text{lb}$, and $a = 5\,\text{ft}$, $b = 7\,\text{ft}$.

***7-16.** Draw the shear and bending-moment diagrams for the cantilever beam.

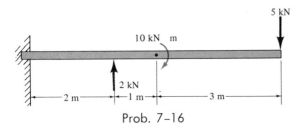

Prob. 7-16

7-17. Draw the shear and bending-moment diagrams for the beam.

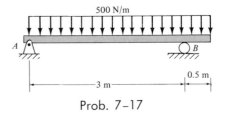

Prob. 7-17

7-18. The work platform supports an 80-kg man having a mass center at G. Draw the shear and bending moment diagrams of the telescopic column AB ($0 \leqslant x \leqslant 4\,\text{m}$) due to this loading.

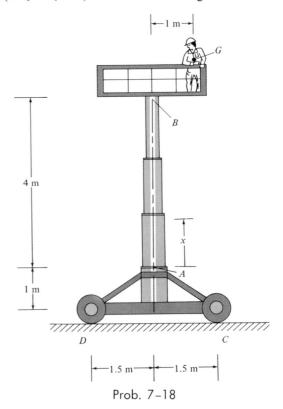

Prob. 7-18

7-19. Draw the shear and bending-moment diagrams for the beam.

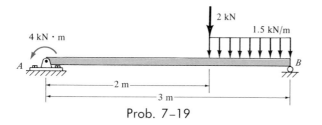

Prob. 7-19

*7-20. Draw the shear and bending-moment diagrams for the beam. Set $w = 400$ N/m and $l = 3$ m.

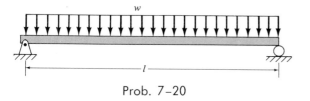

Prob. 7-20

*7-20a. Solve Prob. 7-20 if $w = 300$ lb/ft and $l = 10$ ft.

7-21. Draw the shear and bending-moment diagrams for the beam.

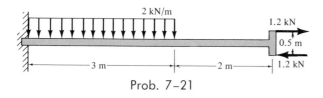

Prob. 7-21

7-22. Draw the shear and bending-moment diagrams for the beam.

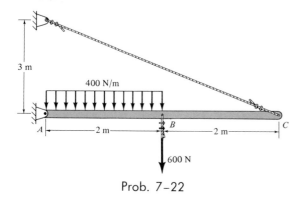

Prob. 7-22

7-23. Draw the shear and bending-moment diagrams for the beam. Set $w = 500$ N/m, $F = 750$ N, and $a = 0.5$ m.

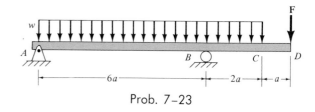

Prob. 7-23

*7-24. Draw the shear and bending-moment diagrams for the beam.

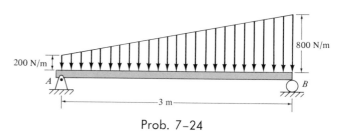

Prob. 7-24

7-25. Draw the shear and bending-moment diagrams for the beam. Set $w = 500$ N/m and $a = 3$ m.

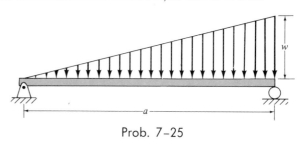

Prob. 7-25

7-25a. Solve Prob. 7-25 if $w = 120$ lb/ft and $a = 18$ ft.

7-26. Draw the shear and bending-moment diagrams for the beam.

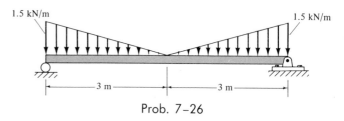

Prob. 7-26

7-27. Draw the shear and bending-moment diagrams for the beam.

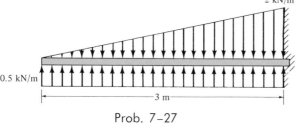

Prob. 7-27

7-29. Determine the internal axial force, shear force, and the bending moment as a function of θ $(0° \leqslant \theta \leqslant 180°)$ for the semicircular cantilever arch. The arch has a total mass of 80 kg.

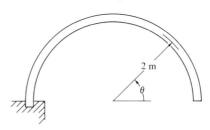

Prob. 7-29

***7-28.** Determine the internal axial force, shear force, and the bending moment as a function of $0° \leqslant \theta \leqslant 180°$ and $0 \leqslant x \leqslant 1$ m for the beam loaded as shown.

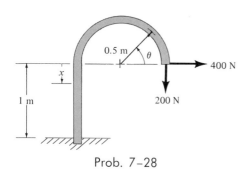

Prob. 7-28

*7-3. Relations Among Distributed Load, Shear, and Bending Moment

In cases where a beam is subjected to several concentrated forces, couples, and distributed loads, the method of constructing the shear and bending-moment diagrams discussed in Sec. 7-2 may become quite tedious. In this section, a simpler method for constructing these diagrams will be discussed—a method based upon differential relations that exist between the load, shear, and bending moment.

Consider the beam AD shown in Fig. 7-9a, which is subjected to an arbitrary distributed loading $w = w(x)$ and a series of concentrated forces and couples. In the following discussion, *the distributed load will be considered positive when the loading acts upward*, as shown. The free-body

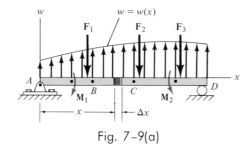

Fig. 7–9(a)

diagram for a small segment of the beam having a length Δx is shown in Fig. 7–9b. Since this segment has been chosen at a point x along the beam, which is *not* subjected to a concentrated force or couple, any results obtained will not apply at points of concentrated loading. The internal shear force and bending moment shown on the free-body diagram are assumed to act in the *positive direction* according to the established sign convention, Fig. 7–5. Note that both the shear force and moment acting on the right-hand face must be increased by a small, finite amount in order to keep the segment in equilibrium. The distributed loading has been replaced by a concentrated force $w(x)\,\Delta x$ that acts at a fractional distance $\epsilon(\Delta x)$ from the right end, where $0 < \epsilon < 1$ (for example, if $w(x)$ is *uniform*, $\epsilon = \frac{1}{2}$). Applying the equations of equilibrium,

$$+\uparrow \Sigma F_y = 0; \qquad V + w(x)\,\Delta x - (V + \Delta V) = 0$$
$$\Delta V = w(x)\,\Delta x$$

and

$$\zeta + \Sigma M_O = 0; \quad -V\,\Delta x - M - w(x)\,\Delta x\,\epsilon(\Delta x) + (M + \Delta M) = 0$$
$$\Delta M = V\,\Delta x + w(x)\epsilon(\Delta x)^2$$

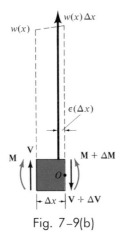

Fig. 7–9(b)

Dividing by Δx and taking the limit as $\Delta x \to 0$ the above equations become

$$\frac{dV}{dx} = w(x) \qquad (7\text{--}1)$$

and

$$\frac{dM}{dx} = V \qquad (7\text{--}2)$$

These two equations provide a convenient means for plotting the shear-force and bending-moment diagrams. For a specific point in the beam, Eq. 7–1 states that the *slope* of the *shear diagram* is equal to the intensity of the *distributed loading* at the point, while Eq. 7–2 states that the *slope* of the *moment diagram* is equal to the intensity of *shear force* at the point. In particular, if the shear is equal to zero, $dM/dx = 0$, and, therefore, points of zero shear correspond to points of maximum (or possibly minimum) moment.

Equations 7–1 and 7–2 may be integrated between points B and C of the beam, Fig. 7–9*a*, in which case

$$\Delta V_{BC} = \int_{x_B}^{x_C} w(x)\, dx \qquad (7\text{--}3)$$

and

$$\Delta M_{BC} = \int_{x_B}^{x_C} V\, dx \qquad (7\text{--}4)$$

Equation 7–3 states that the *change in shear* between B and C is equal to the *area* under the *distributed loading curve* between these points. Similarly, from Eq. 7–4, the *change in moment* between B and C is equal to the *area* under the *shear diagram* from B to C.

From the above derivation and the discussion in Sec. 7–2, it should be noted that Eqs. 7–1 and 7–3 cannot be used at points where a concentrated force acts, since these equations do not account for the sudden change in shear at these points. Similarly, because of a discontinuity of moment, Eqs. 7–2 and 7–4 cannot be used at points where a couple is applied.

The following two examples illustrate the method used to apply these equations for the construction of the shear and bending-moment diagrams. After working through these examples, it is recommended that Examples 7–3 and 7–4 be solved using this method.

Example 7-5

Sketch the shear and bending-moment diagrams for the cantilever beam shown in Fig. 7–10a.

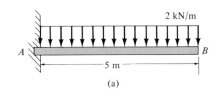

(a)

Solution

The reactions at A are first calculated by using the equations of equilibrium and are indicated on the free-body diagram shown in Fig. 7–10b. It is convenient to plot both the shear and bending-moment diagrams directly below the free-body diagram of the beam, as shown in the figure.

Shear Diagram. When constructing the shear diagram, the shear forces at points A and B are plotted first, since their values can be taken directly from the free-body diagram of the beam. In accordance with the positive sign convention, Fig. 7–5, the shear force at A is $V_A = +10$ kN and at B, $V_B = 0$. At an intermediate point C, the intensity of load acting on the beam is -2 kN/m (a negative value since the load acts downward). In accordance with Eq. 7–1, the slope of the shear diagram at this point is $dV/dx|_C = -2$ kN/m. In a similar manner, the slope at point D is $dV/dx|_D = -2$ kN/m. Since the distributed load is uniform, for other points along the beam this same slope is obtained. Thus, the shear diagram has a *constant* negative slope of -2 kN/m, which extends from end A to end B.

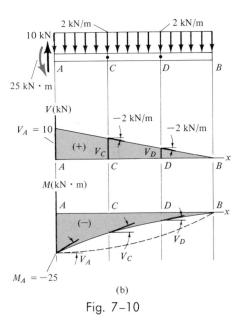

(b)

Fig. 7–10

Moment Diagram. From the free-body diagram, the moment diagram must start at a value of $M_A = -25$ kN · m and terminate at a value of $M_B = 0$. Do you see why M_A is negative? (Refer to Fig. 7–5.) Using Eq. 7–2, the slope of the moment diagram at point A, as shown in Fig. 7–10b, is $dM/dx|_A = V_A = +10$ kN · m/m. The *slope* of the moment diagram will always be positive along the entire axis of the beam since the *shear* is always positive; however, the *slope* will *decrease* as the distance x increases because the *shear decreases*. The resulting moment function is thus a parabola, as shown in Fig. 7–10b. One may ask why this curve is not concave upward as shown by the dashed line. This cannot be the case since the slope of the moment diagram at A is required to be greater than the slope at C; i.e., $V_A > V_C$, as shown in the shear diagram. Similarly, $V_C > V_D$ and, furthermore, the slope at B must be zero.

From these diagrams it is seen that when the distributed loading is constant, the shear diagram will vary linearly with x and the moment diagram will be proportional to x^2. The actual equations of these curves can be obtained using the methods discussed in Sec. 7–2. In a similar manner, if the distributed loading is triangular, that is, varies linearly with x, the shear diagram will be proportional to x^2 and the bending-moment diagram will be a function of x^3. This increase in the exponential power of x in going from the loading to the shear diagram to the bending-moment diagram is due to the integration expressed by Eqs. 7–3 and 7–4.

281

Example 7-6

Sketch the shear and bending-moment diagrams for the beam shown in Fig. 7–11a. Give values at the points where the slopes of the shear and bending-moment diagrams are discontinuous.

$$\curvearrowleft \ \Sigma M_E = 0$$

$$-3(9) + F_{B_y}(8) - 1\left(4\right) + 7kN\cdot m = 0$$

$$F_{B_y} = +3 kN$$

$$\uparrow + \Sigma F_y = 0$$

$$-3kN + 3kN - 1kN + F_{E_y} = 0$$

$$F_{E_y} = 1 kN$$

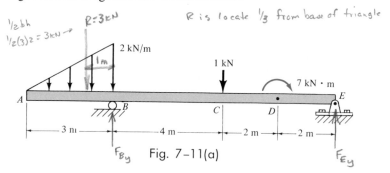

Fig. 7–11(a)

Solution

The free-body diagram of the beam is shown in Fig. 7–11b. For equilibrium, the vertical reactions at B and E are 3 kN and 1 kN, respectively. Why is it necessary to compute these external reactions?

Shear Diagram. Since the shear force at point A is zero, the shear diagram starts at zero. The triangular *load* acts *downward* and *increases* in magnitude from points A to B; hence, from Eq. 7–1, the *slope* of the *shear diagram is negatively increasing* from points A to B. For example, the slopes at points F and H are $dV/dx|_F = -w_F$ and $dV/dx|_H = -w_H$, respectively, where $|w_H| > |w_F|$. Since the distributed loading within region AB is triangular (varies with x), the shear diagram within this region is parabolic (varies with x^2). The total resultant force of the distributed loading is $\frac{1}{2}(3$ m$)(2$ kN/m$) = 3$ kN; hence, the internal shear force just to the *left* of point B is -3 kN. This follows from Eq. 7–3; that is, the change in shear $(V_B - 0)$ is equal to the area under the loading diagram between points A and B. A free-body diagram of segment AB of the beam extending just to the *right* of the concentrated force at point B, Fig. 7–11c, reveals that the shear force just to the *right* of support B must be zero for equilibrium. Since no load acts between points B and C, $dV/dx|_{BC} = 0$, and therefore the shear force remains constant (or zero) within this region, Fig. 7–11b. A free-body diagram for a segment of the beam extending just to the right of the concentrated force at point C, Fig. 7–11d, indicates a jump in the shear force to -1 kN. (Why is it negative?) Within region CDE of the beam the shear force again remains constant, since no concentrated force or distributed load acts within this region, Fig. 7–11b.

Fig. 7–11(b)

Moment Diagram. The moment acting at point A on the free-body diagram of the beam is zero, and therefore the moment diagram begins at zero. From Eq. 7–2, the *slope* of the moment diagram between points A and B becomes increasingly *negative,* since the *shear* within region AB is increasingly *negative.* The curve within this region varies with x^3. As shown on the free-body diagram in Fig. 7–11c, the required internal moment for equilibrium acting at point B is $M_B = -3$ kN $\cdot$ m. Since the shear jumps from a value of -3 kN to zero at point B, the slope of the moment diagram at B makes a corresponding change. The slope within region BC remains at a constant value of $dM/dx |_{BC} = 0$. Why? The moment at point C can be verified by isolating segment ABC of the beam and determining the internal moment at C, using the equations of equilibrium; see Fig. 7–11d. However, another method for determining this internal moment consists of using Eq. 7–4, in which case the change in moment between B and C is $\Delta M_{BC} = (4$ m$)(0$ kN$) = 0$ (the area under the shear diagram within region BC). The moment acting at C is therefore $M_C = M_B + \Delta M_{BC} = -3$ kN $\cdot$ m $+ 0 = -3$ kN $\cdot$ m. The jump made in the shear diagram at C causes the moment diagram to change slope from 0 to -1 kN $\cdot$ m/m. This slope, like the shear, remains constant to D, as shown on the moment diagram. The internal moment acting at the left of D is $M_D = M_C + \Delta M_{CD} = -3$ kN $\cdot$ m $+ (-1$ kN$)(2$ m$) = -5$ kN $\cdot$ m. Using this value, a free-body diagram for a segment of the beam about point D, Fig. 7–11e, indicates that the internal moment acting to the *right* of the concentrated 7-kN $\cdot$ m couple must be $+2$ kN $\cdot$ m for equilibrium. Therefore, there is a jump in the moment diagram at D, Fig. 7–11b. The moment diagram has a constant slope of -1 kN $\cdot$ m/m within region DE and closes to zero at E. The moment must be zero at E, since there is no moment on the free-body diagram at this point.

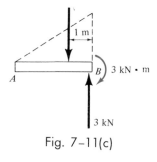

Fig. 7–11(c)

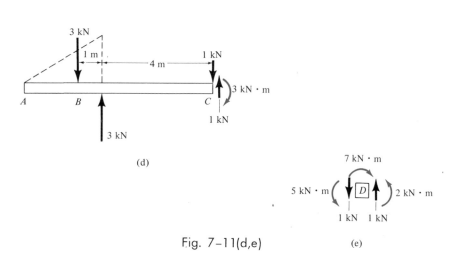

(d)

7 kN $\cdot$ m

5 kN $\cdot$ m D 2 kN $\cdot$ m

1 kN   1 kN

(e)

Fig. 7–11(d,e)

7–30. Sketch the shear and bending-moment diagrams for the beam in Prob. 7–20.

7–30a. Sketch the shear and bending-moment diagrams for the beam in Prob. 7–20a.

7–31. Sketch the shear and bending-moment diagrams for the beam in Prob. 7–21.

*7–32. Sketch the shear and bending-moment diagrams for the beam in Prob. 7–22.

7–33. Sketch the shear and bending-moment diagrams for the beam in Prob. 7–23.

7–34. Sketch the shear and bending-moment diagrams for the beam in Prob. 7–24.

7–35. Sketch the shear and bending-moment diagrams for the beam in Prob. 7–25.

7–35a. Sketch the shear and bending-moment diagrams for the beam in Prob. 7–25a.

*7–36. Sketch the shear and bending-moment diagrams for the beam in Prob. 7–26.

7–37. Sketch the shear and bending-moment diagrams for the beam in Prob. 7–27.

7–38. Sketch the shear and bending-moment diagrams for the axle AB of the train wheels. The load transferred to the axle is 40 kN, distributed equally as shown. Assume that vertical force reactions of the rails occur at C and D.

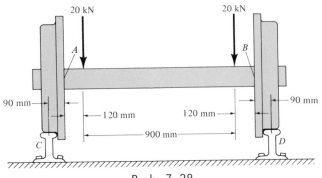

Prob. 7–38

7–39. Sketch the shear and bending-moment diagrams for the beam.

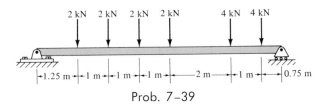

Prob. 7–39

*7–40. Sketch the shear and bending-moment diagrams for the beam. Set $F_1 = 1.5$ kN, $F_2 = 2$ kN, $F_3 = 1.25$ kN, and $a = 2$ m.

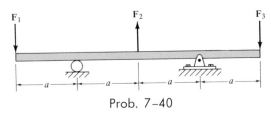

Prob. 7–40

*7–40a. Solve Prob. 7–40 if $F_1 = 100$ lb, $F_2 = 200$ lb, $F_3 = 100$ lb, and $a = 5$ ft.

7–41. Sketch the shear and bending-moment diagrams for the lathe spindle if it is subjected to the concentrated loads shown. The vertical equilibrium forces at the bearings A and B are to be calculated.

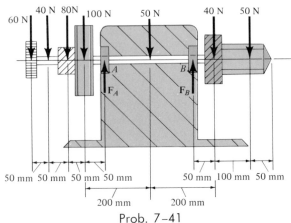

Prob. 7–41

7-42. Sketch the shear and bending-moment diagrams for the beam.

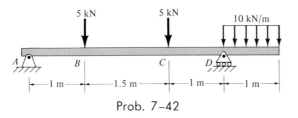

Prob. 7–42

7-43. Sketch the shear and bending-moment diagrams for the beam.

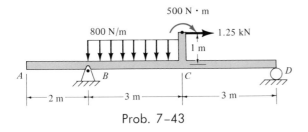

Prob. 7–43

*** 7-44.** Sketch the shear and bending-moment diagrams for the beam.

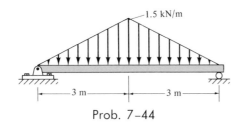

Prob. 7–44

7-45. Sketch the shear and bending-moment diagrams for the beam. Set $w_1 = 3 \text{ kN/m}$, $w_2 = 1.25 \text{ kN/m}$, $F = 2 \text{ kN}$, and $a = 3 \text{ m}$.

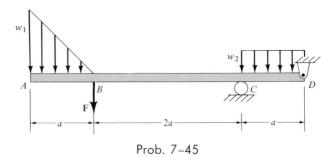

Prob. 7–45

7-45a. Solve Prob. 7–45 if $w_1 = 3000 \text{ lb/ft}$, $w_2 = 1000 \text{ lb/ft}$, $F = 2000 \text{ lb}$, and $a = 3 \text{ ft}$.

*7–4. Cables

Flexible cables and chains are often used in engineering structures for support and to transport loads from one member to another. When used for supporting suspension bridges and trolley wheels, cables form the main load-carrying element of the structure. In the force analysis of such systems, the weight of the cable itself may frequently be neglected. However, when cables are used for transmission lines and guys for radio antennas and derricks, the cable weight may become important and must be included in the structural analysis. Regardless of which loading conditions are present, provided this loading is coplanar with the cable, the

285

requirements for equilibrium are formulated in an identical manner. Two cases will be considered in the analysis that follows: (1) a cable subjected to an external distributed load, and (2) a cable subjected to its own weight.

In deriving the necessary relations between the tensile force, sag, length, and span of a cable, we will make the assumption that the cable is *perfectly flexible* and *inextensible*. Due to flexibility, the cable offers no resistance to bending, and therefore, the tensile force acting in the cable is always tangent to the cable at points along its length. Being inextensible, the cable has a constant length both before and after the load is applied. The derivations that follow are based on knowing the cable's *final configuration*, i.e., the geometry of the cable *after* the loading is applied. With the geometry known, a cable segment is then treated the same as a rigid body.

Cable Subjected to an External Distributed Load.   Consider the weightless cable shown in Fig. 7–12a, which is subjected to a loading function $w = w(x)$, *as measured in the x direction*. The free-body diagram of a small segment of the cable having a length Δs is shown in Fig. 7–12b. Since the tensile force in the cable changes continuously in both magnitude and direction along the length of the cable, then on the free-body diagram this change in cable tension is denoted by $\Delta \mathbf{T}$. The distributed load is represented by its resultant force $w(x)(\Delta x)$, which acts at a fractional distance $\epsilon(\Delta x)$ from point O, where $0 < \epsilon < 1$. Applying the equations of equilibrium yields

$$\xrightarrow{+} \Sigma F_x = 0; \qquad -T \cos \theta + (T + \Delta T) \cos (\theta + \Delta \theta) = 0$$
$$+\uparrow \Sigma F_y = 0; \qquad -T \sin \theta - w(x)(\Delta x) + (T + \Delta T) \sin (\theta + \Delta \theta) = 0$$
$$\zeta + \Sigma M_O = 0; \qquad w(x)(\Delta x)\epsilon(\Delta x) - T \cos \theta \, \Delta y + T \sin \theta \, \Delta x = 0$$

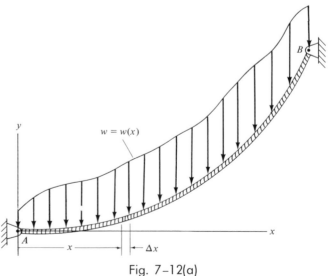

Fig. 7–12(a)

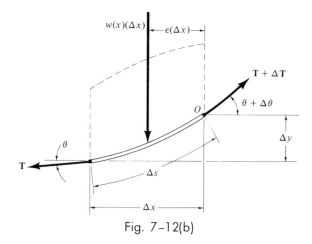

Fig. 7–12(b)

Dividing each of the preceding equations by Δx and taking the limit as $\Delta x \to 0$, and hence $\Delta y \to 0$, $\Delta \theta \to 0$, and $\Delta T \to 0$, we obtain

$$\frac{d(T \cos \theta)}{dx} = 0 \qquad (7\text{–}5)$$

$$\frac{d(T \sin \theta)}{dx} - w(x) = 0 \qquad (7\text{–}6)$$

$$\frac{dy}{dx} = \tan \theta \qquad (7\text{–}7)$$

Integrating Eq. 7–5, we have

$$T \cos \theta = \text{constant} = F_H \qquad (7\text{–}8)$$

where F_H represents the horizontal component of tensile force at any point along the cable.

Integrating Eq. 7–6 gives

$$T \sin \theta = \int w(x)\, dx \qquad (7\text{–}9)$$

Dividing Eq. 7–9 by Eq. 7–8 eliminates T. Then, using Eq. 7–7, we obtain

$$\tan \theta = \frac{dy}{dx} = \frac{1}{F_H} \int w(x)\, dx$$

Performing a second integration yields

$$y = \frac{1}{F_H} \int \left(\int w(x)\, dx \right) dx \qquad (7\text{–}10)$$

This equation is used to determine the deflection curve for the cable, $y = f(x)$. The force F_H and the two constants, say C_1 and C_2, resulting from the integration of Eq. 7–10 are determined by applying the boundary conditions for the cable.

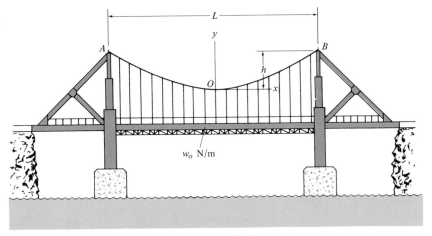

Fig. 7-13(a)

Example 7-7

The cable of a suspension bridge supports half of the uniform road surface between the two columns at A and B, as shown in Fig. 7-13a. If this loading is w_o N/m, determine the maximum force developed in the cable and the required length of the complete cable. The span length L and sag h are known.

Solution

For reasons of symmetry, the origin of coordinates has been placed at the cable's center. The deflection curve of the cable can be found by using Eq. 7-10. Noting that $w(x) = w_o$,

$$y = \frac{1}{F_H} \int \left(\int w_o \, dx \right) dx$$

Integrating this equation twice gives

$$y = \frac{1}{F_H} \left(\frac{w_o x^2}{2} + C_1 x + C_2 \right) \tag{1}$$

The constants of integration may be determined by using the boundary conditions $y = 0$ at $x = 0$, and $dy/dx = 0$ at $x = 0$. Substituting into Eq. (1) yields $C_1 = C_2 = 0$. The deflection curve then becomes

$$y = \frac{w_o}{2F_H} x^2 \tag{2}$$

This is the equation of a *parabola*. The constant F_H may be obtained by using the boundary condition $y = h$ at $x = L/2$. Thus,

$$F_H = \frac{w_o L^2}{8h} \tag{3}$$

Therefore, Eq. (2) becomes

$$y = \frac{4h}{L^2}x^2 \qquad (4)$$

The maximum tension in the cable may be determined using Eq. 7–8. Since

$$T = \frac{F_H}{\cos\theta}, \qquad \text{for} \qquad 0 \leqslant \theta < \frac{\pi}{2}$$

T_{max} will occur when θ is *maximum*, i.e., at point B, Fig. 7–13a. From Eq. (2) this occurs when

$$\left.\frac{dy}{dx}\right|_{max} = \tan\theta_{max} = \left.\frac{w_o}{F_H}x\right|_{x=L/2}$$

or

$$\theta_{max} = \tan^{-1}\left(\frac{w_o L}{2F_H}\right) \qquad (5)$$

Therefore,

$$T_{max} = \frac{F_H}{\cos(\theta_{max})} \qquad (6)$$

Using the triangular relationship shown in Fig. 7–13b, which is based on Eq. (5), Eq. (6) may be written as

$$T_{max} = \frac{\sqrt{4F_H^2 + w_o^2 L^2}}{2}$$

Substituting Eq. (3) into the above equation yields

$$T_{max} = \frac{w_o L}{2}\sqrt{1 + \left(\frac{L}{4h}\right)^2} \qquad \textit{Ans.}$$

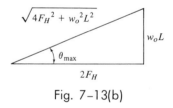

Fig. 7–13(b)

For a differential segment of cable length,

$$ds = \sqrt{(dx)^2 + (dy)^2} = \sqrt{1 + \left(\frac{dy}{dx}\right)^2}\,dx$$

Hence, the total length of the cable, $\mathcal{L}$, can be determined by integration, i.e.,

$$\mathcal{L} = \int ds = 2\int_0^{L/2}\sqrt{1 + \left(\frac{w_o x}{F_H}\right)^2}\,dx \qquad (7)$$

Integrating and substituting the limits yields

$$\mathcal{L} = \frac{L}{2}\sqrt{1 + \left(\frac{w_o L}{2F_H}\right)^2} + \frac{F_H}{w_o}\sinh^{-1}\left(\frac{w_o L}{2F_H}\right)$$

Using Eq. (3) and rearranging terms, the final result can be written as

$$\mathcal{L} = \frac{L}{2}\left[\sqrt{1 + \left(\frac{4h}{L}\right)^2} + \frac{L}{4h}\sinh^{-1}\left(\frac{4h}{L}\right)\right] \qquad \textit{Ans.}$$

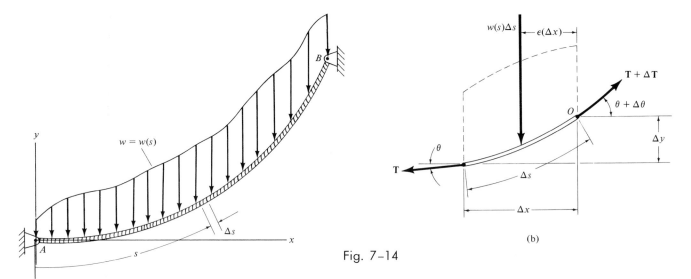

Fig. 7-14

Cable Subjected to Its Own Weight. When the weight of the cable becomes important in the force analysis, the loading function along the cable becomes a function of the arc length s, rather than the projected length x. A generalized loading function $w = w(s)$ acting along the cable is shown in Fig. 7-14a. The free-body diagram for a segment of the cable is shown in Fig. 7-14b. Applying the equilibrium equations to the force system on this diagram, one obtains identical relationships as those given by Eqs. 7-5, 7-6, and 7-7, but with ds replacing dx. Therefore, it may be shown that

$$T \cos \theta = F_H$$
$$T \sin \theta = \int w(s) \, ds$$
$$\frac{dy}{dx} = \frac{1}{F_H} \int w(s) \, ds \qquad (7\text{-}11)$$

To perform a direct integration of Eq. 7-11, it is necessary to replace dy/dx by ds/dx. Since

$$\frac{dy}{dx} = \sqrt{\left(\frac{ds}{dx}\right)^2 - 1}$$

we obtain

$$\frac{ds}{dx} = \left\{ 1 + \frac{1}{F_H^2} \left(\int w(s) \, ds \right)^2 \right\}^{1/2}$$

Separating the variables and integrating yields

$$x = \int \frac{ds}{\left\{ 1 + \frac{1}{F_H^2} \left(\int w(s) \, ds \right)^2 \right\}^{1/2}} \qquad (7\text{-}12)$$

The two constants of integration, say C_1 and C_2, are found using the boundary conditions for the cable.

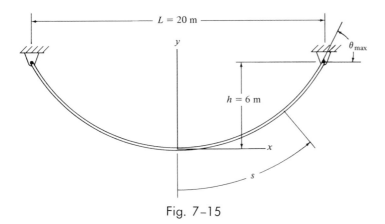

Fig. 7-15

Example 7-8

Determine the deflection curve, the length, and the maximum tension in the uniform cable shown in Fig. 7-15. The cable weighs $w_o = 5 \text{ N/m}$.

Solution

For reasons of symmetry, the origin of coordinates is located at the center of the cable. Applying Eq. 7-12 where $w(s) = w_o$ yields

$$x = \int \frac{ds}{\left\{1 + \dfrac{1}{F_H^2}\left(\int w_o \, ds\right)^2\right\}^{1/2}}$$

Integrating the term under the integral sign in the denominator, we have

$$x = \int \frac{ds}{\left\{1 + \dfrac{1}{F_H^2}\left(w_o s + C_1\right)^2\right\}^{1/2}}$$

Substituting $u = (1/F_H)(w_o s + C_1)$ so that $du = (w_o/F_H) \, ds$, a second integration yields

$$x = \frac{F_H}{w_o}(\sinh^{-1} u + C_2)$$

or

$$x = \frac{F_H}{w_o}\left\{\sinh^{-1}\left[\frac{1}{F_H}(w_o s + C_1)\right] + C_2\right\} \tag{1}$$

From Eq. 7-11,

$$\frac{dy}{dx} = \frac{1}{F_H}\int w_o \, ds$$

or

$$\frac{dy}{dx} = \frac{1}{F_H}(w_o s + C_1)$$

Since $dy/dx = 0$ at $s = 0$, $C_1 = 0$ and

$$\frac{dy}{dx} = \tan \theta = \frac{w_o s}{F_H} \tag{2}$$

where θ is the angle of the cable at any point. The constant C_2 may be evaluated by using the condition $s = 0$ at $x = 0$ in Eq. (1), in which case $C_2 = 0$. To obtain the deflection curve, $y = f(x)$, solve for s in Eq. (1), which yields

$$s = \frac{F_H}{w_o} \sinh \left(\frac{w_o}{F_H} x\right) \tag{3}$$

Now, substitute Eq. (3) into Eq. (2), in which case

$$\frac{dy}{dx} = \sinh \left(\frac{w_o}{F_H} x\right)$$

Hence,

$$y = \frac{F_H}{w_o} \cosh \left(\frac{w_o}{F_H} x\right) + C_3 \tag{4}$$

If the boundary condition $y = 0$ at $x = 0$ is applied, the constant $C_3 = -F_H/w_o$, and therefore the deflection curve becomes

$$y = \frac{F_H}{w_o} \left[\cosh \left(\frac{w_o}{F_H} x\right) - 1\right]$$

This equation defines the shape of a *catenary curve*. The constant F_H is obtained by using the boundary condition that $y = h$ at $x = L/2$, in which case

$$h = \frac{F_H}{w_o} \left[\cosh \left(\frac{w_o L}{2F_H}\right) - 1\right] \tag{5}$$

Since $w_o = 5$ N/m, $h = 6$ m, and $L = 20$ m, Eqs. (4) and (5) become

$$y = \frac{F_H}{5\,\text{N/m}} \left[\cosh \left(\frac{5\,\text{N/m}}{F_H} x\right) - 1\right] \tag{6}$$

$$6\,\text{m} = \frac{F_H}{5\,\text{N/m}} \left[\cosh \left(\frac{50\,\text{N}}{F_H}\right) - 1\right] \tag{7}$$

Equation (7) can be solved for F_H by using a trial-and-error procedure. The result is

$$F_H = 45.8\,\text{N}$$

and therefore Eq. (6) becomes

$$y = 9.16[\cosh (0.109x) - 1]\,\text{m} \qquad\qquad Ans.$$

Using Eq. (3), the half-length of the cable is ($x = 10$ m)

$$\frac{\mathcal{L}}{2} = \frac{45.8\,\text{N}}{5\,\text{N/m}} \sinh \left[\frac{5\,\text{N/m}}{45.8\,\text{N}}(10\,\text{m})\right] = 12.1\,\text{m}$$

Hence,

$$\mathcal{L} = 24.2 \text{ m} \qquad\qquad Ans.$$

Since $T = F_H/\cos\theta$, the maximum tension occurs when θ is maximum, i.e., at $s = \mathcal{L}/2 = 12.1$ m. Using Eq. (2) yields

$$\left.\frac{dy}{dx}\right|_{\text{max}} = \tan\theta_{\text{max}} = \frac{5 \text{ N/m}(12.1 \text{ m})}{45.8 \text{ N}} = 1.32$$

$$\theta_{\text{max}} = 52.9°$$

Thus,

$$T_{\text{max}} = \frac{45.8 \text{ N}}{\cos 52.9°} = 75.9 \text{ N} \qquad\qquad Ans.$$

Problems

Neglect the weight of the cable in the following problems, *unless* specified.

7-46. Determine the maximum loading w_o N/m that the cable can support if it is capable of sustaining a maximum tension of 80 kN before it will break.

* 7-48. The platform has a total mass of 5 Mg and is supported uniformly between points A and B by means of *two* cables. Determine the maximum force developed in each cable.

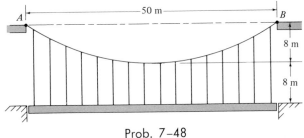

Prob. 7-48

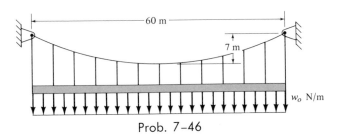

Prob. 7-46

7-47. Determine the maximum tension developed in the cable if it is subjected to a uniform load of 600 N/m.

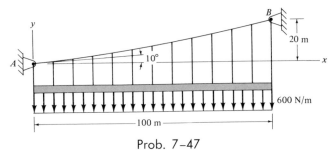

Prob. 7-47

7–49. The cable is subjected to the triangular loading. If the slope of the cable at point A is zero, determine the equation of the curve $y = f(x)$ which defines the cable AB, and the maximum tension developed in the cable. L, h, and w_o are known.

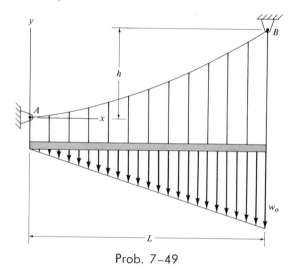

Prob. 7–49

7–50. The cable will break when the maximum tension reaches $T_{max} = 8$ kN. Determine the minimum length of the cable and the sag h if it carries a uniform distributed load of $w = 400$ N/m of horizontal projection. The span is $L = 25$ m.

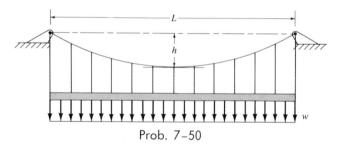

Prob. 7–50

7–50a. Solve Prob. 7–50 if $T_{max} = 2000$ lb, $w = 30$ lb/ft, and $L = 60$ ft.

7–51. A 40-m-long chain has a total mass of 100 kg and is suspended between two points 10 m apart, at the same elevation. Determine the maximum tension and the sag in the chain.

*7–52. A uniform cord is suspended between two points having the same elevation. Determine the sag-to-span ratio so that the maximum tension in the cord equals the cord's total weight.

7–53. If the cable has a mass of 3kg/m, determine the maximum and minimum tension in the cable. What is the cable's length?

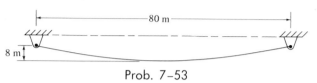

Prob. 7–53

7–54. The telephone cable has a mass of 500 g/m. If the cable sags 1.5 m, determine the maximum tension in the cable and its length between the poles.

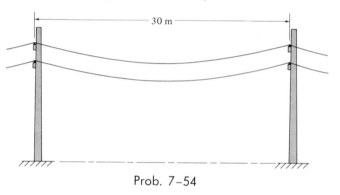

Prob. 7–54

7–55. The cable has a mass of $m = 3$ kg/m and a length of $\mathcal{L} = 150$ m. Determine the sag h when it spans $L = 75$ m. Also, compute the minimum tension in the cable.

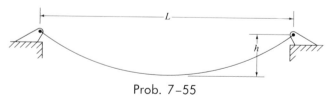

Prob. 7–55

7–55a. Solve Prob. 7–55 if the cable has a weight of $w = 6$ lb/ft, and $\mathcal{L} = 150$ ft, $L = 100$ ft.

*7–56. The yacht is anchored with a chain that has a total length of 40 m and a mass of 18 kg/m. If the chain makes a 60° angle with the horizontal and the tension in the chain at point A is 7 kN, determine the length of chain l_d, which is lying at the bottom of the sea. What is the distance d? Assume that buoyancy effects of the water on the chain are negligible. *Hint:* Establish the origin of the coordinate system at B as shown in order to compute the chain length BA.

7–57. Show that the deflection curve of the cable discussed in Example 7–8 reduces to Eq. (4) in Example 7–7 when the *hyperbolic cosine function* is expanded in terms of a series and only the first two terms are retained. (The answer indicates that the *catenary* may be replaced by a *parabola* in the analysis of problems in which the sag is small. In this case, the cable weight is assumed to be uniformly distributed along the horizontal.)

7–58. The buoyant (or vertical) component of force acting on the weather balloon is 500 N. If cable OB is 70 m in length and it has a mass of 500 g/m, determine the height h of the balloon.

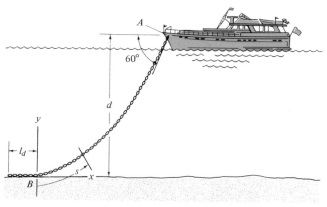

Prob. 7–56

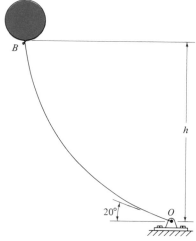

Prob. 7–58

8–1. Characteristics of Dry Friction

In the previous chapters the surface of contact between two bodies was considered to be perfectly *smooth*. Because of this, the force of interaction between the bodies always acts *normal* to the surface at points of contact. In reality, however, all surfaces are *rough,* and depending upon the nature of the problem, the ability of a body to support a *tangential* as well as a *normal* force at its contacting surface must be considered. The tangential force developed at a surface of contact is caused by friction. Specifically, *friction* may be defined as a force of resistance acting on a body which prevents or inhibits any possible slipping of the body. This force always acts *tangent* to the surface at points of contact with other bodies, and is directed so as to oppose the possible or existing motion of the body at that point.

The effects of friction are both advantageous and detrimental. Without friction one would be unable to walk or to make use of any sort of transportation vehicle, such as an automobile. Furthermore, many engineering devices, such as belt drives and brakes, require frictional forces in order to function. The adverse effects of friction caused by the rubbing and wearing away of surfaces are reduced when bearings, rollers, and lubricants are used in machines.

Types of Friction. In general, two types of friction can occur between surfaces. *Fluid friction* exists when the contacting surfaces are separated by a film of fluid such as oil. When relative motion between the bodies occurs, frictional forces are developed between thin layers of the fluid. These forces depend upon the relative velocity of the fluid layers and the viscosity or shearing resistance of the fluid. For this reason problems involving fluid friction are generally studied in fluid mechanics. In this book only the effects of *dry friction* will be presented. This type of friction

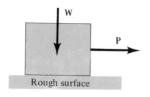

(a)

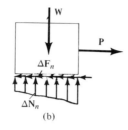

(b)

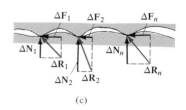

(c)

Fig. 8–1(a–c)

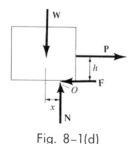

Fig. 8–1(d)

is often called *Coulomb friction,* since its characteristics were extensively studied by C. A. Coulomb in 1781. Specifically, dry friction occurs between the contacting surfaces of rigid bodies in the absence of a lubricating fluid.

Theory of Dry Friction. The theory of dry friction can best be explained by considering what effects are caused by pulling horizontally on a block of uniform weight $\mathbf{W}$ which is resting on a rough horizontal floor, Fig. 8–1a. As shown on the free-body diagram of the block, Fig. 8–1b, the floor exerts a distribution of both *normal force* $\Delta\mathbf{N}_n$ and *frictional force* $\Delta\mathbf{F}_n$ along the contacting surface. For equilibrium, the normal forces must act *upward* to balance the block's weight $\mathbf{W}$ and the frictional forces act to the left, to prevent the applied force $\mathbf{P}$ from moving the block to the right. Close examination between the contacting surfaces of the floor and block reveals how these frictional and normal forces develop, Fig. 8–1c. It can be seen that many microscopic irregularities exist between the two surfaces and, as a result, reactive forces $\Delta\mathbf{R}_n$ are developed at each of the protuberances.* These forces act at all points of contact and, as shown, each reactive force contributes both a frictional component $\Delta\mathbf{F}_n$ and a normal component $\Delta\mathbf{N}_n$.

For simplicity in the foregoing analysis, the effect of these distributed loadings will be indicated by their *resultants* $\mathbf{N}$ and $\mathbf{F}$ and then represented on the free-body diagram of the block as shown in Fig. 8–1d. Clearly, $\mathbf{F}$ always acts *tangent to the contacting surface, opposite* to the direction of $\mathbf{P}$. The normal force $\mathbf{N}$ is directed upward to balance the weight $\mathbf{W}$ and acts on the bottom of the block, a distance x to the right of the line of action of $\mathbf{W}$. This location for $\mathbf{N}$ is necessary to balance the "tipping effect" caused by $\mathbf{P}$. For example, if $\mathbf{P}$ is applied at a height h from the surface, Fig. 8–1d, then moment equilibrium about point O is satisfied if $Wx = Ph$ or $x = Ph/W$.

---

*Besides mechanical interactions as explained here, a detailed treatment of the nature of frictional forces might also include the effects of temperature, density, cleanliness, and electrical attraction between the contacting surfaces.

Tipping. Provided the block does *not slip,* any increase in P causes a corresponding increase in x and, as a result, this tends to concentrate the distribution of normal force farther toward the block's right corner, thereby increasing the chance for tipping, Fig. 8–1e. Indeed, *tipping* occurs if the contacting surface is "rough" enough to hold the block from slipping and the applied force $P = P_t \geqslant (W/h)(a/2)$, where $x = a/2$, Fig. 8–1f.

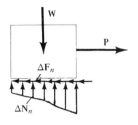

(e)

Impending Motion. In cases where h is small or the surfaces of contact are rather "slippery," the frictional force **F** may *not* be great enough to balance the magnitude of **P,** and consequently the block will tend to slip *before* it can tip. In other words, as the magnitude of **P** is slowly increased, the magnitude of **F** correspondingly increases until it attains a certain *maximum value* F_s, called the *limiting static frictional force,* Fig. 8–1g. When this value is reached, the block is in *unstable equilibrium,* since any further increase in P will cause motion. Experimentally, it has been determined that the magnitude of the limiting static frictional force **F**$_s$, is *directly proportional* to the magnitude of the resultant normal force **N.** This may be expressed mathematically as

$$F_s = \mu_s N \tag{8-1}$$

where the constant of proportionality, μ_s (mu "sub" s), is called the *coefficient of static friction.*

Typical values for μ_s, found in most engineering handbooks, are given in Table 8–1. It should be noted that μ_s is dimensionless and depends only upon the characteristics of the two surfaces in contact. Furthermore, a wide range of values is given for each value of μ_s, since experimental testing was done under variable conditions of roughness and cleanliness of the contacting surfaces. For applications, therefore, it is important that both caution and judgment be exercised when selecting a coefficient of friction for a given set of conditions. When an exact calculation of F_s is required, the coefficient of friction should be determined directly by an experiment that involves the two materials to be used.

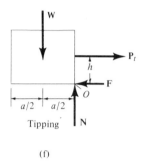

Tipping

(f)

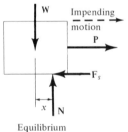

Equilibrium

(g)

Fig. 8–1(e–g)

Table 8–1

Contact Materials	Coefficient of Static Friction (μ_s)
Metal on ice	0.03–0.05
Wood on wood	0.30–0.70
Leather on wood	0.20–0.50
Leather on metal	0.30–0.60
Aluminum on aluminum	1.10–1.70

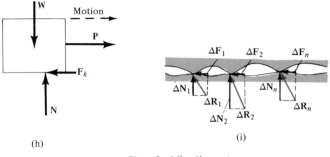

(h)

(i)

Fig. 8–1(h, i)

Motion. If the magnitude of **P** is increased so that it becomes greater than F_s, the frictional force at the contacting surfaces drops slightly to a smaller value F_k, called the *kinetic frictional force*. The block will *not* be held in equilibrium $(P > F_k)$; instead, it begins to slide with increasing speed, Fig. 8–1*h*. The drop made in the frictional force magnitude, from F_s (static) to F_k (kinetic), can be explained by again examining the surfaces of contact, Fig. 8–1*i*. Here it is seen that when $P > F_s$, then **P** essentially "lifts" the block out of its settled position and causes it to "ride" on *top* of the peaks at the contacting surface. Consequently, the resultant contact forces $\Delta\mathbf{R}_n$ are aligned slightly more in the vertical direction than before (Fig. 8–1*c*), and hence contribute *smaller* frictional components, $\Delta\mathbf{F}_n$, as when the irregularities are meshed.

Experiments with sliding blocks indicate that the magnitude of the resultant frictional force (F_k) is directly proportional to the magnitude of the resultant normal force **N**. This may be expressed mathematically as

$$F_k = \mu_k N \qquad\qquad (8\text{–}2)$$

where the constant of proportionality, μ_k, is called the *coefficient of kinetic friction*. Typical values for μ_k are approximately 25 per cent *smaller* than those listed in Table 8–1 for μ_s.

Rules of Dry Friction. As a result of *experiments* that pertain to the above discussion, the following rules, which apply to bodies subjected to dry friction, may be stated:

1. The frictional force acts *tangent* to the contacting surfaces in a direction *opposed* to the *relative motion* or tendency for motion of one surface against another.

2. The magnitude of the limiting static frictional force $\mathbf{F}_s$ that can be developed is independent of the area of contact provided the normal pressure is not great enough to severely deform or crush the contacting surfaces of the bodies.
3. The magnitude of the limiting static frictional force $\mathbf{F}_s$ is generally greater than the magnitude of the kinetic frictional force $\mathbf{F}_k$ for any two surfaces of contact. However, if one of the bodies is moving with a *very low velocity* over the surface of another, F_k becomes approximately equal to F_s.
4. When *slipping* at the point of contact is *about to occur,* the magnitude of the limiting static frictional force $\mathbf{F}_s$ is proportional to the magnitude of the normal force $\mathbf{N}$ at the point of contact, such that $F_s = \mu_s N$ (Eq. 8–1).
5. When *slipping* at the point of contact is *occurring,* the magnitude of the kinetic frictional force $\mathbf{F}_k$ is proportional to the magnitude of the normal force $\mathbf{N}$ at the point of contact, such that $F_k = \mu_k N$ (Eq. 8–2).

Angle of Friction. It should be observed that Eqs. 8–1 and 8–2 have a specific, yet *limited,* use in the solution of friction problems. In particular, the frictional force acting at a contacting surface is determined from $F = \mu_k N$ only if *relative motion* is occurring between the two surfaces. Furthermore, if two bodies are *stationary,* the magnitude of frictional force, F, *does not necessarily* equal $\mu_s N$; instead, F must satisfy the inequality $F \leqslant \mu_s N$. Only when *impending motion* occurs does F reach its upper limit, $F = F_s = \mu_s N$. This situation may be better understood by considering the block shown in Fig. 8–2a, which is acted upon by a force $\mathbf{P}$. In this case consider $P = F_s$, so that the block is on the *verge of sliding.* For equilibrium, the normal force $\mathbf{N}$ and frictional force $\mathbf{F}_s$ combine to create a resultant force $\mathbf{R}_s$. The angle ϕ_s that $\mathbf{R}_s$ makes with $\mathbf{N}$ is called the *angle of static friction* or the *angle of repose.* From the figure,

$$\phi_s = \tan^{-1}\left(\frac{F_s}{N}\right) = \tan^{-1}\left(\frac{\mu_s N}{N}\right) = \tan^{-1}\mu_s$$

Provided the block is *not in motion,* any horizontal force $P < F_s$ causes a resultant force $\mathbf{R}$, which has a line of action directed at an angle ϕ from the vertical such that $\phi \leqslant \phi_s$. If the applied force $\mathbf{P}$ creates *motion* of the block, then $P = F_k$. In this case, the resultant $\mathbf{R}_k$ has a line of action defined by ϕ_k, Fig. 8–2b. This angle is referred to as the *angle of kinetic friction,* where

$$\phi_k = \tan^{-1}\left(\frac{F_k}{N}\right) = \tan^{-1}\left(\frac{\mu_k N}{N}\right) = \tan^{-1}\mu_k$$

By comparison, $\phi_s > \phi_k$.

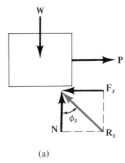

(a)

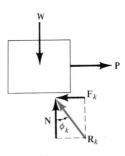

(b)

Fig. 8–2

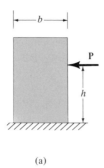

(a)

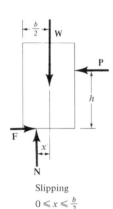

Slipping

$$0 \leqslant x \leqslant \frac{b}{2}$$

$$(F = \mu N)$$

(b)

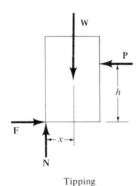

Tipping

$$x = \frac{b}{2}$$

$$(F \leqslant \mu N)$$

(c)

Fig. 8–3

8–2. Problems Involving Dry Friction

If a rigid body is in equilibrium when it is subjected to a system of forces, which includes the effect of friction, the force system must not only satisfy the equations of equilibrium but *also* the laws that govern the frictional forces.

Types of Frictional Problems. In general, there are two types of mechanics problems involving dry friction.

In the "first type" of problem, one has to determine the force needed to either cause impending motion or slipping of a body along *all* its contacting surfaces. If impending motion occurs, the static frictional forces reach their limiting value, i.e., $F = \mu_s N$; whereas if the body is slipping or sliding, kinetic frictional forces are developed, $F = \mu_k N$ (see Examples 8–1, 8–2, and 8–3).

In the "second type" of problem, the applied loads acting on the body (or system of bodies) are either known or have to be determined; and for the analysis one has to *check* if the equilibrium frictional force developed at each surface of contact is less than or equal to the maximum static frictional force that can be developed at these points, $F \leqslant \mu_s N$ (see Examples 8–4, 8–5, and 8–6).

In some cases, this second type of problem may involve several ways in which a body (or system of bodies) may move out of equilibrium and, as a result, each situation will have to be *investigated separately*. For example, a block resting on a rough surface, Fig. 8–3a, may either slip, Fig. 8–3b, or tip, Fig. 8–3c, as the force **P** is increased in magnitude. The actual situation is determined by calculating P for each case, and then choosing the case for which P is the smallest. If in *both cases* the *same value* for P is calculated, then both slipping and tipping occur simultaneously. In cases where a curved body is in contact with another surface, *rolling* rather than tipping may be possible. For example, consider the cylinder wedged between the floor and the lever, Fig. 8–4a. As the magnitude of **P** is increased, two possibilities for motion exist: the cylinder can roll without slipping at A, and as a result it would have to slip at B, Fig. 8–4b; or it could roll along the lever and slip at A, Fig. 8–4c. If the analysis of each of these cases reveals the *same magnitude* for **P**, then slipping at both surfaces, with no rolling, occurs.

Equilibrium Versus Frictional Equations. As stated above, if a body can be subjected to various types of movement, the frictional forces at some of the points of contact may *not* be equal to their maximum static value; instead, $F < \mu_s N$. In this case, F must be determined from the equations of equilibrium, and since F is an "equilibrium force," the sense of direction of **F** can be *assumed*. The correct direction of **F** is made known *after* the equations of equilibrium are applied and the magnitude of **F** is deter-

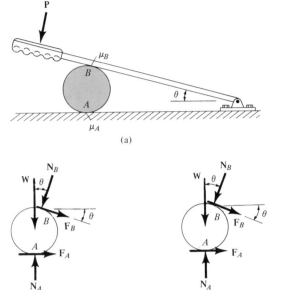

(a)

Rolling at A $(F_A \leqslant \mu_A N_A)$
Slipping at B $(F_B = \mu_B N_B)$

(b)

Rolling at B $(F_B \leqslant \mu_B N_B)$
Slipping at A $(F_A = \mu_A N_A)$

(c)

Fig. 8-4

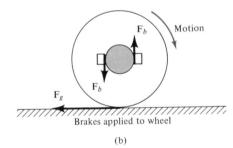

Wheel rolling due to force of friction

(a)

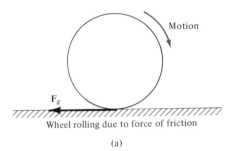

Brakes applied to wheel

(b)

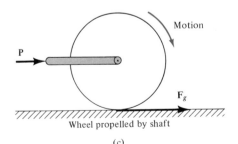

Wheel propelled by shaft

(c)

Fig. 8-5

mined. In particular, if the magnitude is negative, the directional sense of
F is the reverse of that which was assumed. This convenience of *assuming*
the sense of direction of **F** is possible because the equilibrium equations
are *vector* equations; i.e., the scalar equilibrium equations equate to zero
the *components of vectors* acting in the *same direction*.

In cases where the scalar frictional equation $F = \mu_s N$ is used in the
solution of a problem, the convenience of *assuming* the directional sense of
F is *lost*, since this equation relates the magnitudes of two vectors that act
in *perpendicular directions* ($\mathbf{F} = \mathbf{F}_s$ is always perpendicular to **N**). Consequently, **F** *must* be shown acting with its *correct direction* on the free-body
diagram. In this regard, the direction of the frictional force will *always* be
such as to either *oppose the motion or impend the motion of the body* over
its contacting surface. As an example, consider a wheel rolling freely on
the ground, Fig. 8-5a. In order to give the wheel a forward clockwise
motion, the frictional force at the ground, $\mathbf{F}_g$, acts to the left. If brakes are
applied to the wheel, Fig. 8-5b, the braking frictional forces $\mathbf{F}_b$, acting on
the wheel, oppose the rolling motion of the wheel. Finally, if the wheel's
rolling motion is propelled by a shaft that exerts a driving force **P** on the
wheel, Fig. 8-5c, the ground frictional force $\mathbf{F}_g$ acts to the right to prevent
the wheel from slipping.

The following two-step procedure should be used when solving equilibrium problems involving dry friction.

Step 1: Draw the necessary free-body diagrams and determine the number of equations required for a complete solution. Recall that only three equations of coplanar equilibrium can be written for each body. Consequently, if there are more unknowns than equations of equilibrium, it will be necessary to apply the frictional equation, $F = \mu_s N$, at some, if not all, points of contact to obtain the required number of equations.*

Step 2: Apply the necessary frictional equations and the equations of equilibrium, and solve for the unknowns. If the problem involves a three-dimensional force system such that it becomes difficult to obtain the force components or the necessary moment arms, apply the equations of equilibrium using Cartesian vectors.

The following example problems numerically illustrate this two-step procedure.

Example 8-1

The crate shown in Fig. 8–6a has a mass of 20 kg and is subjected to a force of 100 N. Determine the angle θ such that the crate is on the verge of moving up the plane. The coefficient of static friction is $\mu_s = 0.3$.

Solution

Step 1: As shown on the free-body diagram, Fig. 8–6b, the resultant $\mathbf{N}_C$ must act a distance x from the centerline of the crate in order to counteract the tipping effect caused by **P**. There are *four* unknowns: θ, F, N_C, and x. Since *three* equations of equilibrium are available, the frictional equation applies at the contacting surface and provides the necessary fourth equation for the solution. (Use of this equation is also implied by the problem statement, since if the crate is on the verge of slipping up the plane, the maximum static frictional force must be developed on the crate.) Note that **F** must act *down* the plane to prevent upward motion of the crate. It is important that the correct direction of **F** be specified on the free-body diagram since the frictional equation is to be used in the solution.

---

*In cases where the problem is of the "second type," several points of application of the frictional equation may be possible. For example, consider again the cylinder in Fig. 8–4a. If the weight **W** is known, the cylinder is subjected to four unknown force magnitudes: N_A, F_A, N_B, and F_B. The solution for these unknowns will be *unique* if it satisfies the *three* equilibrium equations *and only one* frictional equation. Consequently, the possibilities are either $F_B = \mu_B N_B$ (in which case $F_A \leqslant \mu_A N_A$), Fig. 8–4b, or $F_A = \mu_A N_A$ (in which case $F_B \leqslant \mu_B N_B$), Fig. 8–4c.

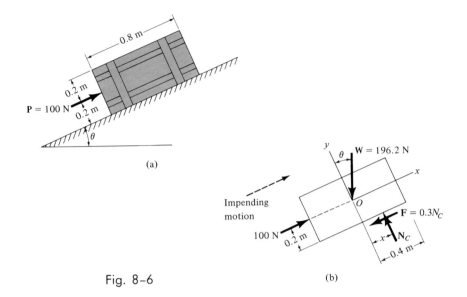

Fig. 8-6

(a)

(b)

Step 2: Writing the frictional equation yields

$$F_s = \mu_s N; \qquad\qquad F = 0.3N_C$$

Using this result and applying the equations of equilibrium, we have

$$+\nearrow \Sigma F_x = 0; \qquad 100 - 0.3N_C - 196.2 \sin\theta = 0 \qquad\qquad (1)$$
$$+\nwarrow \Sigma F_y = 0; \qquad N_C - 196.2 \cos\theta = 0 \qquad\qquad (2)$$
$$\curvearrowright +\Sigma M_O = 0; \qquad -N_C(x) + 0.3N_C(0.2) = 0 \qquad\qquad (3)$$

From Eq. (3), N_C cancels, so that $x = 0.3(0.2) = 0.06$ m. Since the answer is positive, the direction of x was assumed correctly. Furthermore, $x < 0.4$ m, so that the crate will not tip over. Eliminating N_C from Eqs. (1) and (2) yields the following trigonometric equation:

$$100 - 58.9 \cos\theta - 196.2 \sin\theta = 0 \qquad\qquad (4)$$

The angle θ can be obtained by using a trial-and-error solution, i.e., by assuming successive values of θ until the equation is satisfied. A more direct approach would involve using the trigonometric identity $\sin^2\theta + \cos^2\theta = 1$. Solving for the $\cos\theta$ in Eq. (4) and substituting into the identity yields a quadratic equation in terms of $\sin\theta = x'$. When the two roots $\sin\theta_1 = x_1'$ and $\sin\theta_2 = x_2'$ have been determined, the *least* of the two values for θ is chosen since it represents the physical situation. Finally, N_C is determined from Eq. (1) or (2). The results are

$$\theta = 12.5° \qquad N_C = 191.5 \text{ N} \qquad\qquad Ans.$$

Example 8-2

The uniform plank shown in Fig. 8–7a has a mass of 15 kg and rests against a floor and wall for which the coefficients of static friction are $(\mu_s)_A = 0.30$ and $(\mu_s)_B = 0.20$, respectively. Determine the distance s to which a man having a mass of 70 kg can climb without causing the plank to slip.

Solution

Step 1: The free-body diagram for the plank is shown in Fig. 8–7b. The *five* unknowns, N_B, F_B, N_A, F_A, and s, can be determined from the *three* equations of equilibrium and *two* frictional equations applied at points A and B. The frictional forces $\mathbf{F}_A$ and $\mathbf{F}_B$ must be drawn in their correct direction, so that they oppose the tendency for motion of the plank, Fig. 8–7b. Why?

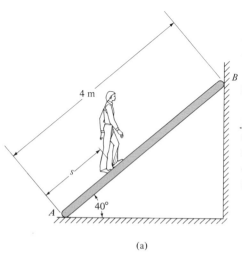

(a)

Fig. 8–7

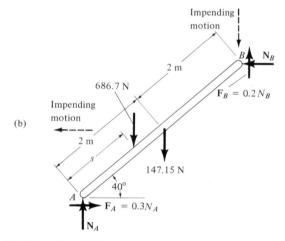

(b)

Step 2: Writing the frictional equations,

$$F = \mu_s N; \qquad\qquad F_A = 0.3 N_A$$
$$F_B = 0.2 N_B$$

Using these results and applying the equations of equilibrium,

$\xrightarrow{+} \Sigma F_x = 0; \qquad\qquad 0.3 N_A - N_B = 0$

$+\uparrow \Sigma F_y = 0; \qquad N_A - 686.7 - 147.15 + 0.2 N_B = 0$

$\zeta + \Sigma M_A = 0; \qquad -686.7(s \cos 40°) - 147.15(2 \cos 40°)$
$$+ 0.2 N_B(4 \cos 40°) + N_B(4 \sin 40°) = 0$$

Solving these equations simultaneously yields

$$N_A = 786.7 \text{ N}$$
$$N_B = 236.0 \text{ N}$$
$$s = 1.00 \text{ m} \qquad\qquad\qquad Ans.$$

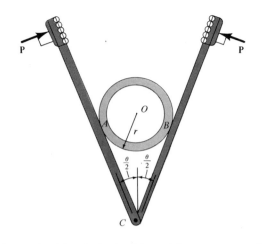

Fig. 8-8(a)

Example 8-3

The pipe shown in Fig. 8-8a is gripped between two levers that are pinned at C. If the coefficient of friction between the levers and the pipe is $\mu = 0.3$, determine the maximum angle θ so that the pipe is gripped without slipping. Neglect the weight of the pipe.

Solution

Step 1: A free-body diagram of the pipe is shown in Fig. 8-8b. There are five unknowns: N_A, F_A, N_B, F_B, and θ. The *three* equations of equilibrium and *two* frictional equations at A and B apply. The frictional forces act downward to prevent the upward motion of the pipe.

Step 2: The frictional equations are

$$F_s = \mu_s N; \qquad\qquad F_A = \mu N_A$$
$$F_B = \mu N_B$$

Using these results, and applying the equations of equilibrium, yields

$$\xrightarrow{+}\Sigma F_x = 0; \quad N_A \cos(\theta/2) + \mu N_A \sin(\theta/2) - N_B \cos(\theta/2) \qquad (1)$$
$$- \mu N_B \sin(\theta/2) = 0$$

$$\zeta + \Sigma M_O = 0; \qquad \mu N_B(r) - \mu N_A(r) = 0 \qquad\qquad (2)$$

$$+\uparrow\Sigma F_y = 0; \quad N_A \sin(\theta/2) - \mu N_A \cos(\theta/2) + N_B \sin(\theta/2) \qquad (3)$$
$$- \mu N_B \cos(\theta/2) = 0$$

From either one of Eqs. (1) and (2) it is seen that $N_A = N_B$, which could also have been determined directly from the symmetry of both geometry and loading. Substituting this result into Eq. (3), we obtain

$$\sin(\theta/2) - \mu \cos(\theta/2) = 0$$

so that

$$\tan(\theta/2) = \frac{\sin(\theta/2)}{\cos(\theta/2)} = \mu = 0.3$$

$$\theta = 2 \tan^{-1} 0.3 = 33.4° \qquad\qquad Ans.$$

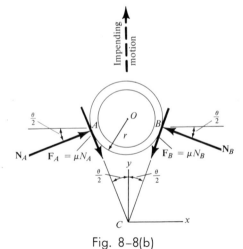

Fig. 8-8(b)

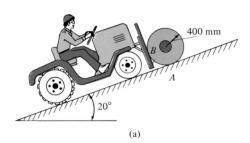

(a)

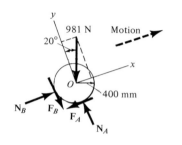

(b)

Fig. 8-9

Example 8-4

Determine the normal force that must be exerted on the 100-kg spool shown in Fig. 8-9a to push it up the 20° incline at constant velocity. The coefficients of static and kinetic friction at the points of contact are $(\mu_s)_A = 0.18$, $(\mu_k)_A = 0.15$ and $(\mu_s)_B = 0.45$, $(\mu_k)_B = 0.4$.

Solution

Step 1: The free-body diagram of the spool is shown in Fig. 8-9b. The four unknowns, N_A, F_A, N_B, and F_B, can be determined from the *three* equations of equilibrium and *one* frictional equation, which applies either at A or B. If slipping occurs at A, the spool will *slide* up the incline; whereas if slipping occurs at B, the spool *rolls* up the incline. The problem requires determination of N_B.

Step 2: Applying the equations of equilibrium,

$$+\nearrow \Sigma F_x = 0; \qquad -F_A + N_B - 981 \sin 20° = 0 \qquad (1)$$

$$+\nwarrow \Sigma F_y = 0; \qquad N_A - F_B - 981 \cos 20° = 0 \qquad (2)$$

$$\zeta + \Sigma M_O = 0; \qquad -F_B(400 \text{ mm}) + F_A(400 \text{ mm}) = 0 \qquad (3)$$

Spool Rolls Up Incline. In this case slipping occurs at B. Hence,

$$F = \mu_k N; \qquad F_B = 0.40 N_B \qquad (4)$$

$$F \leqslant \mu_s N; \qquad F_A \leqslant 0.18 N_A \qquad (5)$$

The direction of the frictional force at B must be correctly specified. Why? Since the spool is being forced up the plane, $\mathbf{F}_B$ acts downward to prevent the clockwise rolling motion of the spool, Fig. 8-9b. Solving Eqs. (1) through (4), we have

$$N_A = 1145.5 \text{ N}$$
$$F_A = 223.7 \text{ N}$$
$$N_B = 559.2 \text{ N}$$
$$F_B = 223.7 \text{ N}$$

The assumption regarding no slipping at A should be checked using Eq. (5).

$$\overset{?}{223.7 \text{ N} \leqslant 0.18(1145.5)} = 206.2 \text{ N}$$

The inequality does *not apply,* and therefore slipping occurs at A and not at B. Hence, the other case of motion must be investigated.

Spool Slides Up Incline. In this case,

$$F = \mu_k N; \qquad F_A = 0.15 N_A \qquad (6)$$

$$F = \mu_s N; \qquad F_B \leqslant 0.45 N_B \qquad (7)$$

Solving Eqs. (1) through (3) and (6) yields

$$N_A = 1084.5 \text{ N}$$
$$F_A = 162.7 \text{ N}$$
$$N_B = 498.2 \text{ N}$$
$$F_B = 162.7 \text{ N} \qquad \text{Ans.}$$

The validity of the solution can be checked by testing the assumption that no slipping occurs at B, Eq. (7).

$$162.7 \text{ N} < 0.45(498.2 \text{ N}) = 224.2 \text{ N} \qquad \text{(check)}$$

Example 8-5

The brake shown in Fig. 8–10a bears against the outer surface of a 20-kg wheel for which $\mu_s = 0.45$ and $\mu_k = 0.40$. Determine the horizontal and vertical components of force acting at the pin O if a force of 200 N is applied to the handle.

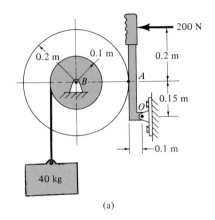

(a)

Solution

Step 1: The free-body diagrams for both the wheel and brake are shown in Fig. 8–10b. All six unknowns, N_A, F_A, O_x, O_y, B_x, and B_y, can be obtained from the equilibrium equations, three for the brake and three for the wheel. For the analysis $\mathbf{F}_A$ is therefore treated as an equilibrium force that is not necessarily equal to the *maximum frictional force* developed by the brake; i.e., F_A must satisfy the inequality $F_A \leqslant \mu_s N_A$ *if the wheel is to remain at rest.*

Step 2: Summing moments about point B on the wheel yields a direct solution for the magnitude of $\mathbf{F}_A$.

$$\zeta + \Sigma M_B = 0; \qquad -392.4(0.1) + F_A(0.2) = 0$$
$$F_A = 196.2 \text{ N}$$

The "positive" answer indicates that $\mathbf{F}_A$ is shown having the correct sense of direction on the free-body diagram. Using this result and summing moments about O on the brake to obtain N_A yields

$$\zeta + \Sigma M_O = 0; \quad -200(0.35) + N_A(0.15) + 196.2(0.1) = 0$$
$$N_A = 335.9 \text{ N}$$

Since $\mu_s = 0.45$, the maximum static frictional force that can be developed at A is

$$F_s = \mu_s N; \qquad (F_s)_A = 0.45(335.9) = 151.2 \text{ N}$$

By comparison, the frictional force *required* for equilibrium, $F_A = 196.2$ N, is *greater* than the maximum static frictional force $(F_s)_A$ developed at the brake, i.e., $F_A > \mu_s N_A$. Consequently, *the wheel is rotating* and the problem must be solved on this basis.

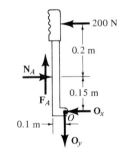

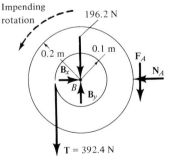

(b)

Fig. 8–10

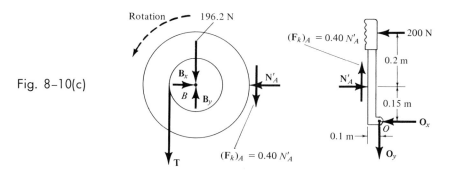

Fig. 8–10(c)

Since slipping occurs at A, the kinetic frictional force developed by the brake is related to the normal force by the equation

$$F_k = \mu_k N; \qquad\qquad (F_k)_A = 0.40 N'_A$$

Here it is important that the correct direction of $(\mathbf{F}_k)_A$ be specified on the free-body diagram. Why? In this case $(\mathbf{F}_k)_A$ acts downward to resist the rotating motion of the wheel, Fig. 8–10c.

Applying the three equations of equilibrium to the brake, since it does not move,

$$\xrightarrow{+}\Sigma F_x = 0; \qquad\qquad -O_x + N'_A - 200 = 0$$
$$+\uparrow\Sigma F_y = 0; \qquad\qquad -O_y + 0.40 N'_A = 0$$
$$\zeta+\Sigma M_O = 0; \quad -N'_A(0.15) - 0.40 N'_A(0.1) + 200(0.35) = 0$$

Hence,

$$N'_A = 368.4 \text{ N}$$
$$O_x = 168.4 \text{ N} \qquad\qquad Ans.$$
$$O_y = 147.4 \text{ N} \qquad\qquad Ans.$$

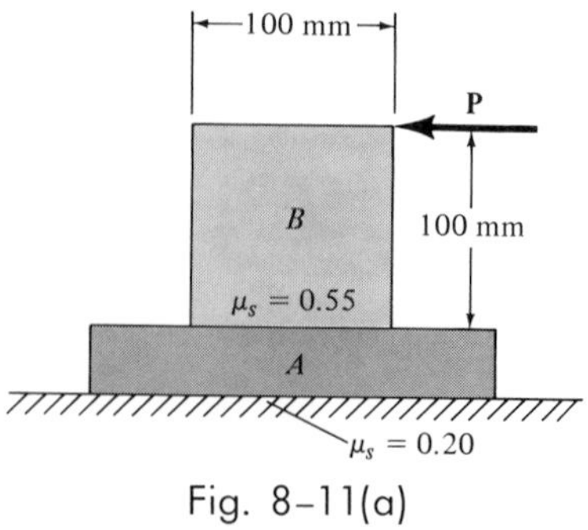

Example 8–6

The homogeneous blocks A and B shown in Fig. 8–11a have masses of 10 kg and 6 kg, respectively. Determine the greatest magnitude of force $\mathbf{P}$ that can be applied without causing the blocks to move. The static coefficients of friction between the contacting surfaces are given in the figure.

Solution

Free-body diagrams of the blocks are shown in Fig. 8–11b. In this problem tipping of block B will have to be investigated. Consequently, the

——100 mm——

P

B   100 mm

$\mu_s = 0.55$

A

$\mu_s = 0.20$

Fig. 8–11(a)

dimension x is used to locate the position of N_B under block B. The possibility of block A tipping will not be investigated, since it is long and flat. There are six unknowns, P, N_B, F_B, x, N_A, and F_A. *Five* equations of equilibrium are available, three for block B and two for block A. (The location of N_A is not specified, so moment equilibrium of A will not be considered.) Only *one* other equation must be used to obtain the necessary *sixth* equation for a complete solution. Three possibilities for motion therefore exist: (1) impending motion of A, $F_A = 0.20N_A$; (2) impending motion of B on A, $F_B = 0.55N_B$; or (3) tipping of B on A, $x = 50$ mm. The magnitude of $\mathbf{P}$ needed to cause each of these motions will be computed. The least value of P is the required answer. Why?

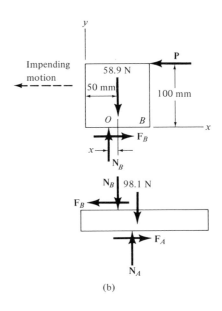

(b)

Impending Motion of A. The free-body diagram of both blocks treated as one system is shown in Fig. 8–11c. The three unknowns, N_A, F_A, and P, are determined from the two force equations of equilibrium and the law of static friction applied at A.

$$F = \mu_s N; \qquad\qquad F_A = 0.20N_A$$

so that,

$$\xrightarrow{+}\Sigma F_x = 0; \qquad\qquad 0.20N_A - P = 0$$
$$+\uparrow\Sigma F_y = 0; \qquad\qquad N_A - 98.1 - 58.9 = 0$$

Hence,

$$P = 31.4\text{ N}$$
$$N_A = 157.0\text{ N}$$
$$F_A = 31.4\text{ N}$$

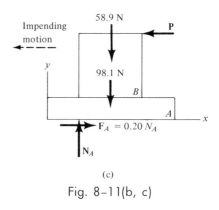

(c)

Fig. 8–11(b, c)

Impending Motion of B. The free-body diagram of block B, when it is about to slide, is shown in Fig. 8–11b. The four unknowns, P, N_B, F_B, and x, are obtained by applying the three equations of equilibrium to the block and the law of static friction,

$$F = \mu_s N; \qquad\qquad F_B = 0.55N_B$$

so that,

$$\xrightarrow{+}\Sigma F_x = 0; \qquad\qquad 0.55N_B - P = 0$$
$$+\uparrow\Sigma F_y = 0; \qquad\qquad N_B - 58.9 = 0$$
$$\zeta+\Sigma M_O = 0; \qquad\qquad 58.9(x) - P(100) = 0$$

Hence,

$$P = 32.4\text{ N}$$
$$N_B = 58.9\text{ N}$$
$$x = 55.0\text{ mm}$$

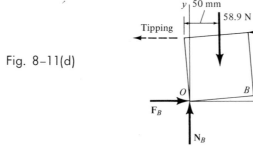

Fig. 8–11(d)

Tipping of B.   It can be seen from the above solution that since $x > 50$ mm, which is half the width of the block, tipping will occur *before* the block can slip. Using the free-body diagram shown in Fig. 8–11d to investigate this situation, we have

$$\xrightarrow{+} \Sigma F_x = 0; \qquad\qquad F_B - P = 0$$
$$+\uparrow \Sigma F_y = 0; \qquad\qquad N_B - 58.9 = 0$$
$$\zeta + \Sigma M_O = 0; \qquad\quad -P(100) + 58.9(50) = 0$$

so that

$$P = 29.5 \text{ N}$$
$$N_B = 58.9 \text{ N}$$
$$F_B = 29.5 \text{ N}$$

The smallest magnitude of force **P** calculated is $P = 29.5$ N. As a result, when $P > 29.5$ N, block B will begin to tip.

Problems

Assume that $\mu = \mu_s = \mu_k$ in the following problems where specified.

8–1. The crate has a mass of 200 kg and is subjected to a towing force **P** acting at an angle $\theta = 20°$ with the horizontal. If $\mu_s = 0.5$, determine the magnitude of **P** to just start the crate moving down the plane.

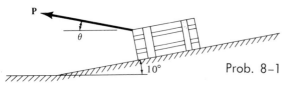

Prob. 8–1

8–2. If the coefficient of kinetic friction between the inclined surface and the crate in Prob. 8–1 is $\mu_k = 0.3$,

determine the angle θ of the force $P = 600$ N that will allow the 200-kg crate to move down the plane with constant velocity.

8–3. Specify the angle θ and the magnitude P of the smallest force needed to tow the 50-kg crate up the incline. The coefficient of friction between the surface of contact and the crate is $\mu = 0.5$.

Prob. 8–3

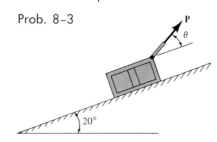

*8-4. A cable is wrapped around the inner core of a spool. Determine the magnitude of the vertical force **P** that must be applied to the end of the cable to rotate the spool. The coefficient of friction between the spool and the contacting surfaces at A and B is $\mu = 0.6$. The spool has a mass of 150 kg.

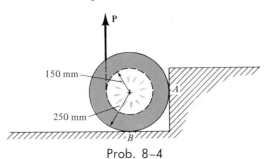

Prob. 8-4

8-5. The coiled belt spring is used as a feeding device for small boxes of merchandise, each box having a mass of $m = 1.5$ kg. If the coil exerts a *constant* force of $P = 30$ N on the box at A, determine the smallest number of boxes that can be kept under constant pressure by the device without any movement of the belt. The coefficient of friction between the belt and the ground is $\mu_g = 0.4$, and between the belt and each box $\mu_b = 0.35$. Neglect tipping of the boxes, the weight of the belt, and any friction that it exerts on A.

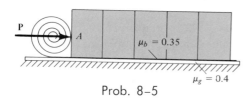

Prob. 8-5

8-5a. Solve Prob. 8-5 if each box has a weight of $W = 4$ lb and $P = 10$ lb, $\mu_g = 0.4$, and $\mu_b = 0.35$.

8-6. A winch is mounted at E on the front of a 2.5-Mg pickup truck. As the cable is drawn in, it loads a crate onto the truck. Determine the largest mass m of the crate that can be loaded without causing the truck to move. The truck is braked only at its rear wheels and has a center of mass at G. The coefficient of static friction between the wheels and the ground is $\mu_w = 0.4$, and

between the crate and the ground $\mu_c = 0.5$. The pulley at D is frictionless.

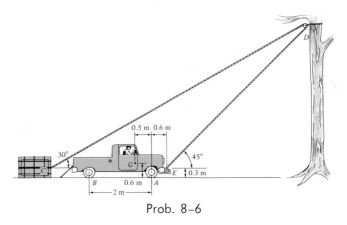

Prob. 8-6

8-7. Using the data in Prob. 8-6, solve the problem assuming that all four wheels of the truck are braked. $\mu_w = 0.4$ between the wheels and the ground.

*8-8. The mine car and its contents have a total mass of 6 Mg and a center of gravity at G. If the coefficient of static friction between the wheels and the tracks is $\mu_s = 0.4$ when the wheels are locked, find the normal force acting on the front wheels at B and the rear wheels at A when (a) only the brakes at A are locked, and (b) the brakes at both A and B are locked. Does the car move when only the brakes at A are locked?

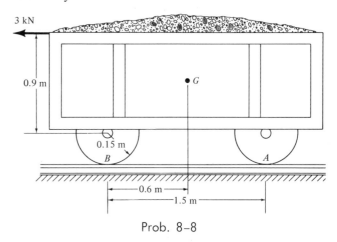

Prob. 8-8

8–9. A uniform beam has a mass of 15 kg and rests on two surfaces at points A and B. Determine the maximum distance x to which the girl can slowly walk up the beam before it begins to slip. The girl has a mass of 53 kg and walks up the beam with a constant velocity.

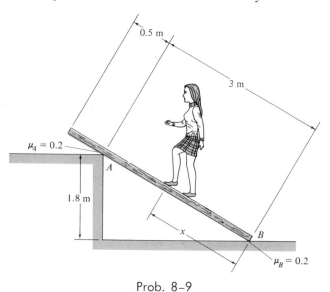

Prob. 8–9

8–10a. Solve Prob. 8–10 if each of the links has a weight of $W = 15$ lb and $l = 2$ ft.

8–11. A 20-kg ladder has a center of mass at G. If the coefficients of friction at A and B are $\mu_A = 0.3$ and $\mu_B = 0.2$, respectively, determine the smallest horizontal force that the man must exert on the ladder at point C in order to push the ladder forward.

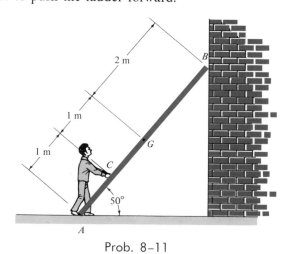

Prob. 8–11

8–10. The two rods each have a mass of $m = 10$ kg, and they are pinned together at B. If the coefficient of static friction at C is $\mu_s = 0.5$, determine the maximum angle θ for equilibrium. Set $l = 0.5$ m.

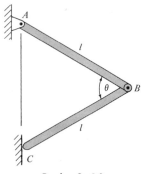

Prob. 8–10

***8–12.** Solve Prob. 8–11 if the wall at B is smooth, i.e., $\mu_B = 0$.

8–13. The spool has a mass of 20 kg. If a cord is wrapped around its inner core and is attached to the wall, and the coefficient of friction at A is $\mu = 0.15$, determine if the spool remains in equilibrium when it is released.

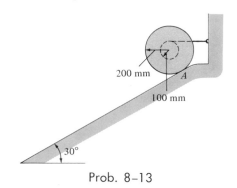

Prob. 8–13

8-14. A 20-kg disk rests on an inclined surface for which $\mu_s = 0.2$. Determine the maximum vertical force **P** that may be applied to link AB without causing the disk to slip at C.

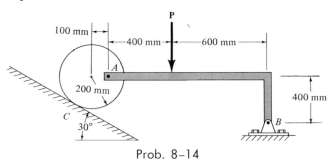

Prob. 8-14

8-15. A single force **P** is applied to the handle of the drawer. If friction is neglected at the bottom side and the coefficient of static friction along the sides is $\mu_s = 0.4$, determine the largest spacing s between the handles so that the drawer does not bind at the corners A and B when the force **P** is applied to one of the handles. Set $a = 1.25$ m and $b = 0.3$ m.

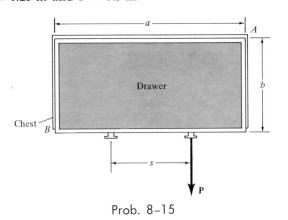

Prob. 8-15

8-15a. Solve Prob. 8-15 if $a = 4$ ft, $b = 1.5$ ft, and $\mu_s = 0.6$.

*8-16. Two blocks A and B, each having a mass of 10 kg, are connected by the linkage shown. If the coefficient of static friction at the contacting surfaces is $\mu_s = 0.5$, determine the largest vertical force **P** that may

be applied to pin C of the linkage without causing the blocks to move. Neglect the weight of the links.

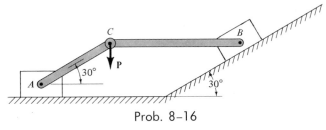

Prob. 8-16

8-17. Determine the minimum force **F** needed to push the two 100-kg cylinders up the plane. The force acts parallel to the plane and the coefficients of friction at the contacting surfaces are $\mu_A = 0.3$, $\mu_B = 0.25$, and $\mu_C = 0.4$. Each cylinder has a radius of 150 mm.

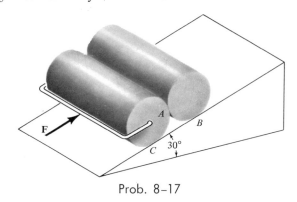

Prob. 8-17

8-18. Determine the compressive force acting on A if a 100-N force is applied to the wedge at C. Neglect friction at B and consider the coefficient of friction at the contacting surfaces of the wedge to be $\mu = 0.3$.

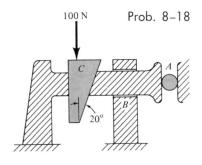

Prob. 8-18

8–19. Determine the force **P** that must be applied to the 20-kg block *B* to lift the 100-kg block *A*. The coefficient of friction for all contacting surfaces is $\mu = 0.3$.

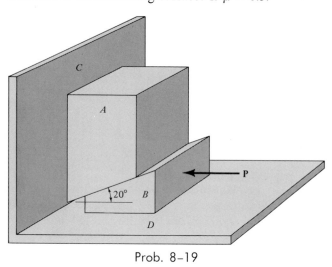

Prob. 8–19

****8–20.** The compound beam is adjusted into the horizontal position by means of a wedge located at its right support. If the coefficient of friction between the wedge and the surfaces of contact is $\mu = 0.25$, determine the horizontal force **P** required to push the wedge forward. Set $F = 10$ kN, $a = 7$ m, and $b = 3$ m.

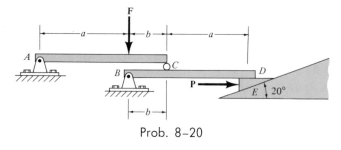

Prob. 8–20

****8–20a.** Solve Prob. 8–20 if $F = 300$ lb, $a = 6$ ft, $b = 2.5$ ft, and $\mu = 0.3$.

8–21. If the vertical force acting on the bracket at *A* is 500 N, determine the horizontal force **P** required to push the wedge to the left. Neglect the weight of the wedge

and the bracket. The coefficient of friction between all the contacting surfaces of the wedge is $\mu = 0.2$.

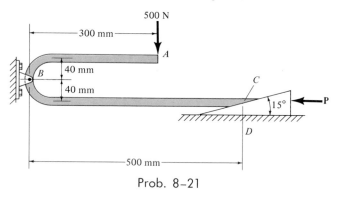

Prob. 8–21

8–22. A 5-kg wedge is placed in the grooved slot of an inclined plane. Determine the maximum angle θ for the incline without causing the wedge to slip. The coefficient of static friction between the wedge and the surfaces of contact is $\mu_s = 0.2$.

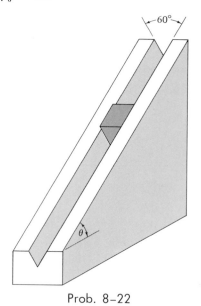

Prob. 8–22

8–23. Determine the minimum applied force **P** required to move wedge *A*. The spring is compressed a distance of 200 mm. Neglect the weight of *A* and *B*. The coefficient

316

of static friction for all contacting surfaces is $\mu_s = 0.35$. Neglect friction at the rollers.

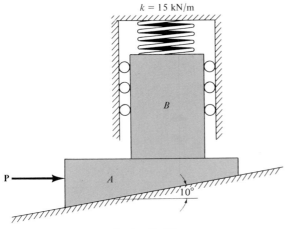

Prob. 8-23

*8-24. If the coefficient of static friction between the drum and brake mechanism is $\mu_s = 0.4$, determine the horizontal and vertical components of reaction at the pin O. Does the 300-N force prevent the drum from rotating? Neglect the weight and thickness of the brake. The drum has a mass of 15 kg.

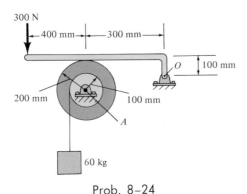

Prob. 8-24

8-25. Gravel is stored in a conical pile at a materials yard. If the height of the pile is $h = 10$ m and the coefficient of static friction between the gravel particles is $\mu_s = 0.4$, determine the approximate diameter d of the

pile. For the calculation, neglect the "irregularities" of the particles, and first determine the angle θ of the pile by considering one of the particles to be represented by a *block* resting on an inclined plane of angle θ, where $\mu = 0.4$.

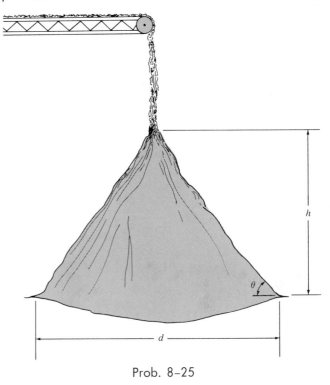

Prob. 8-25

8-25a. Solve Prob. 8-25 if $\mu = 0.3$ and $h = 20$ ft.

8-26. A man attempts to support a stack of books horizontally by applying a compressive force of $F = 225$ N on the ends of the stack with his hands. If each book has a mass of 0.95 kg, determine the greatest number of books that can be supported in the stack. The coefficient of friction between the man's hands and a book is $\mu_h = 0.45$ and between any two books, $\mu_b = 0.4$.

Prob. 8-26

8-27. The beam is supported by a pin at A and a roller having a radius of 20 mm at B. If the coefficient of static friction between the roller and the beam and inclined plane is $\mu_B = \mu_C = 0.2$, determine the largest angle θ of the incline so that the roller does not slip for any force P applied to the beam.

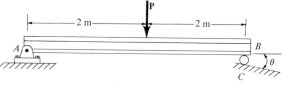

Prob. 8-27

* **8-28.** A roofer, having a mass of 70 kg, walks slowly in an upright position down along the surface of a dome that has a radius of curvature of $r = 20$ m. If the coefficient of friction between his shoes and the dome is $\mu = 0.7$, determine the angle θ at which he first begins to slip.

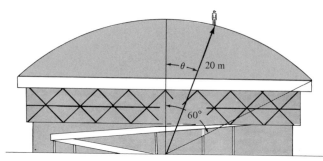

Prob. 8-28

8-29. The wedge has a negligible weight and a coefficient of static friction $\mu_s = 0.2$ with all contacting surfaces. Determine the angle θ so that it is "self-locking." This requires no slipping for any magnitude of the force P applied to the joint.

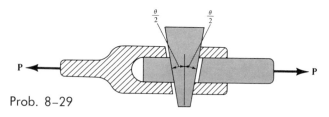

Prob. 8-29

8-30. To prevent clockwise rotation of the wheel A, a small cylinder of negligible weight is placed between the wheel and the wall. If the coefficient of static friction at the points of contact B and C is $\mu_s = 0.3$, determine the largest radius r of the cylinder, which is so placed to lock the wheel and keep it from rotating when any moment M is applied to the wheel. Set $a = 225$ mm and $R = 200$ mm.

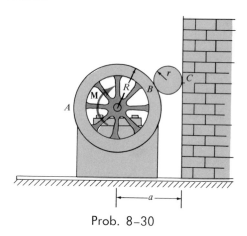

Prob. 8-30

8-30a. Solve Prob. 8-30 if $a = 1.5$ ft, $R = 1.25$ ft, and $\mu_s = 0.25$.

8-31. An ax is driven into the tree stump. If the coefficient of static friction between the ax and the wood is $\mu_s = 0.15$, determine the smallest angle θ of the blade that will cause the ax to be "self-locking," i.e., so it will not slip out. Neglect the weight of the ax.

Prob. 8-31

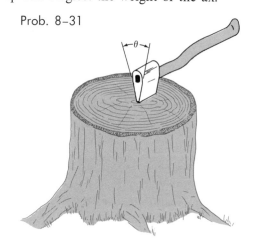

*8–32. The friction grip is used to prevent the plate p from moving horizontally to the right by placing the loosely fitting cylinder C of negligible weight up against the plate. If the coefficient of static friction between all points of contact is $\mu_s = 0.3$, determine the design angle θ of the *jaws* that will prevent movement of the plate for any force **P** applied to the plate.

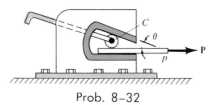

Prob. 8–32

8–33. The uniform 60-kg crate C rests uniformly on a 10-kg dolly D. If the front casters of the dolly at A are locked to prevent rolling, determine the maximum force **P** that may be applied without causing motion of the crate. The coefficient of static friction between the casters and the floor is $\mu_f = 0.35$ and between the dolly and the crate, $\mu_d = 0.5$.

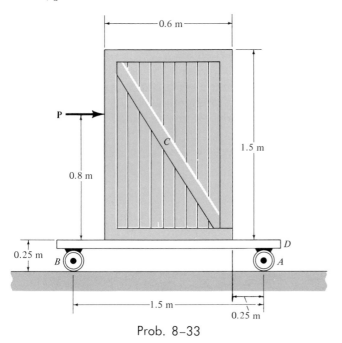

Prob. 8–33

8–34. Using the data in Prob. 8–33, solve the problem assuming that all four casters of the dolly are braked, where $\mu_f = 0.35$.

8–35. The block has a mass of $m = 5$ kg and a height $h = 100$ mm. If it is subjected to a horizontal force **P**, determine the width w so that the block slips and tips at the same time. What is the magnitude of **P** required to do this?

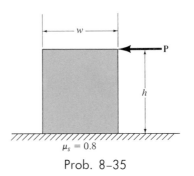

Prob. 8–35

8–35a. Solve Prob. 8–35 if the block has a weight of $W = 25$ lb and $h = 5$ in.

*8–36. Block C has a mass of 60 kg and is confined between two walls by smooth rollers. If the block rests on top of a spool that has a mass of 50 kg, determine the minimum cable force **P** needed to move the spool. The cable is wrapped around the spool's inner core. The coefficients of friction at A and B are $\mu_A = 0.3$ and $\mu_B = 0.5$.

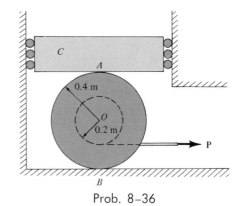

Prob. 8–36

8-37. The beam AB has a mass of 20 kg and is subjected to a force of 250 N. It is supported at one end by a pin and at the other end by a spool having a mass of 35 kg. If a cable is wrapped around the inner core of the spool, determine the minimum cable force **P** needed to move the spool from under the beam. The coefficients of friction at B and D are $\mu_B = 0.4$ and $\mu_D = 0.2$, respectively.

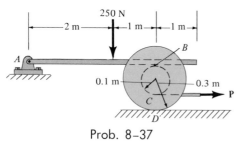

Prob. 8–37

8-38. The block has a mass of $m = 5$ kg and is attached to a light rod AD that pivots at pin A. If the coefficient of static friction between the plane and the block is $\mu_s = 0.4$, determine the minimum angle θ at which the block may be placed on the plane without slipping. Neglect the *size* of the block in the calculation. Set $a = 200$ mm, $b = 300$ mm, and $c = 400$ mm.

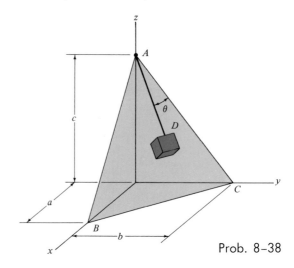

Prob. 8–38

*8–3. Frictional Forces on Screws

A *screw* may be thought of simply as an inclined plane wrapped around a cylinder. A nut initially at position A on the screw shown in Fig. 8–12 will move up to B when rotated 360° around the screw. This rotation is equivalent to translating the nut up an inclined plane of height p and length $l = 2\pi r$, where r is the mean radius of the thread. The rise p is often referred to as the *pitch* of the screw, where the *pitch angle* is given by $\theta_p = \tan^{-1}(p/2\pi r)$.

In most cases screws are used as fasteners; however, in many types of

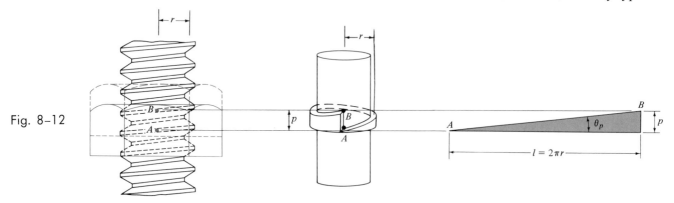

Fig. 8–12

machines screws are incorporated to transmit power or motion from one part of the machine to another. A *square-threaded screw* is most commonly used for the latter purpose, especially when large forces are applied along the axis of the screw.

Frictional Analysis. When a screw is subjected to large axial loads, the frictional forces developed at the thread become important in the analysis to determine the force needed to turn the screw. Consider, for example, the square-threaded jack screw shown in Fig. 8–13, which supports the vertical load **W** and twisting moment **M**.* If moments are summed about the axis of the screw, **M** can be thought of as being equivalent to the moment of a horizontal force **S** acting at the mean radius r of the thread, so that $M = Sr$. The reactive forces of the jack to loads S and W are actually distributed over the circumference of the screw thread in contact with the screw hole in the jack, that is, within region h shown in Fig. 8–13. For simplicity, this portion of thread can be imagined as being unwound from the screw and represented as a simple block resting on an inclined plane having the screw's pitch angle θ_p, Fig. 8–14a. The inclined plane represents the inside *supporting thread* of the jack base. The block is subjected to the total axial load **W** acting on the jack, the horizontal force **S** (which is related to the applied moment **M**), and the *resultant force* **R**, which the plane exerts on the block. As shown, force **R** has components acting normal, **N**, and tangent, **F**, to the contacting surface.

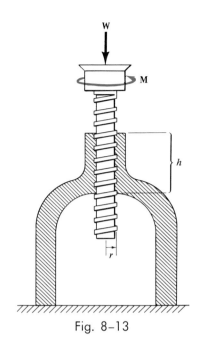

Fig. 8–13

Upward Screw Motion. Provided M is great enough, the screw (and hence the block) is either on the verge of upward impending motion, or motion is occurring. Under these conditions, **R** acts at an angle $(\phi + \theta_p)$ from the vertical as shown in the figure, where $\phi = \tan^{-1}(F/N) = \tan^{-1}(\mu N/N) = \tan^{-1}\mu$. Applying the two force equations of equilibrium to the block, we obtain

$$\xrightarrow{+}\Sigma F_x = 0; \qquad S - R\sin(\phi + \theta_p) = 0$$
$$+\uparrow\Sigma F_y = 0; \qquad R\cos(\phi + \theta_p) - W = 0$$

Eliminating R and solving for S, then substituting this value into the equation $M = Sr$, yields

$$\boxed{M = Wr\tan(\phi + \theta_p)} \qquad (8\text{–}3)$$

This equation gives the required value M necessary to cause upward impending motion of the screw when $\phi = \phi_s = \tan^{-1}\mu_s$ (the angle of static friction). If ϕ is replaced by $\phi_k = \tan^{-1}\mu_k$ (the angle of kinetic friction), Eq. 8–3 would give a smaller value M necessary to maintain uniform upward motion of the screw.

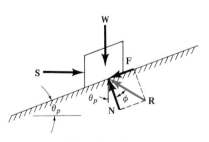

Fig. 8–14(a)

*For applications, **M** is developed by applying a horizontal force **P** at right angles to the end of a lever that would be fixed to the screw.

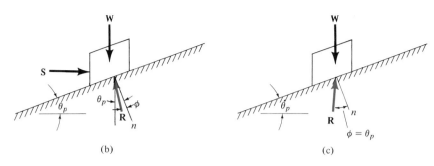

Fig. 8–14(b, c)

(b)   (c)

Downward Screw Motion. When the load **W** is to be *lowered*, the direction of **M** is reversed, in which case the angle ϕ (ϕ_s or ϕ_k) lies on the opposite side of the normal *n* to the plane supporting the block. This case is shown in Fig. 8–14b for $\phi < \theta_p$. Thus, Eq. 8–3 becomes

$$M = Wr \tan(\phi - \theta_p) \qquad (8\text{–}4)$$

Self-Locking Screw. If the moment **M** is *removed*, the screw will remain *self-locking;* i.e., it will support the weight **W** *by friction forces alone* provided $\phi > \theta_p$. However, if $\phi = \phi_s = \theta_p$, Fig. 8–14c, the screw will be on the verge of rotating downward. When $\theta_p > \phi_s$, a restraining load or moment is needed to prevent the screw from rotating downward.

Example 8–7

The turnbuckle shown in Fig. 8–15 has a square thread with a mean radius of 5 mm and a pitch of 2 mm. If the coefficient of friction between the screw and the turnbuckle is $\mu_s = 0.25$, determine the moment **M** that must be applied to draw the end screws closer together. Is the turnbuckle "self-locking"?

Solution

The moment *M* may be obtained by using Eq. 8–3. Since friction at two screws must be overcome, this requires

$$M = 2[Wr \tan(\phi_s + \theta_p)] \qquad (1)$$

Here, $W = 2000$ N, $r = 5$ mm, $\phi_s = \tan^{-1}\mu_s = \tan^{-1}(0.25) = 14.04°$, and $\theta_p = \tan^{-1}(p/2\pi r) = \tan^{-1}\{2 \text{ mm}/[2\pi(5 \text{ mm})]\} = 3.64°$. Substituting these values into Eq. (1) and solving gives

$$M = 2[(2000 \text{ N})(5 \text{ mm}) \tan(14.04° + 3.64°)]$$
$$M = 6375.1 \text{ N} \cdot \text{mm} = 6.38 \text{ N} \cdot \text{m} \qquad \qquad Ans.$$

When the moment *M* is *removed*, the turnbuckle will be self-locking; i.e., it will not unscrew, since $\phi_s > \theta_p$.

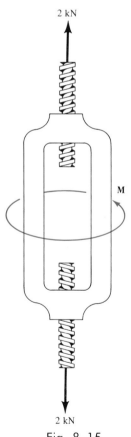

2 kN

M

2 kN

Fig. 8–15

*8–4. Frictional Forces on Collar Bearings, Pivot Bearings, and Disks

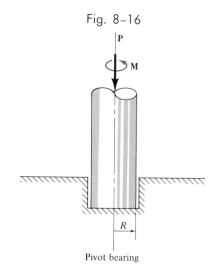

Fig. 8–16

Pivot or *collar bearings* are commonly used to support the *axial* or *normal loads* of a rotating shaft. These two types of support are shown in Fig. 8–16. Provided the bearings are not lubricated, or only partially lubricated, the laws of dry friction may be applied to determine the required moment **M** needed to turn the shaft when the shaft supports an axial force **P**.

Frictional Analysis. Consider, for example, the end of a collar-bearing shaft shown in Fig. 8–17, which is subjected to an axial force **P** and has a total bearing or contact area $\pi(R_2^2 - R_1^2)$. In the following analysis the normal pressure p is considered to be *uniformly distributed* over this area—a reasonable assumption provided the bearing is new and evenly supported. Since $\Sigma F_z = 0$, p, measured as a force per unit area, is

$$p = \frac{P}{\pi(R_2^2 - R_1^2)}$$

The moment **M,** needed to cause impending rotation of the shaft, can be determined from moment equilibrium of the frictional forces $d\mathbf{F}$ developed at the bearing surface by applying $\Sigma M_z = 0$. A small area element $dA = (r\, d\theta)(dr)$, shown in Fig. 8–17, is subjected to a normal force

$$dN = p\, dA = \frac{P}{\pi(R_2^2 - R_1^2)} dA$$

Pivot bearing

(a)

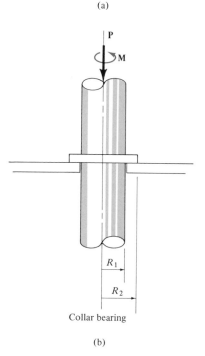

Collar bearing

(b)

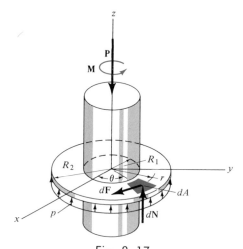

Fig. 8–17

and a frictional force

$$dF = \mu_s \, dN = \frac{\mu_s P}{\pi(R_2^2 - R_1^2)} dA$$

The normal force dN does not create a moment about the z axis of the shaft; however, the frictional force does, namely $dM = r \, dF$. Integration is needed to compute the total moment created by all the frictional forces acting on differential areas dA. Therefore, for impending rotational motion,

$$\Sigma M_z = 0; \qquad\qquad M - \int_A r \, dF = 0$$

Using $dA = (r \, d\theta)(dr)$ and integrating over the entire bearing area yields

$$M = \int_{R_1}^{R_2} \int_0^{2\pi} r \left[\frac{\mu_s P}{\pi(R_2^2 - R_1^2)} \right] (r \, d\theta \, dr) = \frac{\mu_s P}{\pi(R_2^2 - R_1^2)} \int_{R_1}^{R_2} r^2 \, dr \int_0^{2\pi} d\theta$$

or

$$M = \tfrac{2}{3}\mu_s P \left(\frac{R_2^3 - R_1^3}{R_2^2 - R_1^2} \right) \qquad\qquad (8\text{-}5)$$

This equation gives the magnitude of moment $\mathbf{M}$ required for impending rotation of the shaft. The frictional moment developed at the end of the shaft, when it is *rotating* at constant speed, can be found by substituting μ_k for μ_s in Eq. 8–5.

When $R_2 = R$ and $R_1 = 0$, as in the case of a pivot bearing, Fig. 8–16a, Eq. 8–5 reduces to

$$M = \tfrac{2}{3}\mu_s P R \qquad\qquad (8\text{-}6)$$

Recall from the initial assumption, that both Eqs. 8–5 and 8–6 apply only for bearing surfaces subjected to *constant pressure*. If the pressure is not uniform, a variation of the pressure as a function of the bearing area must be determined before integrating to obtain the moment M. The following example illustrates this concept.

Example 8–8

The uniform bar shown in Fig. 8–18a has a total mass m. If it is assumed that the normal pressure acting at the contacting surface varies linearly along the length of the bar as shown, determine the couple $\mathbf{M}$ required to rotate the bar. Assume that the bar's width s is negligible in comparison to its length l. The coefficient of friction is equal to μ_s.

Solution

A free-body diagram of the bar is shown in Fig. 8–18b. Since the bar has a total weight of $W = mg$, the intensity w_o of the distributed load at the

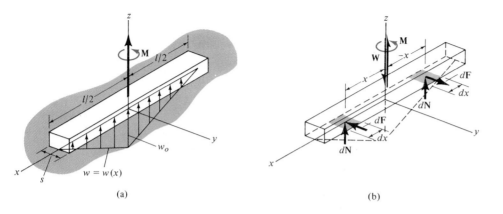

(a)

(b)

Fig. 8–18

center ($x = 0$) is determined from vertical force equilibrium.

$$+\uparrow \Sigma F_z = 0; \quad -mg + 2\left[\frac{1}{2}\left(\frac{l}{2}\right)w_o\right] = 0 \quad w_o = \frac{2mg}{l}$$

Since $w = 0$ at $x = l/2$, the distributed load expressed as a function of x is

$$w = w_o\left(1 - \frac{2x}{l}\right) = \frac{2mg}{l}\left(1 - \frac{2x}{l}\right)$$

The magnitude of the force dN acting on a segment of area having a length dx is therefore

$$dN = w\,dx = \frac{2mg}{l}\left(1 - \frac{2x}{l}\right)dx$$

The magnitude of the frictional force dF acting on the same element of area is

$$dF = \mu\,dN = \frac{2\mu mg}{l}\left(1 - \frac{2x}{l}\right)dx$$

Hence, the moment created by this force about the z axis is

$$dM = x\,dF = \frac{2\mu mg}{l}x\left(1 - \frac{2x}{l}\right)dx$$

The summation of moments about the z axis of the bar is determined by integration, which yields

$$\Sigma M_z = 0; \qquad M - 2\int_0^{l/2} \frac{2\mu mg}{l}x\left(1 - \frac{2x}{l}\right)dx = 0$$

$$M = \frac{4\mu mg}{l}\left(\frac{x^2}{2} - \frac{2x^3}{3l}\right)\bigg|_0^{l/2}$$

$$M = \frac{\mu mgl}{6} \qquad\qquad\qquad Ans.$$

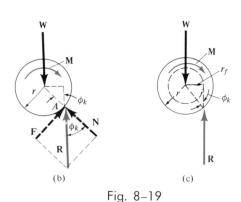

(a)

(b)   (c)

Fig. 8–19

When a shaft or axle is subjected to lateral loads, a *journal bearing* is commonly used for support. Well-lubricated journal bearings are subjected to the laws of fluid mechanics, in which the viscosity of the lubricant, the speed of rotation, and the amount of clearance between the shaft and bearing are needed to determine the frictional resistance of the bearing. When the bearing is not lubricated, or only partially lubricated, however, a reasonable analysis of the frictional resistance can be based on the laws of dry friction.

Frictional Analysis. A typical journal-bearing support is shown in Fig. 8–19a. As the shaft rotates in the direction shown in the figure, it rolls up against the wall of the bearing to some point A where slipping occurs. If the lateral load acting at the end of the shaft is **W**, it is necessary that the bearing reactive force **R** acting at A be equal and opposite to **W**, Fig. 8–19b. The moment **M** needed to maintain rotation of the shaft can be found by summing moments about the z axis of the shaft; i.e.,

$$\Sigma M_z = 0; \qquad -M + (R \sin \phi_k)r = 0$$

or

$$M = Rr \sin \phi_k$$

where ϕ_k is the angle of kinetic friction defined by $\tan \phi_k = F/N = \mu_k N/N = \mu_k$. In Fig. 8–19c, it is seen that $r \sin \phi_k = r_f$. The dashed circle with radius r_f is called the *friction circle,* and as the shaft rotates, the reaction **R** will always be tangent to it. If the bearing is partially lubricated, μ_k is small, and therefore $\mu_k = \tan \phi_k \approx \sin \phi_k \approx \phi_k$. Under these conditions, a reasonable approximation to the moment needed to overcome the frictional resistance becomes

$$M \approx Rr\mu_k \tag{8–7}$$

The following example illustrates a common application of this equation.

Example 8–9

The 100-mm-diameter pulley shown in Fig. 8–20a fits loosely on a 10-mm-diameter shaft for which the coefficients of static and kinetic friction are $\mu = 0.4$. Determine the minimum tension **T** in the belt to (a) raise the 100-kg block at constant velocity, and (b) lower the block at constant velocity. Assume no slipping occurs between the belt and pulley and neglect the weight of the pulley.

Solution
Part (a). A free-body diagram of the pulley is shown in Fig. 8–20b. When

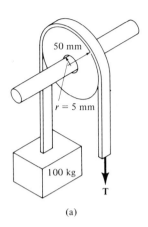

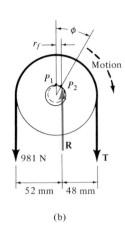

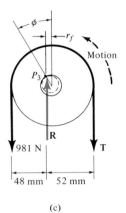

Fig. 8-20

(a)   (b)   (c)

the pulley is subjected to cable tensions of 981 N each, the pulley makes contact with the shaft at point P_1. As the tension **T** is *increased,* the pulley will roll around the shaft to point P_2 before motion impends. From the figure, the friction circle has a radius $r_f = r \sin \phi$. Using the simplification $\sin \phi \approx \mu$, $r_f \approx r\mu = (5 \text{ mm})(0.4) = 2 \text{ mm}$, so that summing moments about P_2 gives

$$\zeta + \Sigma M_{P_2} = 0; \qquad 981 \text{ N}(52 \text{ mm}) - T(48 \text{ mm}) = 0$$
$$T = 1062.8 \text{ N} \qquad\qquad Ans.$$

Part (b). When the block is lowered, the resultant force **R** acting on the shaft passes through point P_3, as shown in Fig. 8-20c. Summing moments about this point yields

$$\zeta + \Sigma M_{P_3} = 0; \qquad 981 \text{ N}(48 \text{ mm}) - T(52 \text{ mm}) = 0$$
$$T = 905.5 \text{ N} \qquad\qquad Ans.$$

Problems

8-39. Prove that the pitch p must be less than $2\pi r\mu_s$ for the jack screw shown in Fig. 8-13 to be "self-locking."

*** 8-40.** The square-threaded bolt is used to join two plates together. If the bolt has a mean diameter of $d = 20$ mm and a pitch of $p = 3$ mm, determine the smallest torque **M** required to loosen the bolt if the tension in the bolt is $T = 40$ kN. The coefficient of static friction between the threads and the bolt is $\mu_s = 0.15$.

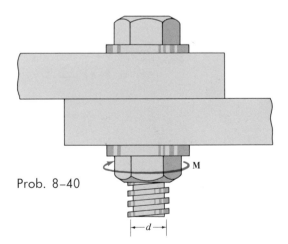

Prob. 8-40

*8-40a. Solve Prob. 8-40 if $d = 0.75$ in., $p = 0.2$ in., $T = 3000$ lb, and $\mu_s = 0.3$.

8-41. The square-threaded screw of the vise clamps has a mean diameter of 20 mm and a pitch of 4 mm. If $\mu_s = 0.25$ for the threads, and the turning force applied perpendicular to the handle is $P = 100$ N, determine the compressive force in the block.

8-42. The automobile jack is subjected to a vertical load of $F = 8$ kN. If a square-threaded screw, having a pitch of 5 mm and a mean diameter of 10 mm, is used in the jack, determine the force that must be applied perpendicular to the handle to (a) raise the load, and (b) lower the load; $\mu = 0.2$. The supporting plate exerts *only* vertical forces at A and B, and each cross link has a total length of 200 mm.

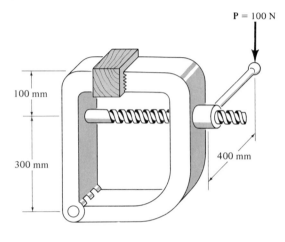

P = 100 N

100 mm

300 mm

400 mm

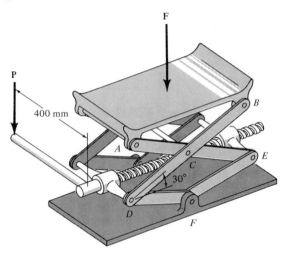

F

P

400 mm

B

A

E

C

30°

D

F

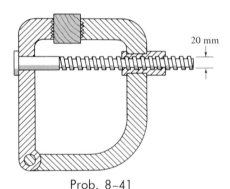

20 mm

Prob. 8-41

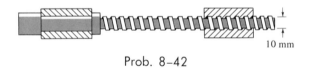

10 mm

Prob. 8-42

8-43. The *universal testing machine* is used to measure the strength of a specimen S when the specimen is subjected to large tensile or compressive loads. These forces are applied by raising or lowering the movable head H of the machine. The head has a mass of 200 kg and rests on two square-threaded screws, each with a mean diameter of 100 mm and a pitch of 10 mm. The coefficient of static friction between each screw and the head is $\mu_s = 0.2$. If the specimen is compressed in the machine with a force of 12 kN, which is indicated on the dial, determine the moment (N · m) that must be applied along the axis of each screw by the motor of the machine to increase the load. Neglect friction along the four guide posts G. The head H can slide freely along the four guide posts, whereas the block C cannot.

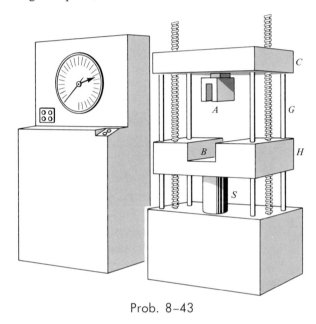

Prob. 8–43

*8–44. If a tensile specimen is attached at A and B in the testing machine of Prob. 8–43, determine the moment that must be applied along the axis of each screw by the machine to create a tensile force in the specimen of 24 kN.

8–45. The collar bearing uniformly supports an axial force of $P = 5$ kN. If a moment of $M = 40$ N · m is required to overcome friction developed by the bearing,

determine the coefficient of friction μ acting at the bearing surface. Set $d_1 = 40$ mm and $d_2 = 60$ mm.

Prob. 8–45

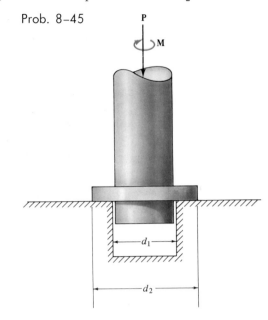

8–45a. Solve Prob. 8–45 if $P = 1500$ lb, $M = 160$ lb · ft, $d_1 = 3$ in., and $d_2 = 5$ in.

8–46. The *double-collar bearing* is subjected to an axial force $P = 4$ kN. Assuming that collar A supports $0.75P$ and collar B supports $0.25P$, both with a uniform distribution of pressure, determine the maximum frictional moment **M** that may be resisted by the bearing. $\mu_s = 0.2$ for both collars.

Prob. 8–46

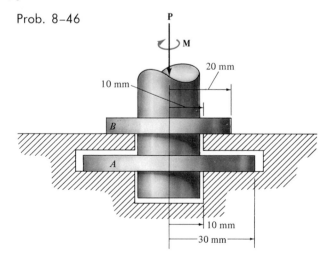

8-47. Determine the moment **M** that must be applied to the collar bearing in Prob. 8–46 to cause impending motion of the shaft. Assume that the force $P = 4$ kN is distributed equally to each collar. The coefficient of friction for collar A is $\mu_s = 0.2$, and for collar B, $\mu_s = 0.3$.

****8-48.** The *disk clutch* is used in standard automobile transmissions. If the four springs force the two plates A and B together, determine the force in each spring required to transmit a moment of $M = 600$ N · m across the plates. Assume the pressure created is uniformly distributed between the plates. The coefficient of friction is $\mu = 0.7$.

8-50. A tube has a total mass of $m = 80$ kg, length $l = 3$ m, and radius $R = 0.75$ m. If it rests in sand for which the coefficient of friction is $\mu = 0.23$, determine the moment **M** needed to turn it. Assume that the pressure distribution along the length of the tube is defined by $p = p_o \sin \theta$. For the solution it is necessary to determine p_o, the peak pressure, in terms of the mass and tube dimensions.

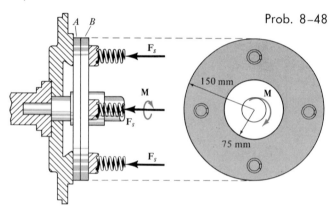

Prob. 8–48

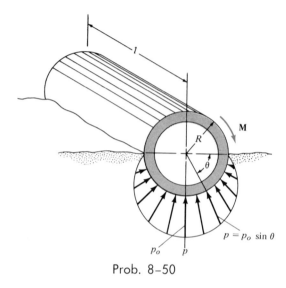

Prob. 8–50

8-49. The plate A is engaged to plate B using three springs. If each spring exerts a force of 750 N on plate B and $\mu = 0.35$ for the contacting surfaces, determine the maximum moment **M** that may be transmitted across the plates. Assume the pressure created is uniformly distributed between the plates.

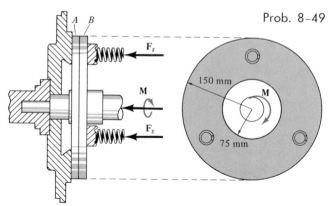

Prob. 8–49

8-50a. Solve Prob. 8–50 if the tube has a weight $W = 150$ lb, and $l = 10$ ft, $R = 1.5$ ft. Set $\mu = 0.3$ and define p_o in terms of the weight and tube dimensions.

8-51. Because of wearing at the edges, the pivot bearing is subjected to a conical pressure distribution at its surface of contact. Determine the moment **M** required to turn the shaft which supports an axial force **P.** The coefficient of friction is μ. For the solution, it is necessary to determine the peak pressure p_o in terms of P and the bearing radius R.

Prob. 8-51

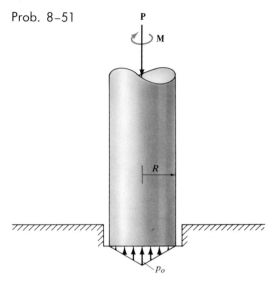

*8-52. Assuming that the variation of pressure at the bottom of the pivot bearing is defined as $p = p_o(R_2/r)$, determine the moment **M** needed to rotate the shaft if the applied axial force is **P**. The coefficient of friction is μ. For the solution, it is necessary to determine p_o in terms of P and the bearing dimensions R_1 and R_2.

Prob. 8-52

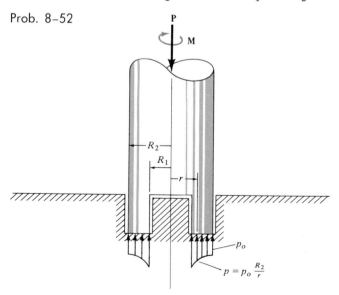

8-53. A 200-mm-diameter post is driven 3 m into sand for which $\mu = 0.15$. If the normal pressure acting *completely around the post* varies linearly with depth as shown, determine the frictional moment **M** that must be overcome to rotate the post.

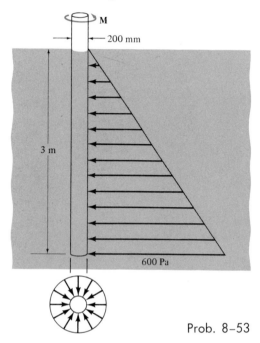

Prob. 8-53

8-54. The collar bushing B fits loosely over the fixed shaft S, which has a radius of 20 mm. Determine the smallest force $\mathbf{T}_A$ needed to pull the belt downward at A. Also determine the resultant normal and frictional components of force developed on the collar bushing if $\mu = 0.2$ between the collar and the shaft. Assume that the belt does not slip on the collar; rather, the collar slips on the shaft.

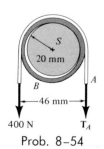

Prob. 8-54

8-55. A disk having an outer diameter of $d = 120$ mm fits loosely over a fixed shaft having a diameter of $d_s = 30$ mm. If the coefficient of friction between the disk and the shaft is $\mu = 0.15$, and the disk has a mass of $m = 50$ kg, determine the smallest vertical force **F** acting on the rim, which must be applied to the disk to cause it to slip over the shaft.

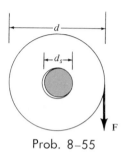

Prob. 8–55

8-55a. Solve Prob. 8–55 if the disk weighs $W = 100$ lb, and $d = 12$ in., $d_s = 3$ in., $\mu = 0.25$.

*8-56. If the smallest tension force $T_A = 480$ N is required to pull the belt downward at A over the shaft S, determine the coefficient of friction between the loosely fitting collar bushing and the shaft. Assume the belt does not slip on the collar; rather, the collar slips on the shaft.

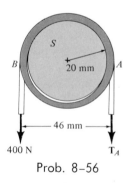

Prob. 8–56

8-57. The axle of the 200-mm-diameter pulley fits loosely in a 50-mm-diameter pin hole. If $\mu = 0.30$, determine the minimum tension **T** required to raise the

40-kg load. Neglect the weight of the pulley and assume that the cord does not slip on the pulley.

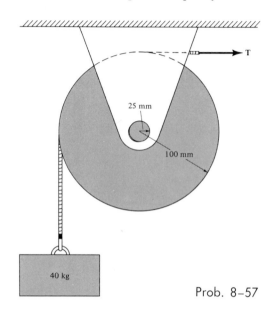

Prob. 8–57

8-58. The collar fits *loosely* around a fixed shaft that has a radius of 50 mm. If the coefficient of friction between the shaft and the collar is $\mu = 0.3$, determine the largest tension **T** in the horizontal segment of the belt so that the belt can be lowered in the direction of the 400-N force with a constant speed. Assume that the belt does not slip on the collar; rather, the collar slips on the shaft. Neglect the weight and thickness of the belt and collar. The radius, measured from the center of the collar to the mean thickness of the belt, is 54 mm.

Prob. 8–58

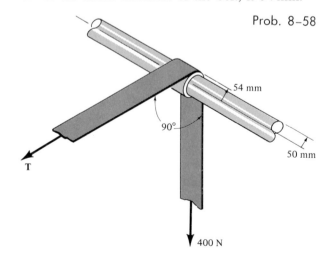

*8-6. Frictional Forces on Flat Belts

In the design of belt drives or band brakes it is necessary to determine the frictional forces developed between a belt and its contacting surface. Consider the flat belt passing over a fixed drum of radius r, as shown in Fig. 8-21a. The *total* angle of belt contact in radians is β, and the coefficient of friction between the two surfaces is μ. If it is known that the tension acting in the belt on the right of the drum is $\mathbf{T_1}$, let it be required to find the tension $\mathbf{T_2}$ needed to pull the belt counterclockwise over the surface of the drum. *Obviously, T_2 must be greater than T_1 since the belt must overcome the resistance of friction at the surface of contact.*

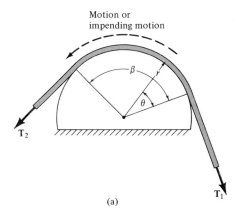

(a)

Frictional Analysis. A free-body diagram of the *entire belt* is shown in Fig. 8-21b. It is seen that the normal force $\mathbf{N}$ and the frictional force $\mathbf{F}$, acting at different points along the contacting surface of the belt, will vary both in magnitude and direction. Due to this unknown force distribution, the analysis of the problem will proceed on the basis of initially studying the forces acting on a differential element of the belt.

A free-body diagram of an element having a length $ds = r\,d\theta$ is shown in Fig. 8-21c. Assuming either impending motion or motion of the belt, the magnitude of frictional force $dF = \mu\,dN$. This force opposes the sliding motion of the belt and thereby increases the magnitude of tensile force acting in the belt by dT. Applying the two force equations of equilibrium, we have

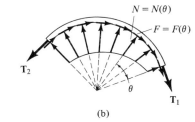

(b)

$$\xrightarrow{+}\Sigma F_x = 0; \quad T\cos\left(\frac{d\theta}{2}\right) + \mu\,dN - (T + dT)\cos\left(\frac{d\theta}{2}\right) = 0$$

$$+\uparrow\Sigma F_y = 0; \quad dN - (T + dT)\sin\left(\frac{d\theta}{2}\right) - T\sin\left(\frac{d\theta}{2}\right) = 0$$

Since $d\theta$ is of *infinitesimal size*, $\sin(d\theta/2)$ and $\cos(d\theta/2)$ can be replaced by $d\theta/2$ and 1, respectively. Also, the *product* of two infinitesimals dT and $d\theta/2$ may be neglected when compared to infinitesimals of the first order. The above two equations therefore reduce to

$$\mu\,dN = dT$$

and

$$dN = T\,d\theta$$

Eliminating dN between these two relations yields

$$\frac{dT}{T} = \mu\,d\theta$$

Integrating this equation between all the points of contact that the belt makes with the drum, and noting that $T = T_1$ at $\theta = 0$, and $T = T_2$ at $\theta = \beta$, yields

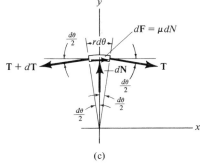

(c)

Fig. 8-21

$$\int_{T_1}^{T_2} \frac{dT}{T} = \mu \int_0^\beta d\theta$$

or

$$\ln \frac{T_2}{T_1} = \mu\beta$$

Solving for T_2, we obtain

$$T_2 = T_1 e^{\mu\beta} \tag{8–8}$$

where μ = coefficient of static or kinetic friction between the belt and the surface of contact

β = angle of belt to surface contact, measured in radians

T_2, T_1 = belt tensions; T_1 opposes the direction of motion (or impending motion) of the belt, while T_2 acts in the direction of belt motion (or impending motion); because of friction, $T_2 > T_1$.

e = 2.718..., base of the natural logarithm.

Note that Eq. 8–8 is *independent* of the *radius* of the drum and instead depends upon the angle of belt contact, β. Furthermore, as indicated by the integration, this equation is valid for flat belts placed on *any shape* of contacting surface.

Example 8–10

The maximum tension that can be developed in the belt, shown in Fig. 8–22a, is 500 N. If the pulley at A is free to rotate and the coefficient of static friction at the fixed drums B and C is $\mu_s = 0.25$, determine the

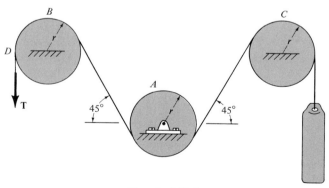

Fig. 8–22(a)

largest mass of the cylinder that can be lifted by the belt. Assume that the force **T** applied at the end of the belt is directed vertically downward, as shown.

Solution

Lifting the cylinder, which has a weight $W = mg$, causes the belt to move counterclockwise over the drums at B and C; hence, the maximum tension T_2 in the belt occurs at D. Thus, $T = 500$ N. A section of the belt passing over the drum at B is shown in Fig. 8–22b. Since $180° = \pi$ rad, the angle of contact between the drum and the belt is $\beta = (135°/180°)\pi = 3\pi/4$ rad. Using Eq. 8–8, we have

$$T_2 = T_1 e^{\mu_s \beta}; \qquad\qquad 500 \text{ N} = T_1 e^{0.25[(3/4)\pi]}$$

Hence,

$$T_1 = \frac{500 \text{ N}}{e^{0.25[(3/4)\pi]}} = \frac{500}{1.80} = 277.4 \text{ N}$$

Since the pulley at A is free to rotate, equilibrium requires that the tension in the belt remains the *same* on both sides of the pulley.

The section of the belt passing over the drum at C is shown in Fig. 8–22c. The load $W < 277.4$ N. Why? Applying Eq. 8–8, we obtain

$$T_2 = T_1 e^{\mu_s \beta}; \qquad\qquad 277.4 = W e^{0.25[(3/4)\pi]}$$
$$W = 153.9 \text{ N}$$

so that

$$m = \frac{W}{g} = \frac{153.9}{9.81}$$
$$= 15.7 \text{ kg} \qquad\qquad Ans.$$

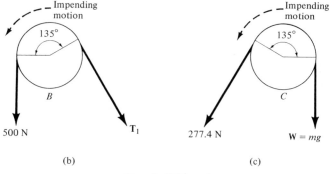

(b)   (c)

Fig. 8–22(b, c)

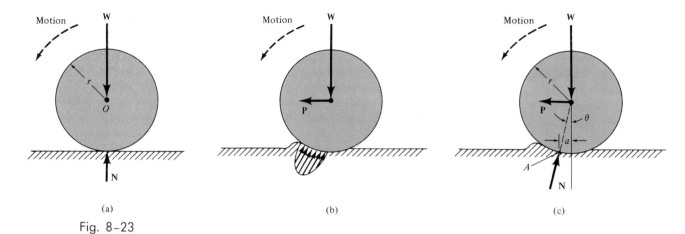

(a) (b) (c)

Fig. 8-23

*8-7. Rolling Resistance

If a *rigid* cylinder of weight **W** rolls at constant velocity along a *rigid* surface, the normal force exerted by the surface on the cylinder acts at the tangent point of contact, as shown in Fig. 8-23a. Under these conditions, provided the cylinder does not encounter frictional resistance from the air, motion would continue indefinitely. Actually, however, no materials are perfectly rigid; and therefore the reaction of the surface on the cylinder consists of a *distribution* of normal pressure. Assuming that the cylinder is much harder than the surface, a small "hill" is formed which is caused by the indentation of the cylinder into the surface, Fig. 8-23b. Since the cylinder can never really "climb" over this "hill," the normal pressure is always present, and consequently, to maintain motion, a driving force **P** must be applied to the cylinder. For analysis, the normal pressure distribution can be replaced by its *resultant* force **N**, which acts at an angle θ from the vertical, Fig. 8-23c. To keep the cylinder in equilibrium, i.e., moving at constant velocity, it is necessary that **N** be *concurrent* with the driving force **P** and the weight **W**. Summing moments about point A gives $Wa = P(r\cos\theta)$. Since the deformations of the cylinder are generally small in relation to the radius, $\cos\theta \approx 1$; hence,

$$Wa \approx Pr$$

or

$$P \approx \frac{Wa}{r} \tag{8-9}$$

The distance a is termed the *coefficient of rolling resistance,* which has the dimension of length. For instance, $a \approx 0.5$ mm for a mild steel wheel rolling on a steel rail. For hardened steel ball bearings on steel $a \approx 0.1$ mm. Experimentally, though, this factor is difficult to measure, since it depends upon such parameters as the rate of rotation of the cylinder, the elastic properties of the contacting surfaces, and the frictional effects of the surfaces. For this reason, little reliance is placed on the data for determining a. The analysis presented here does, however, indicate why a heavy load offers greater resistance to motion than a light load under the same conditions. Furthermore, since P needed to *roll* the cylinder over the surface is much less than that needed to *slide* the cylinder across the surface, the analysis indicates why roller or ball bearings are often used to minimize the frictional resistance between moving parts.

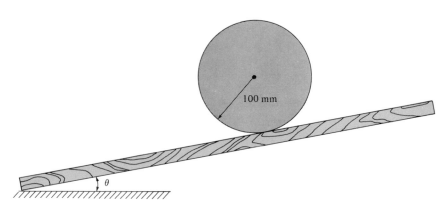

Fig. 8–24(a)

Example 8–11

A 10-kg steel wheel shown in Fig. 8–24a has a radius of 100 mm and rests on an inclined plane made of wood. If θ is increased so that the wheel begins to roll down the incline with constant velocity when $\theta = 1.2°$, determine the coefficient of rolling resistance.

Solution

As shown on the free-body diagram, Fig. 8–24b, when the wheel has impending motion, the normal reaction $\mathbf{N}$ acts at a point A defined by the dimension a. Resolving the weight into rectangular components normal and perpendicular to the incline, and summing moments about point A, yields (approximately)

$$\zeta + \Sigma M_A = 0; \quad 98.1 \cos 1.2°(a) - 98.1 \sin 1.2°(100) = 0$$

Solving,

$$a = 2.1 \text{ mm} \qquad\qquad Ans.$$

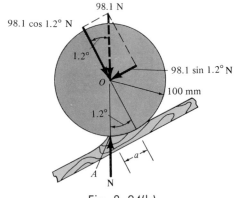

Fig. 8–24(b)

Problems

8-59. Determine the *minimum* tension in the rope at points A and B, which is necessary to maintain equilibrium of the rope. $\mu = 0.3$ between the rope and the fixed post D. The rope is wrapped only once around the post.

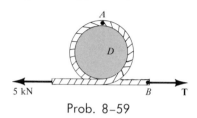

Prob. 8-59

* **8-60.** A cable is attached to plate B, passes over a fixed peg at C, and is attached to the block at A. Using the coefficients of friction shown in the figure, determine the smallest mass of block A that will prevent sliding motion of B down the plane. Block B has a mass of $m_B = 20$ kg.

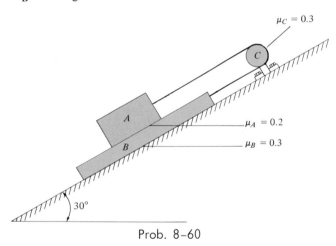

Prob. 8-60

* **8-60a.** Determine the smallest weight of block A if the plate in Prob. 8-60 has a weight of $W_B = 50$ lb.

8-61. The uniform bar AB is supported by a rope that passes over a frictionless pulley at C and a fixed peg at D. If the coefficient of friction between the rope and the peg is $\mu_D = 0.3$, determine the smallest distance x from the end of the bar at which a 20-N force may be placed and not cause the bar to move.

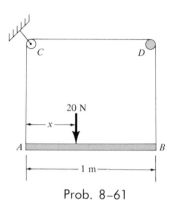

Prob. 8-61

8-62. If a moment $M = 200$ N · m is applied to the disk, determine the minimum vertical force $\mathbf{P}$ that must be applied to the band brake to prevent the disk from rotating. The coefficient of friction between the band and the disk is $\mu = 0.45$. The disk has a mass of 40 kg.

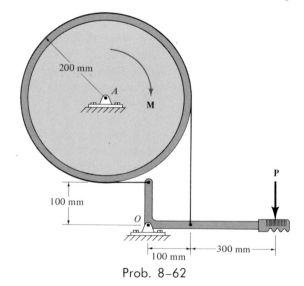

Prob. 8-62

8-63. Granular material, having a density of 1.2 Mg/m³, is transported on a conveyor belt that slides over the fixed surface s, where $\mu = 0.3$. Operation of the belt is provided by a motor that supplies a twist or torque **M** to wheel A. The wheel at B is free to rotate and $\mu_A = 0.4$ between the wheel at A and the belt. If the belt is subjected to a pretension of 300 N when no load is on the belt, determine the greatest volume V of material that is permitted on the belt at any time, without allowing the belt to stop. What is the required torque to drive the belt when it is subjected to this maximum load?

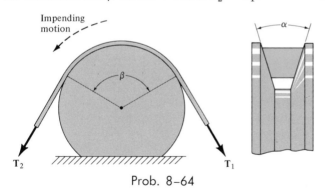

Prob. 8-63

* **8-64.** Show that the frictional relation between the belt tensions, the coefficient of friction μ, and the angular contacts α and β for the V-belt is $T_2 = T_1 e^{[\mu\beta/\sin(\alpha/2)]}$.

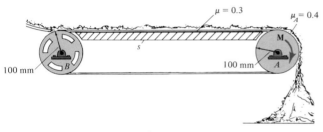

Prob. 8-64

8-65. The 60° V-fan belt of an automobile engine passes around the hub H of a generator G and over the housing F to a fan. If the generator locks, causing the axle of the hub to become fixed, determine the maximum possible moment **M** resisted by the axle as the belt slips over the hub. The belt can sustain a maximum tension of $T_{max} = 800$ N. Set $r = 50$ mm. Assume that slipping of the belt only occurs at H and that the coefficient of friction for the hub is $\mu = 0.30$. Use the result of Prob. 8-64.

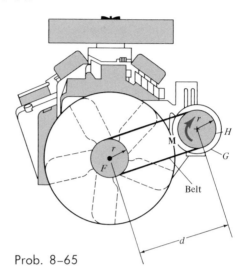

Prob. 8-65

8-65a. Solve Prob. 8-65 if $T_{max} = 175$ lb, $r = 2$ in., and $\mu = 0.25$.

8-66. Two 6-kg blocks are attached to a cord that passes over two fixed drums. If $\mu = 0.3$ at the drums, determine the angle θ the cord makes with the horizontal when a vertical force $P = 200$ N is applied to the cord as shown. The tension in the cord is zero when the blocks just touch the *ground,* i.e., $\theta = 0°$ and $P = 0$.

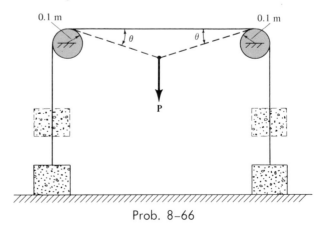

Prob. 8-66

8–67. The 20-kg motor has a center of gravity at G and is pin-connected at C to maintain a tension in the drive belt. Determine the smallest counterclockwise twist or torque $\mathbf{M}$ that must be supplied by the motor to turn the disk B if wheel A locks and causes the belt to slip over the disk. The coefficient of friction between the belt and the disk is $\mu = 0.3$.

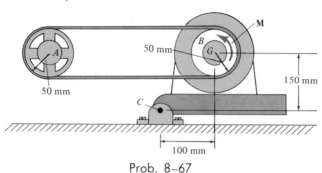

Prob. 8–67

*8–68.** A motor supplies a constant twist or torque of $M = 30$ N · m to wheel B. Power is transmitted to wheel C using a belt having a coeficient of friction of $\mu = 0.3$ between the belt and both wheels B and C. To maintain tension in the belt, the axle at D is free to slide horizontally, and a second belt is used that wraps around both the inner hub, which is attached to B, and the idle pulley at A. If the coefficient of friction between this belt and the hub is $\mu' = 0.4$, determine the largest mass E that can be suspended from the rope attached to A such that wheel B continues to turn if wheel C becomes fixed from rotating.

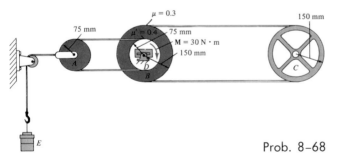

Prob. 8–68

8–69. A ring is tied to the end of a rope at B, and the rope is wrapped around a frictionless pulley A and passes through the ring as shown. If the rope is subjected to a tension $\mathbf{T}$ and the coefficient of friction between the rope and the ring is $\mu = 0.2$, determine the angle θ for equilibrium.

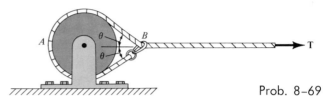

Prob. 8–69

8–70. A tank car has a mass of $m = 30$ Mg and is supported by eight wheels, each of which has a diameter of $d = 500$ mm. If the coefficient of rolling resistance is $a = 0.4$ mm between the tracks and each wheel, determine the magnitude of horizontal force $\mathbf{P}$ required to overcome the rolling resistance of the wheels.

8–70a. Solve Prob. 8–70 if the tank car has a weight of $W = 60(10^3)$ lb, $d = 30$ in., and $a = 0.016$ in.

8–71. Experimentally, it is found that a cylinder having a diameter of 150 mm rolls with a constant speed down an inclined plane having a slope of 18 mm/m. Determine the coefficient of rolling resistance for the cylinder.

*8–72.** The lawn roller has a mass of 100 kg. If the arm BA is held at an angle of $30°$ from the horizontal and the coefficient of rolling resistance for the roller is 25 mm, determine the force $\mathbf{F}$ needed to push the roller at constant speed. Neglect friction developed at the axle and assume that the resultant force $\mathbf{F}$ acting on the handle is applied along the arm BA.

Prob. 8–72

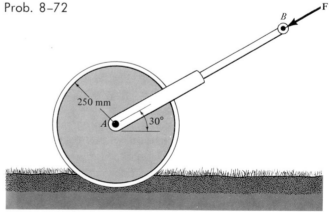

8-73. The pipe is subjected to the load that has a weight **W.** If the coefficients of rolling resistance for the pipe's top and bottom surfaces are a_A and a_B, respectively, show that a force having a magnitude of $P = [W(a_A + a_B)]/2r$ is required to move the load and thereby roll the pipe forward. Neglect the weight of the pipe.

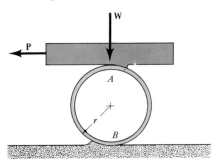

Prob. 8–73

8-74. The 1.5-Mg steel beam is moved over a level surface using a series of 50-mm-diameter rollers for which the coefficient of rolling resistance is 0.4 mm at the ground and 0.2 mm at the bottom surface of the beam. Determine the horizontal force **P** needed to push the beam forward at a constant speed. *Hint:* Use the result of Prob. 8–73.

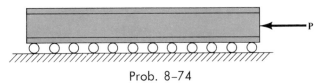

Prob. 8–74

9

Center of Gravity and Centroids

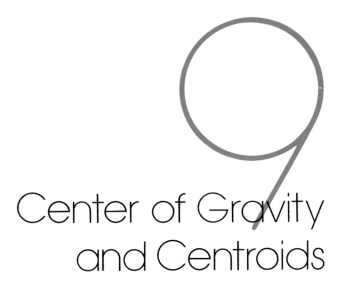

9-1. Center of Gravity and Center of Mass for a System of Particles

The method used for simplifying a distributed surface loading was discussed in Sec. 4–10. In this section, this method will be generalized and applied to finding the center of gravity and center of mass for a system of discrete particles.

Center of Gravity. Consider the system of three different particles shown in Fig. 9–1a, which are distributed across the top surface of an imaginary weightless beam. This parallel system of forces may be reduced to a single resultant force $\mathbf{W}$ acting at point G on the beam as shown in Fig. 9–1b. Applying Eq. 4–14, we require the resultant force $\mathbf{W}$ to have a magnitude of

$$+\downarrow F_R = \Sigma F; \qquad W = W_1 + W_2 + W_3$$

The location of $\mathbf{W}$ is determined by applying the principle of moments, Eq. 4–15, which requires the moment of $\mathbf{W}$ about point O, Fig. 9–1b, to be equivalent to the sum of the moments of the weights of all the particles about O, Fig. 9–1a. Hence,

$$\zeta + M_{R_O} = \Sigma M_O; \qquad \bar{x} W = x_1 W_1 + x_2 W_2 + x_3 W_3$$

Combining the above equations and solving for $\bar{x}$ yields

$$\bar{x} = \frac{\tilde{x}_1 W_1 + \tilde{x}_2 W_2 + \tilde{x}_3 W_3}{W_1 + W_2 + W_3}$$

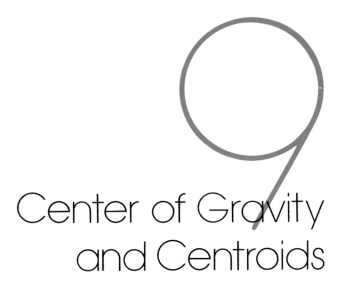

(a)

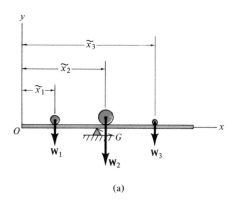

(b)

Fig. 9–1

From this result it may be concluded that the effect caused by a single particle having a weight W, which is equal to the total weight of all the particles, and placed at point G, located a distance $\bar{x}$ to the right of point O, Fig. 9–1b, is *equivalent* to the effect caused by the weights of all three particles shown in Fig. 9–1a. Point G is termed the *center of gravity;* and for this special case, G may be considered as a "balance point" for the beam; i.e., if a fulcrum were placed under the beam at point G as shown in Figs. 9–1a and 9–1b, the beam would remain balanced in the horizontal position.

The preceding equation for $\bar{x}$ may be generalized to include a system of n particles acting on the beam. For this case,

$$\bar{x} = \frac{\sum_{i=1}^{n} \tilde{x}_i W_i}{\sum_{i=1}^{n} W_i}$$

Here W_i is the weight of the arbitrary ith particle of the system and $\tilde{x}_i$ is the *algebraic distance* measured from the origin O to the ith particle.

If n particles, having weights $W_1, W_2, \ldots, W_n$, are all contained in the x-y plane, the location of the arbitrary ith particle may be specified using coordinates $(\tilde{x}_i, \tilde{y}_i)$. Summing moments of the weights of all of the particles about the x and y axes, respectively, in the same manner as above, one obtains the following two equations, which locate the position $G(\bar{x}, \bar{y})$ of the center of gravity for the system of particles:

$$\bar{x} = \frac{\sum_{i=1}^{n} \tilde{x}_i W_i}{\sum_{i=1}^{n} W_i}$$

$$\bar{y} = \frac{\sum_{i=1}^{n} \tilde{y}_i W_i}{\sum_{i=1}^{n} W_i}$$

In the most general case, the ith particle is located at a point $(\tilde{x}_i, \tilde{y}_i, \tilde{z}_i)$ in space. Provided the n particles always maintain their same relative position regardless of how the coordinate axes are oriented, the line of action of the resultant weight will always act through the center of gravity $G(\overline{x}, \overline{y}, \overline{z})$ of the system. The location of this point can be determined by using the principle of moments applied about each of the axes. The results are

9-1. Center of Gravity and 345
Center of Mass for a System
of Particles

$$\overline{x} = \frac{\sum\limits_{i=1}^{n} \tilde{x}_i W_i}{\sum\limits_{i=1}^{n} W_i} \qquad \overline{y} = \frac{\sum\limits_{i=1}^{n} \tilde{y}_i W_i}{\sum\limits_{i=1}^{n} W_i} \qquad \overline{z} = \frac{\sum\limits_{i=1}^{n} \tilde{z}_i W_i}{\sum\limits_{i=1}^{n} W_i} \qquad (9\text{-}1)$$

Center of Mass. To study problems concerning the motion of *matter* under the influence of forces, i.e., dynamics, it is necessary to locate a point called the *center of mass*. In finding the location of this point for a system of particles, one must consider the "weighted" average of the *mass* of each particle. In this way, equations analogous to Eq. 9–1 are formulated, except that m_i is substituted for W_i, i.e.,

$$\overline{x} = \frac{\sum\limits_{i=1}^{n} \tilde{x}_i m_i}{\sum\limits_{i=1}^{n} m_i} \qquad \overline{y} = \frac{\sum\limits_{i=1}^{n} \tilde{y}_i m_i}{\sum\limits_{i=1}^{n} m_i} \qquad \overline{z} = \frac{\sum\limits_{i=1}^{n} \tilde{z}_i m_i}{\sum\limits_{i=1}^{n} m_i} \qquad (9\text{-}2)$$

Since particles have "weight" only when under the influence of a gravitational attraction, the center of gravity depends upon the existence of a gravitational field for its definition. The center of mass, however, is independent of gravitational forces. For example, it would be meaningless to define the center of gravity of particles representing the planets of our solar system, whereas the center of mass of this system is important.

In an exact sense, the center of gravity and the center of mass for a system of particles do *not* coincide. This is because the weight of each particle depends upon its *distance* from the center of the earth, a consequence of Newton's law of gravitation, $F = Gm_1 m_2/r^2$ (Eq. 1–2). Furthermore, gravitational forces are directed toward the center of the earth, and therefore the lines of action of these forces are not actually parallel. For all practical purposes, however, these variations are negligible, and therefore the "moment" computations used to determine the location for the center of gravity and the center of mass are assumed to yield the *same* result. In other words, since it is assumed that the acceleration of gravity g for every particle is *constant*, substituting $W_i = m_i g$ into Eqs. 9–1 yields Eqs. 9–2, since g cancels from both the numerator and denominator of Eqs. 9–1.

Fig. 9–2

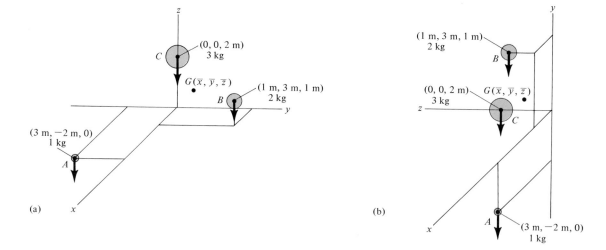

(a)

(b)

Example 9–1

Locate the center of gravity G of the three particles shown in Fig. 9–2a. The mass of each particle is given and each particle's location is specified by its x, y, z coordinates.

Solution

Since the particle weights all act *parallel* to the z axis, Fig. 9–2a, they create *no moment* about this axis; hence, the moment calculations for the positions $\bar{x}$ and $\bar{y}$ of the center of gravity can be "visualized" when the coordinate axis is oriented as shown. In determining the location $\bar{z}$ it may be helpful to imagine the coordinate axis as being rotated 90° about the x axis, Fig. 9–2b, in order to "visualize" how the moments of the weights are developed about this axis.

Since the center of mass and the center of gravity are assumed to coincide, Eqs. 9–2 may be applied for the solution. Using the data listed in either Fig. 9–2a or Fig. 9–2b yields

$$\bar{x} = \frac{\sum \tilde{x}_i m_i}{\sum m_i} = \frac{3(1) + 1(2) + 0(3)}{1 + 2 + 3} = 0.833 \text{ m} \qquad Ans.$$

$$\bar{y} = \frac{\sum \tilde{y}_i m_i}{\sum m_i} = \frac{-2(1) + 3(2) + 0(3)}{1 + 2 + 3} = 0.667 \text{ m} \qquad Ans.$$

$$\bar{z} = \frac{\sum \tilde{z}_i m_i}{\sum m_i} = \frac{0(1) + 1(2) + 2(3)}{1 + 2 + 3} = 1.333 \text{ m} \qquad Ans.$$

Problems

9-1. Locate the center of gravity *G* of the five particles with respect to the origin *O*.

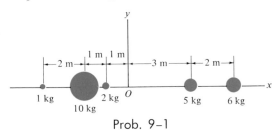

Prob. 9–1

9-2. If the four particles can be replaced by a single 10-kg particle acting at a distance of 2 m to the left of the origin, determine the position $\tilde{x}_p$ and the mass m_p of particle *P*.

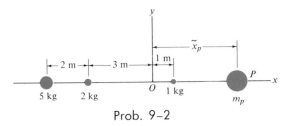

Prob. 9–2

9-3. Locate the coordinates $(\bar{x}, \bar{y})$ of the center of mass for the system of four particles lying in the *x-y* plane.

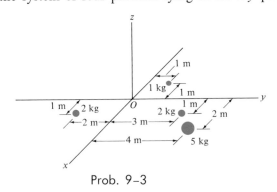

Prob. 9–3

***9-4.** Determine the location (x, y) of a 7-kg particle so that the three particles, which lie in the *x-y* plane, have a center of mass located at the origin *O*.

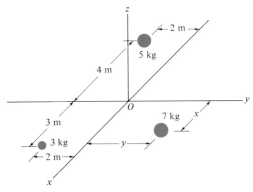

Prob. 9–4

9-5. Locate the center of gravity $(\bar{x}, \bar{y}, \bar{z})$ of the three particles.

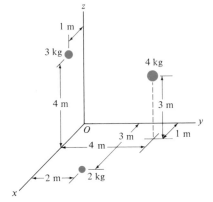

Prob. 9–5

347

9-2. Center of Gravity and Centroid of a Body

Center of Gravity. The principles used to determine the location of the center of gravity for a system of discrete particles may also be used to determine the location of this point for a continuous system, which essentially contains an infinite number of particles. For example, consider the determination of G for a thin wire of length L placed along the x axis, Fig. 9–3a. As an *approximation,* the wire may be divided into n segments or particles, such that the ith segment has a length Δx_i, weight ΔW_i, and is located at $\tilde{x}_i$ from O. Since $W = \sum\limits_{i=1}^{n} \Delta W_i$, then applying the principle of moments, Eq. 4–15, in reference to Figs. 9–3a and 9–3b, yields

$$\curvearrowright + M_{R_O} = \Sigma M_O; \qquad \bar{x} \sum_{i=1}^{n} \Delta W_i = \sum_{i=1}^{n} \tilde{x}_i \Delta W_i$$

Solving for $\bar{x}$, which defines the location of G, gives

$$\bar{x} = \frac{\displaystyle\sum_{i=1}^{n} \tilde{x}_i \Delta W_i}{\displaystyle\sum_{i=1}^{n} \Delta W_i}$$

If smaller and smaller Δx_i segments are chosen until $\Delta x_i \to 0$, and hence $n \to \infty$, the two summations become *integrals* that formulate the *exact* location of the center of gravity. In the limit, each segment of wire becomes a differential element that has a weight $d\mathbf{W}$. Hence,

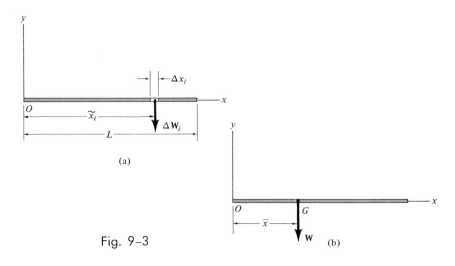

Fig. 9–3

$$\bar{x} = \frac{\displaystyle\lim_{\Delta x_i \to 0}\left[\sum_{i=1}^{n}\tilde{x}_i\,\Delta W_i\right]}{\displaystyle\lim_{\Delta x_i \to 0}\left[\sum_{i=1}^{n}\Delta W_i\right]} = \frac{\displaystyle\int_{W}\tilde{x}\,dW}{\displaystyle\int_{W}dW}$$

The total weight of the wire may therefore be represented by a concentrated force $W = \int_W dW$ acting at a distance $\bar{x}$ from the origin of coordinates, Fig. 9-3b.

In general, a rod or wire may be *curved* such that it must be described in three dimensions as shown in Fig. 9-4. In this case, the location of the center of gravity G is determined using the three formulas

$$\bar{x} = \frac{\displaystyle\int_{W}\tilde{x}\,dW}{\displaystyle\int_{W}dW} \qquad \bar{y} = \frac{\displaystyle\int_{W}\tilde{y}\,dW}{\displaystyle\int_{W}dW} \qquad \bar{z} = \frac{\displaystyle\int_{W}\tilde{z}\,dW}{\displaystyle\int_{W}dW} \qquad (9\text{-}3)$$

The derivations are based upon the principle of moments in the same manner as applied to the straight wire, except in this case "moments" are balanced about *each* of the coordinate axes. In particular the moment balance about the z axis in Fig. 9-4 can be "visualized" by rotating the coordinate axes and the wire 90° about the x axis so that the wire's weight acts parallel to the y axis. At any rate, regardless of how the axes are rotated, the line of action of $\mathbf{W}$ will *always* act through G, a point not necessarily lying on the wire.

Thin Rod or Wire. Since the integration of Eqs. 9-3 is to be performed over the *entire length L* of the wire, it is necessary to express the magnitude of the differential weight $d\mathbf{W}$ in terms of its length dL, Fig. 9-4. Provided the wire has a mass density ρ and constant cross-sectional area A, the elemental volume is $dV = A\,dL$, the differential mass is $dm = \rho\,dV = \rho A\,dL$, and the differential weight is $dW = g\,dm = g\rho A\,dL$, where g is the acceleration of gravity (9.81 m/s²). Substituting dW into Eq. 9-3, the *constants* A and g cancel from the numerator and denominator, leaving

$$\bar{x} = \frac{\displaystyle\int_{L}\tilde{x}\rho\,dL}{\displaystyle\int_{L}\rho\,dL} \qquad \bar{y} = \frac{\displaystyle\int_{L}\tilde{y}\rho\,dL}{\displaystyle\int_{L}\rho\,dL} \qquad \bar{z} = \frac{\displaystyle\int_{L}\tilde{z}\rho\,dL}{\displaystyle\int_{L}\rho\,dL} \qquad (9\text{-}4)$$

In general, ρ is a function of position, $\rho = \rho(x, y, z)$, and must therefore be retained in the integrand of these formulas.

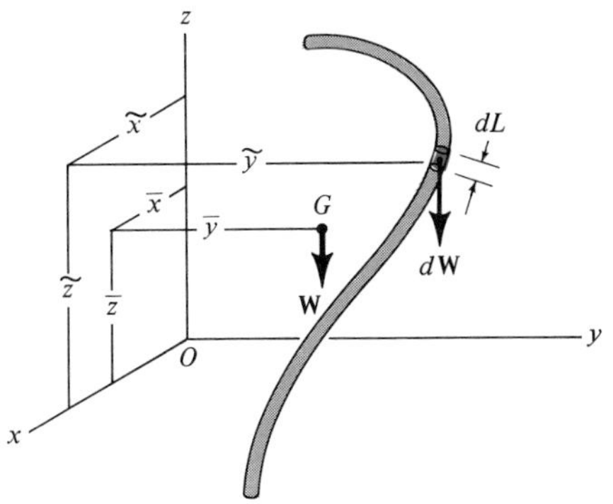

Fig. 9-4

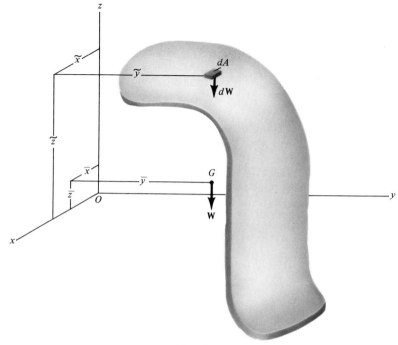

Fig. 9–5

Thin Plate or Shell. Equations 9–3 may also be used for determining the
center of gravity of a plate or shell having a total area A, constant
thickness t, and mass density ρ, Fig. 9–5. In this case, the elemental
volume is $dV = t\, dA$, the differential mass is $dm = \rho t\, dA$, and the differential weight is $dW = g\rho t\, dA$. After substitution, we have

$$\bar{x} = \frac{\int_A \tilde{x}\rho\, dA}{\int_A \rho\, dA} \qquad \bar{y} = \frac{\int_A \tilde{y}\rho\, dA}{\int_A \rho\, dA} \qquad \bar{z} = \frac{\int_A \tilde{z}\rho\, dA}{\int_A \rho\, dA} \qquad (9\text{--}5)$$

Solid Body. For a solid body, having a total volume V and mass density
ρ, Fig. 9–6, the element of volume dV has a weight $dW = g\rho\, dV$ so that
Eqs. 9–3 become

$$\bar{x} = \frac{\int_V \tilde{x}\rho\, dV}{\int_V \rho\, dV} \qquad \bar{y} = \frac{\int_V \tilde{y}\rho\, dV}{\int_V \rho\, dV} \qquad \bar{z} = \frac{\int_V \tilde{z}\rho\, dV}{\int_V \rho\, dV} \qquad (9\text{--}6)$$

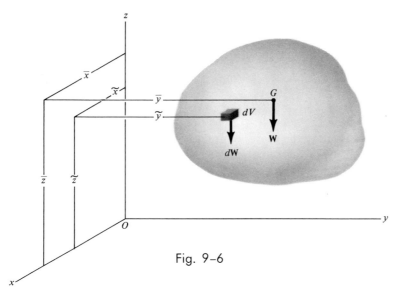

Fig. 9-6

Centroids. If the material composing the body is uniform or *homogeneous*, the mass density ρ will be *constant* throughout the body, and therefore this term will factor out of the integrals and hence *cancel* from both the numerator and denominator in Eqs. 9-4, 9-5, and 9-6. The resulting formulas, which locate the center of gravity or the center of mass of the body, will then depend entirely upon the body's *geometrical properties*. When this is the case, the point $C(\bar{x}, \bar{y}, \bar{z})$ is called the *centroid* of the body.*

Thin Rod or Wire (Line). For *lines* defining the center line of uniform rods or wires having a *constant cross-sectional area,*

$$\bar{x} = \frac{\int_L \tilde{x}\, dL}{\int_L dL} \qquad \bar{y} = \frac{\int_L \tilde{y}\, dL}{\int_L dL} \qquad \bar{z} = \frac{\int_L \tilde{z}\, dL}{\int_L dL} \tag{9-7}$$

Thin Plate or Shell (Area). For *areas* defining the surface of thin plates and shells having a *constant thickness,*

$$\bar{x} = \frac{\int_A \tilde{x}\, dA}{\int_A dA} \qquad \bar{y} = \frac{\int_A \tilde{y}\, dA}{\int_A dA} \qquad \bar{z} = \frac{\int_A \tilde{z}\, dA}{\int_A dA} \tag{9-8}$$

*Of course, a nonhomogeneous body also has a centroid. In this case, however, the centroid and center of gravity will *not* coincide.

Solid Body (Volume).   For *volumes* defining the space occupied by a body,

$$\bar{x} = \frac{\int_V \tilde{x}\, dV}{\int_V dV} \qquad \bar{y} = \frac{\int_V \tilde{y}\, dV}{\int_V dV} \qquad \bar{z} = \frac{\int_V \tilde{z}\, dV}{\int_V dV} \qquad (9\text{-}9)$$

Besides locating the center of gravity of homogeneous solids, computations for the centroids of volumes and area shapes are important in other fields of mechanics. For example, in structural mechanics, the centroid of the cross-sectional area of a beam or column must be located in order to properly design these members. Furthermore, it was shown in Sec. 4–10 that a distributed surface loading may be reduced to a single resultant force having a line of action that passes through the *centroid* of the *volume* or *area* described by the *loading diagram*.

The *centroids* of some shapes may be partially or completely specified by using *symmetry conditions*. In cases where the shape has an axis of symmetry, the centroid of the shape will lie along that axis. For example, the centroid C for the homogeneous wire shown in Fig. 9–7 must lie along the y axis, since for every elemental length dL at a distance $+\tilde{x}$ to the right of the y axis, there is an identical element at a distance $-\tilde{x}$ to the left of the y axis. The total moment for all the elements about the axis of symmetry will therefore cancel; i.e., $\int_L \tilde{x}\, dL = 0$, Eq. 9–7, so that $\bar{x} = 0$. In cases where a shape has two or three axes of symmetry, it follows that the centroid lies at the intersection of these axes, Fig. 9–8.

PROCEDURE FOR ANALYSIS

In general, when applying Eqs. 9–3 through 9–9 it is best to choose a coordinate system which simplifies the equation that describes the boundary of the object. For example, polar coordinates are generally used

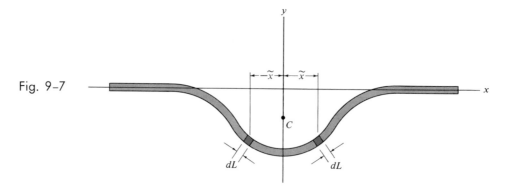

Fig. 9–7

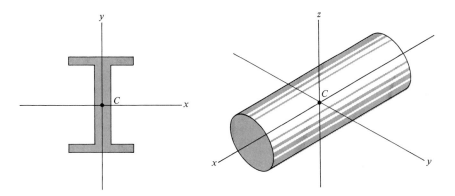

Fig. 9-8

to describe the shape of an object having circular boundaries. The differential element for integration should be chosen so that only *one integration* is required to cover the entire region. This can be done only if the element is of differential size or thickness in one direction. Provided this type of element is selected, the following five-step procedure should be applied when computing the center of gravity or centroid of a solid.

Step 1: Specify the coordinate axes, and choose a differential element for the solid. For lineal elements (wires, rods) this element is represented as a differential line segment; for areas (plates, shells) the element is generally a rectangle, having a finite height and differential width; and for volumes (bodies) the element is either a circular disk, having a finite radius and differential thickness, or a shell having a finite length and radius, and differential thickness.

Step 2: Construct the element so that it intersects the surface boundary of the line, area, or volume at an *arbitrary point*. This fixes the dimensions of the element.

Step 3: Express the length dL, area dA, or volume dV, of the element in terms of the coordinates used to define the boundary of the solid.

Step 4: Determine the "moment arm" or perpendicular distance from each of the coordinate axes to the *centroid or center of gravity of the element*. For an x, y, z coordinate system, these dimensions are represented by $\tilde{x}$, $\tilde{y}$, and $\tilde{z}$.

Step 5: Substitute the data computed in *Steps 3* and *4* into the appropriate equations (Eqs. 9-4 through 9-9) and perform the integration.* Note that integration can be accomplished when the function in the integrand is expressed in terms of the *same variable as the differential thickness of the element*. The limits of the integral are then defined from the two extreme locations of the element's differential thickness, so that when the elements are "summed" or the integration performed, the entire region is covered.

The following examples illustrate this five-step procedure.

*Formulas for integration are given in Appendix A.

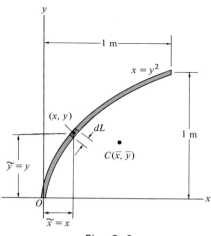

Fig. 9-9

Example 9-2

Locate the centroid of the rod bent into the shape of a parabolic arc, shown in Fig. 9-9.

Solution

Step 1: The differential element is shown in Fig. 9-9.
Step 2: The element intersects the curve at the arbitrary point (x, y).
Step 3: The length of the element is determined using the Pythagorean theorem.

$$dL = \sqrt{(dx)^2 + (dy)^2} = \sqrt{\left(\frac{dx}{dy}\right)^2 + 1}\, dy$$

Since $x = y^2$, then $dx/dy = 2y$. Therefore,

$$dL = \sqrt{(2y)^2 + 1}\, dy$$

Step 4: The centroid of the element is located at $\tilde{x} = x$, $\tilde{y} = y$.
Step 5: Using Eqs. 9-7 and integrating with respect to y yields

$$\bar{x} = \frac{\int_L \tilde{x}\, dL}{\int_L dL} = \frac{\int_0^1 x\sqrt{4y^2 + 1}\, dy}{\int_0^1 \sqrt{4y^2 + 1}\, dy} = \frac{\int_0^1 y^2\sqrt{4y^2 + 1}\, dy}{\int_0^1 \sqrt{4y^2 + 1}\, dy}$$

$$= \frac{0.739}{1.479} = 0.500 \text{ m} \qquad\qquad Ans.$$

$$\bar{y} = \frac{\int_L \tilde{y}\, dL}{\int_L dL} = \frac{\int_0^1 y\sqrt{4y^2 + 1}\, dy}{\int_0^1 \sqrt{4y^2 + 1}\, dy} = \frac{0.848}{1.479} = 0.573 \text{ m} \qquad Ans.$$

Example 9-3

Locate the centroid of the area of the plate shown in Fig. 9-10a.

Solution I

Step 1: The differential element is shown in Fig. 9-10a.
Step 2: The element intersects the curve at the arbitrary point (x, y).
Step 3: The area of the element is $dA = y\, dx$.
Step 4: The centroid of the element is located at $\tilde{x} = x$, $\tilde{y} = y/2$.
Step 5: Using Eqs. 9-8 and integrating with respect to x yields

$$\bar{x} = \frac{\int_A \tilde{x}\, dA}{\int_A dA} = \frac{\int_0^1 xy\, dx}{\int_0^1 y\, dx} = \frac{\int_0^1 x^3\, dx}{\int_0^1 x^2\, dx} = \frac{0.250}{0.333} = 0.75 \text{ m} \qquad Ans.$$

$$\bar{y} = \frac{\int_A \tilde{y}\, dA}{\int_A dA} = \frac{\int_0^1 \left(\frac{y}{2}\right)y\, dx}{\int_0^1 y\, dx} = \frac{\int_0^1 \left(\frac{x^2}{2}\right)x^2\, dx}{\int_0^1 x^2\, dx} = \frac{0.100}{0.333} = 0.3 \text{ m} \qquad Ans.$$

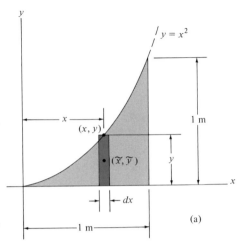

(a)

Solution II

Step 1: The differential element is shown in Fig. 9–10b.

Step 2: The element intersects the curve at the arbitrary point (x, y).

Step 3: The area of the element is $dA = (1 - x)\, dy$.

Step 4: The centroid of the element is located at

$$\tilde{x} = x + \left(\frac{1 - x}{2}\right) = \frac{1 + x}{2}, \quad \tilde{y} = y$$

Step 5: Using Eqs. 9–8, and integrating with respect to y, we obtain

$$\bar{x} = \frac{\int_A \tilde{x}\, dA}{\int_A dA} = \frac{\int_0^1 \left(\frac{1 + x}{2}\right)(1 - x)\, dy}{\int_0^1 (1 - x)\, dy}$$

$$= \frac{\int_0^1 \left(\frac{1 + \sqrt{y}}{2}\right)(1 - \sqrt{y})\, dy}{\int_0^1 (1 - \sqrt{y})\, dy}$$

$$= \frac{\frac{1}{2}\int_0^1 (1 - y)\, dy}{\int_0^1 (1 - \sqrt{y})\, dy} = \frac{0.250}{0.1667} = 0.75 \text{ m} \qquad Ans.$$

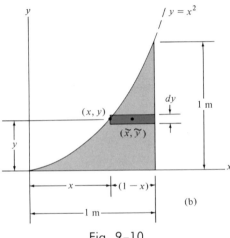

(b)

Fig. 9–10

$$\bar{y} = \frac{\int_A \tilde{y}\, dA}{\int_A dA} = \frac{\int_0^1 y(1 - x)\, dy}{\int_0^1 (1 - x)\, dy} = \frac{\int_0^1 y(1 - \sqrt{y})\, dy}{\int_0^1 (1 - \sqrt{y})\, dy}$$

$$= \frac{\int_0^1 (y - y^{3/2})\, dy}{\int_0^1 (1 - \sqrt{y})\, dy} = \frac{0.0667}{0.1667} = 0.3 \text{ m} \qquad Ans.$$

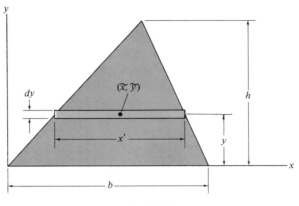

Fig. 9–11

Example 9–4

Determine the distance $\bar{y}$ to the centroid of the area of the triangle shown in Fig. 9–11.

Solution
Step 1: Consider a rectangular element having thickness dy and *variable length* x', Fig. 9–11. By similar triangles, $b/h = x'/(h - y)$ or $x' = (b/h)(h - y)$.
Step 2: The element intersects the sides of the triangle at an arbitrary height y above the x axis.
Step 3: The area of the element is $dA = x' \, dy = (b/h)(h - y) \, dy$.
Step 4: The centroid of the element is located a distance $\tilde{y} = y$ from the x axis.
Step 5: Using the second of Eqs. 9–8, and integrating with respect to y yields

$$\bar{y} = \frac{\int_A \tilde{y} \, dA}{\int_A dA} = \frac{\int_0^h y \frac{b}{h}(h - y) \, dy}{\int_0^h \frac{b}{h}(h - y) \, dy}$$

$$= \frac{\frac{b}{h} \int_0^h (hy - y^2) \, dy}{\frac{b}{h} \int_0^h (h - y) \, dy} = \frac{h}{3} \qquad Ans.$$

Locate the centroid for the paraboloid of revolution, which is generated by revolving the shaded area shown in Fig. 9–12a about the y axis.

Solution I

Since the generated volume is *symmetric* with respect to the y axis,

$$\bar{x} = \bar{z} = 0 \qquad\qquad Ans.$$

Step 1: An element having the shape of a *thin disk* is chosen, Fig. 9–12a. This element has a radius of $r = z$ and a thickness of dy. *Note:* In this "disk" method of analysis, the element of planar area, dA, is always taken *perpendicular* to the axis of revolution.

Step 2: The element intersects the generating curve at the arbitrary point $(0, y, z)$.

Step 3: The volume of the element is $dV = (\pi z^2)\, dy$.

Step 4: The centroid of the element is located at $\tilde{y} = y$.

Step 5: Using the second of Eqs. 9–9, and integrating with respect to y, yields

$$\bar{y} = \frac{\displaystyle\int_V \tilde{y}\, dV}{\displaystyle\int_V dV} = \frac{\displaystyle\int_0^{100} y(\pi z^2)\, dy}{\displaystyle\int_0^{100} (\pi z^2)\, dy} = \frac{100\pi \displaystyle\int_0^{100} y^2\, dy}{100\pi \displaystyle\int_0^{100} y\, dy} = 66.7 \text{ mm} \quad Ans.$$

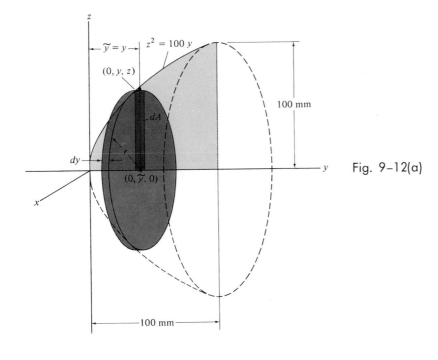

Fig. 9-12(a)

Solution II

Step 1: As shown in Fig. 9–12b, the volume element can be chosen in the form of a *thin cylindrical shell,* where the shell radius is $r = z$ and the thickness is dz. *Note:* In this "shell" method of analysis, the element of planar area, dA, is always taken *parallel* to the axis of revolution.

Step 2: The element intersects the generating curve at point $(0, y, z)$.

Step 3: The volume of the element is $dV = 2\pi r\, dA = 2\pi z(100 - y)\, dz$.

Step 4: The centroid of the element is located at

$$\tilde{y} = y + \frac{100 - y}{2} = \frac{100 + y}{2}$$

Step 5: Using the second of Eqs. 9–9 and integrating with respect to z yields

$$\bar{y} = \frac{\int_V \tilde{y}\, dV}{\int_V dV} = \frac{\int_0^{100} \left(\frac{100 + y}{2}\right) 2\pi z(100 - y)\, dz}{\int_0^{100} 2\pi z(100 - y)\, dz}$$

$$= \frac{\pi \int_0^{100} z(10^4 - 10^{-4}z^4)\, dz}{2\pi \int_0^{100} z(100 - 10^{-2}z^2)\, dz} = 66.7 \text{ mm} \qquad Ans.$$

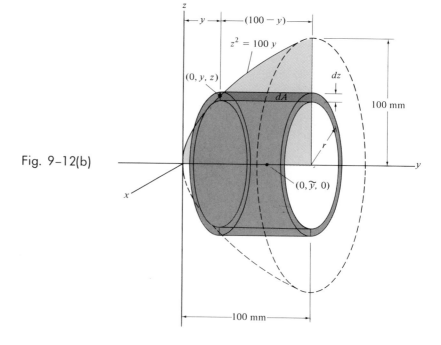

Fig. 9–12(b)

Example 9-6

Locate the center of gravity of the cylinder shown in Fig. 9–13a if the density of the material varies directly with its distance from the base of the cylinder such that $\rho_V = 2000z$ kg/m³.

Solution

For reasons of material symmetry

$$\bar{x} = \bar{y} = 0 \qquad \qquad Ans.$$

Step 1: A disk element is chosen for integration, Fig. 9–13a, since the *density of this entire element is constant* for a given value of z. The element has a constant radius of 0.05 m and a thickness of dz.

Step 2: The element is located along the z axis at the arbitrary point $(0, 0, z)$.

Step 3: The volume of the element is $dV = \pi(0.05)^2\, dz$.

Step 4: The center of gravity of the element is located at $\tilde{z} = z$.

Step 5: Using the third of Eqs. 9–6, and integrating with respect to z, with $\rho = 2000z$ kg/m³, yields

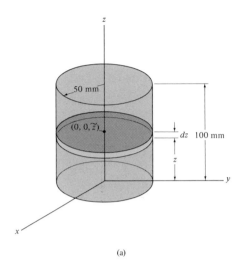

(a)

$$
\bar{z} = \frac{\displaystyle\int_V \tilde{z}\rho\, dV}{\displaystyle\int_V \rho\, dV} = \frac{\displaystyle\int_0^{0.1} (z)(2000z)\pi(0.05)^2\, dz}{\displaystyle\int_0^{0.1} (2000z)\pi(0.05)^2\, dz}
$$

$$
= \frac{5\pi \displaystyle\int_0^{0.1} z^2\, dz}{5\pi \displaystyle\int_0^{0.1} z\, dz} = 0.0667 \text{ m} = 66.7 \text{ mm} \qquad Ans.
$$

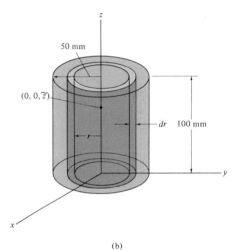

(b)

Fig. 9–13

Note: It is not possible to use a shell element for integration such as shown in Fig. 9–13b, since the density of the material composing the shell would *vary* along the shell's height, and hence the location of $\tilde{z}$ for the element would not be easily determined.

Problems

9-6. Locate the center of mass of the homogeneous wire bent into the shape of a circular arc.

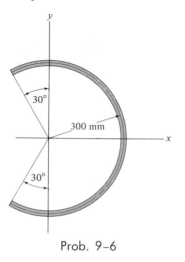

Prob. 9–6

*9-8. Locate the center of mass of the homogeneous rod bent in the form of a parabola.

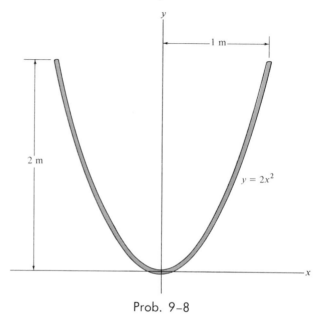

$y = 2x^2$

Prob. 9–8

9-7. Determine the distance $\bar{x}$ to the center of mass of the homogeneous rod bent into the shape shown.

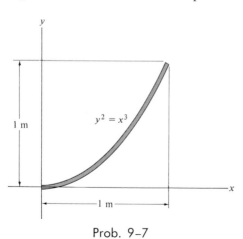

$y^2 = x^3$

Prob. 9–7

9-9. Locate the centroid of the shaded area.

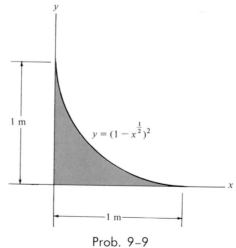

$y = (1 - x^{\frac{1}{2}})^2$

Prob. 9–9

9-10. Locate the centroid $\bar{y}$ of the shaded area if $a = 100$ mm.

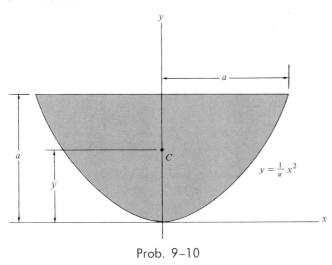

Prob. 9–10

9-10a. Solve Prob. 9–10 if $a = 2$ ft.

9-11. Locate the centroid of the plate area.

Prob. 9–11

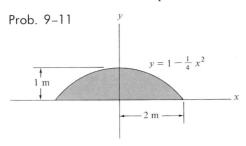

$$y = 1 - \frac{1}{4}x^2$$

*9-12. Locate the centroid for the quarter circular area.

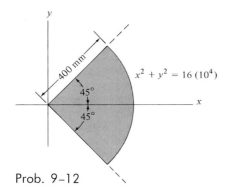

Prob. 9–12

9-13. Locate the centroid of the shaded elliptical sector.

Prob. 9–13

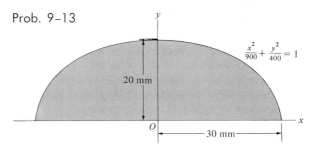

$$\frac{x^2}{900} + \frac{y^2}{400} = 1$$

9-14. Locate the centroid of the shaded area of the plate.

Prob. 9–14

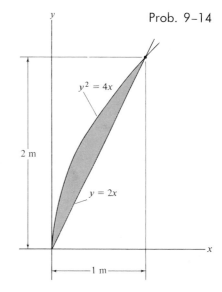

$$y^2 = 4x$$

$$y = 2x$$

9-15. Locate the center of gravity of the homogeneous cantilever beam. Set $a = 2$ m and $b = 0.5$ m.

Prob. 9–15

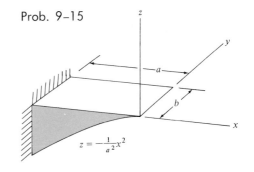

$$z = -\frac{1}{a^2}x^2$$

9–15a. Solve Prob. 9–15 if $a = 3$ ft and $b = 1.5$ ft.

*9–16. Locate the center of mass of the thin homogeneous cylindrical shell.

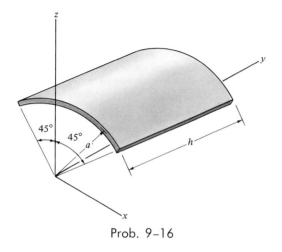

Prob. 9–16

9–17. Locate the center of gravity $\bar{z}$ of the thin homogeneous conical shell.

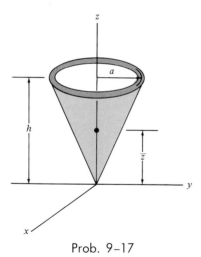

Prob. 9–17

9–18. Locate the center of gravity of the thin homogeneous hemispherical shell. *Suggestion:* Choose a ring element having a center at $(x, 0, 0)$, radius z, and thickness $dL = \sqrt{(dx)^2 + (dz)^2}$.

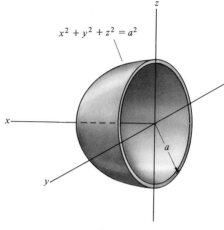

$$x^2 + y^2 + z^2 = a^2$$

Prob. 9–18

9–19. Locate the center of gravity of the hemisphere. The density of the material varies linearly from zero at the origin O to ρ_0 at the surface. *Suggestion:* Choose a hemispherical shell element for integration and use the result of Prob. 9–18.

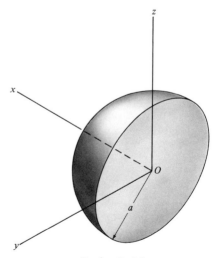

Prob. 9–19

*9-20. Locate the center of mass of the homogeneous "bell-shaped" volume formed by revolving the shaded area about the y axis. Set $a = 1$ m.

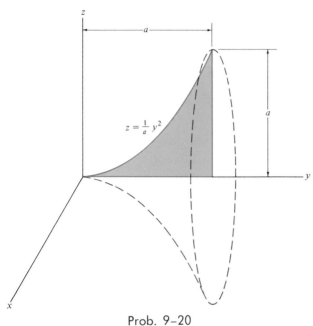

Prob. 9-20

*9-20a. Solve Prob. 9-20 if $a = 2$ ft.

9-21. Locate the center of gravity of the homogeneous cone formed by rotating the shaded area about the y axis.

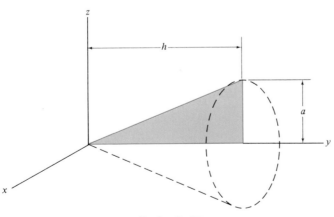

Prob. 9-21

9-22. Locate the center of gravity of the volume generated by revolving the shaded area about the z axis. The material is homogeneous.

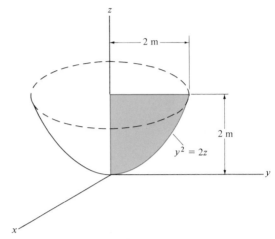

Prob. 9-22

9-23. Locate the center of gravity of the volume generated by revolving the shaded area about the y axis. The material is homogeneous.

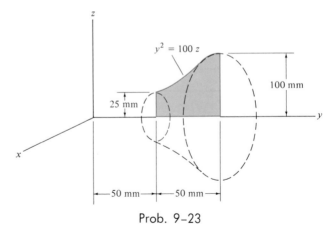

Prob. 9-23

*9-24. Locate the center of gravity for the homogeneous half-cone.

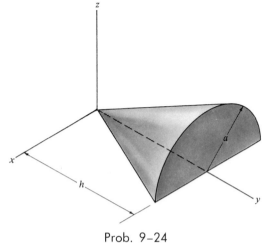

Prob. 9–24

9-25a. Solve Prob. 9–25 if $b = 6$ in., $d = 2$ in., and $c = 2$ in.$^{1/2}$.

9-26. Determine the distance $\bar{z}$ to the centroid of the spherical segment.

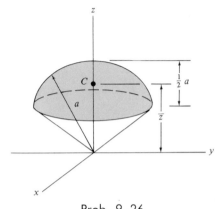

Prob. 9–26

9-25. Locate the center of gravity of the homogeneous solid formed by rotating the shaded area about the *aa* axis. Set $b = 200$ mm, $d = 100$ mm, and $c = 10$ mm$^{1/2}$.

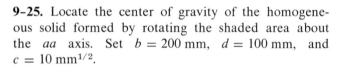

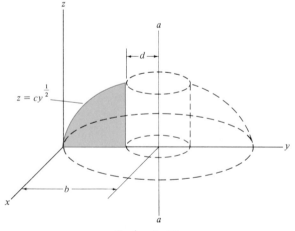

Prob. 9–25

9-27. Locate the center of mass of the homogeneous rectangular pyramid.

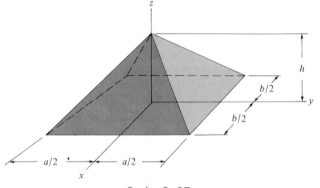

Prob. 9–27

9–3. Composite Bodies

In many cases, a body can be sectioned or divided into several parts having simpler shapes. Provided the weight and location of the center of gravity of each of these "composite parts" are known, one can eliminate the need for integration to determine the center of gravity for the entire body. Application of the principle of moments to each of the composite parts yields formulas analogous to Eqs. 9–1, since there is a finite number of weights to be accounted for. Thus, the necessary formulas for finding the center of gravity for a three-dimensional body become

$$\bar{x} = \frac{\Sigma \tilde{x}W}{\Sigma W} \qquad \bar{y} = \frac{\Sigma \tilde{y}W}{\Sigma W} \qquad \bar{z} = \frac{\Sigma \tilde{z}W}{\Sigma W} \qquad (9\text{--}10)$$

where $\tilde{x}, \tilde{y},$ and $\tilde{z}$ represent the *algebraic distances* from the center of gravity of each composite part to the origin of the coordinates, and ΣW represents the sum of the weights of each of the composite parts or simply the *total weight* of the body.

When the body has a *constant density,* the center of gravity *coincides* with the geometric center or centroid of the body. The centroid for composite lines, areas, and volumes can be found using relations analogous to Eq. 9–10; however, the W's are replaced by L's, A's, and V's, respectively. Centroids for common shapes of lines, areas, shells, and volumes are given in Appendix B.

PROCEDURE FOR ANALYSIS

The following three-step procedure should be used to determine the center of gravity of a body or the centroid of a geometrical object represented by a line, area, or volume:

Step 1: Using a sketch, divide the body or object into a finite number of composite parts. If a *hole,* or geometric region having no material, represents one of the parts, the weight or "size" of the hole is considered a *negative quantity.*

Step 2: Establish the location and orientation of the coordinate axes on the sketch and determine the location $\tilde{x}, \tilde{y}, \tilde{z}$ of the center of gravity or centroid of each part.

Step 3: Determine $\bar{x}, \bar{y}, \bar{z}$ by applying the center of gravity equations (Eq. 9–10) or the analogous centroid equations. If an object is *symmetrical* about an axis, recall that the centroid of the object lies on the axis.

The calculations involved in *Step 3* can be arranged in tabular form, as indicated in the following two examples.

Fig. 9-14

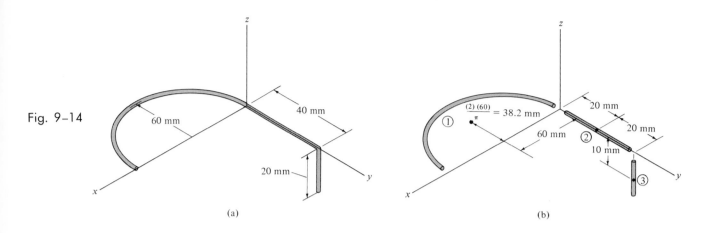

(a)

(b)

Example 9-7

Locate the centroid of the wire shown in Fig. 9–14a.

Solution

Step 1: The wire is divided into three segments as shown in Fig. 9–14b.
Step 2: The location of the centroid for each piece is determined and indicated on the sketch. In particular, the centroid of segment 1 is determined either by integration or using the formula in Appendix B.
Step 3: The calculations are listed in the following table:

Segment	L (mm)	$\tilde{x}$ (mm)	$\tilde{y}$ (mm)	$\tilde{z}$ (mm)	$\tilde{x}L$ (mm²)	$\tilde{y}L$ (mm²)	$\tilde{z}L$ (mm²)
1	$\pi(60) = 188.5$	60	−38.2	0	11 310	−7200	0
2	40	0	20	0	0	800	0
3	20	0	40	−10	0	800	−200
	$\Sigma L = 248.5$				$\Sigma\tilde{x}L = 11\ 310$	$\Sigma\tilde{y}L = -5600$	$\Sigma\tilde{z}L = -200$

Thus,

$$\bar{x} = \frac{\Sigma\ \tilde{x}L}{\Sigma\ L} = \frac{11\ 310}{248.5} = 45.5 \text{ mm} \qquad Ans.$$

$$\bar{y} = \frac{\Sigma\ \tilde{y}L}{\Sigma\ L} = \frac{-5600}{248.5} = -22.5 \text{ mm} \qquad Ans.$$

$$\bar{z} = \frac{\Sigma\ \tilde{z}L}{\Sigma\ L} = \frac{-200}{248.5} = -0.8 \text{ mm} \qquad Ans.$$

Example 9-8

Locate the center of gravity of the composite assembly shown in Fig. 9-15a. The conical frustum has a density of $\rho_c = 8\ \text{Mg/m}^3$ and the hemisphere has a density of $\rho_h = 4\ \text{Mg/m}^3$.

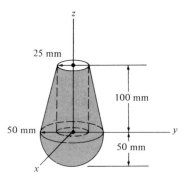

Fig. 9-15(a)

Solution

Step 1: The assembly can be thought of as consisting of four segments as shown in Fig. 9-15b. For the calculations, segments 3 and 4 must be considered as "negative" volumes in order that the four pieces, when added together, yield the total composite shape shown in Fig. 9-15a.

Step 2: Using Appendix B, the computations for the centroid $\tilde{z}$ of each piece are shown in the figure.

Step 3: Because of *symmetry*, note that

$$\bar{x} = \bar{y} = 0 \qquad\qquad Ans.$$

Since $W = mg$ and g is constant, the mass of each piece can be computed from $m = \rho V$ and used for the calculations. Also, $1\ \text{Mg/m}^3 = 10^{-6}\ \text{kg/mm}^3$, so that

Segment	m (kg)	$\tilde{z}$ (mm)	$\tilde{z}m$ (kg·mm)
1	$8(10^{-6})(\frac{1}{3})\pi(50)^2(200) = 4.189$	50	209.440
2	$4(10^{-6})(\frac{2}{3})\pi(50)^3 = 1.047$	-18.75	-19.635
3	$-8(10^{-6})(\frac{1}{3})\pi(25)^2(100) = -0.524$	$100 + 25 = 125$	-65.450
4	$-8(10^{-6})\pi(25)^2(100) = -1.571$	50	-78.540
	$\Sigma m = 3.141$		$\Sigma\tilde{z}m = 45.815$

Applying Eq. 9-10 yields

$$\bar{z} = \frac{\Sigma\tilde{z}W}{\Sigma W} = \frac{\Sigma\tilde{z}m}{\Sigma m} = \frac{45.815}{3.141} = 14.6\ \text{mm} \qquad\qquad Ans.$$

Fig. 9-15(b)

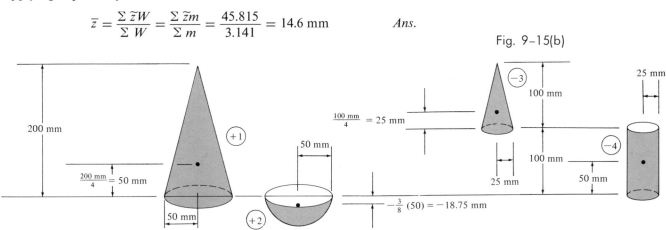

Problems

*9–28. Locate the center of mass of the homogeneous rod.

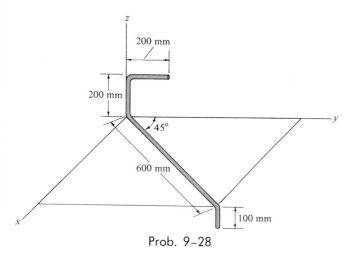

Prob. 9–28

9–29. Locate the center of gravity of the homogeneous rod.

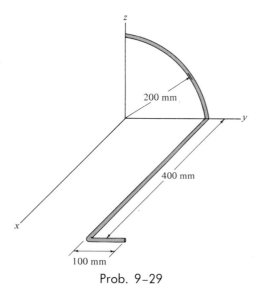

Prob. 9–29

9–30. Locate the center of gravity of the homogeneous wire. Set $a = 150$ mm and $b = 200$ mm.

Prob. 9–30

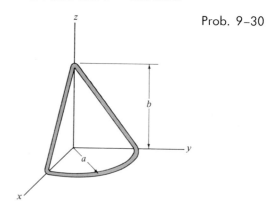

9–30a. Solve Prob. 9–30 if $a = 3$ in. and $b = 4$ in.

9–31. Determine the length of the wire segment AB so that the centroid for the entire composite is located at point C.

Prob. 9–31

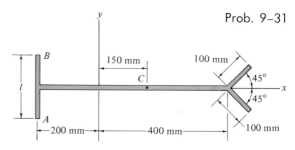

*9–32. Determine the distance $\bar{y}$ to the centroidal axis $\bar{x}\bar{x}$ of the beam's cross-sectional area.

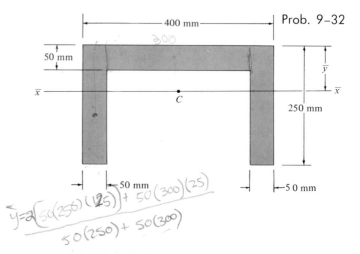

Prob. 9–32

$$y = 2\left(\frac{50(250)(125)}{}\right) + 50(300)(25)$$
$$\overline{50(250) + 50(300)}$$

9–33. Determine the distance $\bar{y}$ to the centroidal axis $\bar{x}\,\bar{x}$ of the beam's cross-sectional area. Neglect the size of the corner welds at A and B for the calculation.

Prob. 9–33

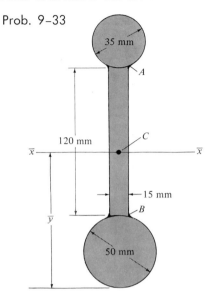

9–34. Determine the distance $\bar{y}$ to the centroidal axis $\bar{x}\,\bar{x}$ of the beam's cross-sectional area. Neglect the size of the corner welds at A and B for the calculation.

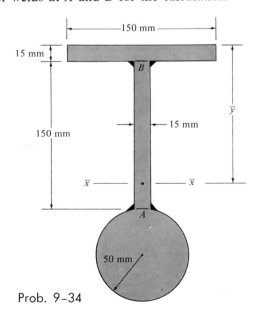

Prob. 9–34

9–35. Locate the centroid for the cross-sectional area of the beam. Set $a = 25$ mm.

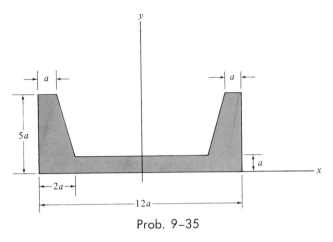

Prob. 9–35

9–35a. Solve Prob. 9–35 with $a = 1.5$ in.

*** 9–36.** Locate the centroid $\bar{y}$ of the cross-sectional area of the beam constructed from a plate, channel, and four angles. Handbook values for the areas, and centroids C_c and C_a of the channel and one of the angles are listed. Neglect the size of all the rivet heads, R, for the calculation.

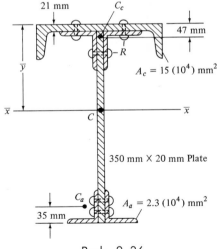

Prob. 9–36

9–37. Locate the centroid of the cross-sectional area of the beam.

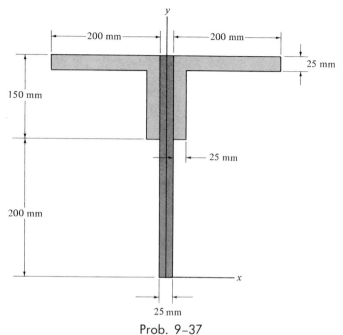

Prob. 9–37

9–38. Determine the distance $\bar{y}$ to the centroidal axis $\bar{x}\,\bar{x}$ of the beam's cross-sectional area.

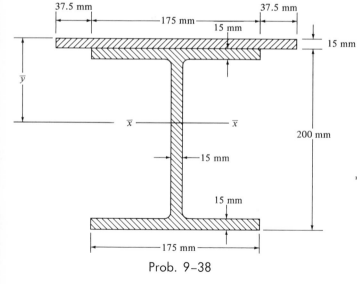

Prob. 9–38

9–39. Locate, approximately, the centroid of the plate.

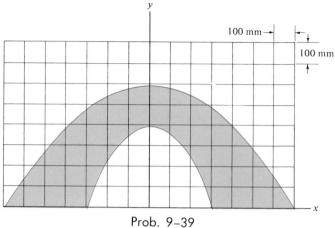

Prob. 9–39

*9–40. Determine the distance $\bar{y}$ to the centroid C for the circular sector having a radius $r = 0.25$ m.

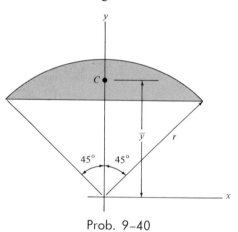

Prob. 9–40

*9–40a. Solve Prob. 9–40 if $r = 5$ in.

9-41. Locate the center of gravity of the solid. The frustum A has a mass density of $\rho_A = 5$ Mg/m³ and the hemisphere B has a density of $\rho_B = 3$ Mg/m³.

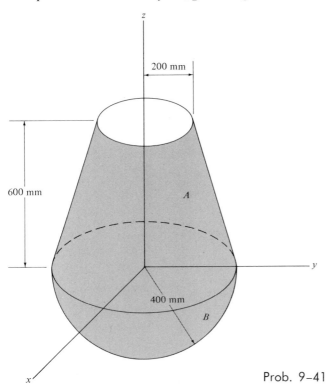

Prob. 9-41

9-42. Determine the center of mass of the airplane. The mass of the various items is shown in the table.

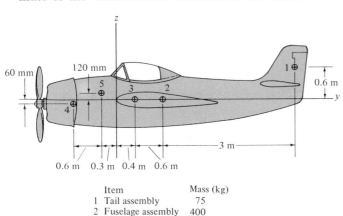

Item	Mass (kg)
1 Tail assembly	75
2 Fuselage assembly	400
3 Wing assembly	425
4 Engine section	60
5 Fixed equipment	450

Prob. 9-42

9-43. The frustum of a right circular cone has a hole drilled along its axis. Locate the center of gravity if the material is homogeneous.

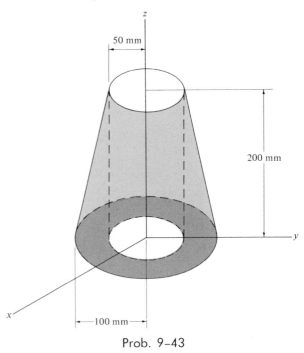

Prob. 9-43

*9-44. A toy skyrocket consists of a conical top, $\rho_t = 600$ kg/m³, a hollow cylinder, $\rho_c = 400$ kg/m³, and a stick having a circular cross section, $\rho_s = 300$ kg/m³. Determine the length of the stick, x, so that the center of gravity G of the skyrocket is located along line aa.

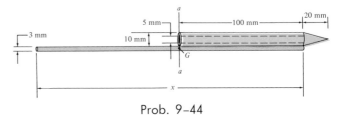

Prob. 9-44

371

9-45. Locate the center of gravity of the two-block assembly. The densities of materials A and B are $\rho_A = 2 \text{ Mg/m}^3$ and $\rho_B = 6 \text{ Mg/m}^3$, respectively. Set $a = 100$ mm.

Prob. 9–45

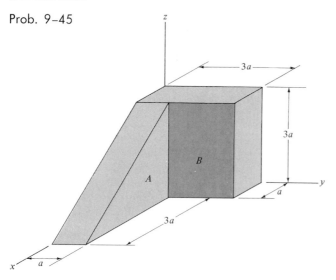

9-45a. Solve Prob. 9-45 if $\rho_A = 150 \text{ lb/ft}^3$, $\rho_B = 400 \text{ lb/ft}^3$, and $a = 1$ in.

9-46. Locate the center of gravity of the sheet-metal bracket if the material is homogeneous and has a constant thickness.

Prob. 9–46

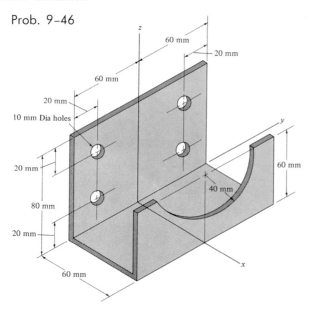

9-47. Locate the center of gravity of the casting that is formed from a hollow cylinder having a density of $\rho = 9 \text{ Mg/m}^3$ and a hemisphere having a density of 3 Mg/m^3.

Prob. 9–47

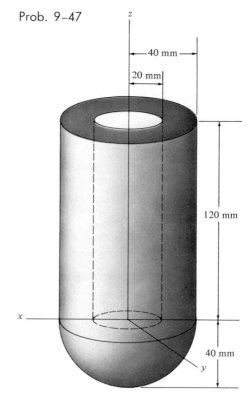

*9-48. Each of the three plates welded to the rod has a density of 7 Mg/m^3 and a thickness of 10 mm. Determine the length l of plate C and the angle of placement, θ, so that the center of gravity of the assembly lies on the y axis. Plates A and B lie in the x-y and z-y planes, respectively.

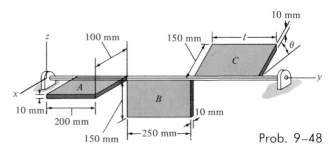

Prob. 9–48

*9–4. Theorems of Pappus and Guldinus

The *theorems of Pappus and Guldinus,* which were first developed by Pappus of Alexandria during the third century A.D. and again at a later time by the Swiss mathematician Paul Guldin or Guldinus (1577–1643), are used to find the surface area or volume of any solid of revolution.

A *surface area of revolution* is generated by revolving a *curve* about a fixed axis; whereas a *volume of revolution* is generated by revolving an *area* about the axis. For example, if the line *AB*, shown in Fig. 9–16, is rotated about the *x* axis, it generates the surface area of a cone. Furthermore, if the triangular area *ABC* shown in Fig. 19–17 is rotated about the *x* axis, it generates the volume of a cone.

The statements and proofs of the theorems of Pappus and Guldinus follow. The proofs require that the generating curves and areas do *not* cross the axis about which they are rotated; otherwise, two sections on either side of the axis would generate areas or volumes having opposite signs and hence would produce a cancellation.

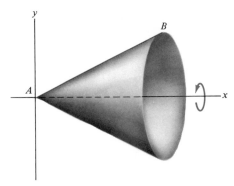

Fig. 9–16

Surface Area. *The area of a surface of revolution equals the product of the length of the generating curve and the distance traveled by the centroid of the curve in generating the surface area.*

Proof: When a differential length dL of the curve shown in Fig. 9–18 is revolved about the *y* axis through a distance of $2\pi x$, it generates a ring having a surface area $dA = 2\pi x \, dL$. The entire surface area, generated by revolving the entire curve about the *y* axis, is therefore $A = 2\pi \int_L x \, dL$. Since $\int_L x \, dL = \bar{x}L$, where $\bar{x}$ locates the centroid *C* of the generating curve *L*, the area becomes $A = 2\pi\bar{x}L$. In general, though,

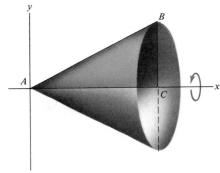

Fig. 9–17

$$A = \theta\bar{x}L \qquad (9\text{–}11)$$

where
A is the surface area of revolution
θ is the angle of revolution, measured in radians, $\theta \leqslant 2\pi$
$\bar{x}$ is the distance between the centroid of the generating curve and the axis
L is the length of the generating curve

Fig. 9–18

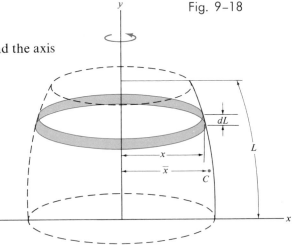

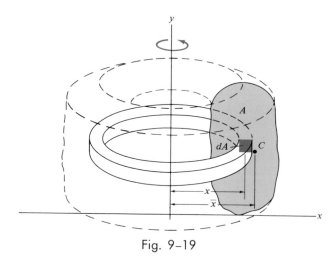

Fig. 9–19

Volume. *The volume of a surface of revolution equals the product of the generating area and the distance traveled by the centroid of the area in generating the volume.*

Proof: When the differential area dA, shown in Fig. 9–19, is revolved about the y axis through a distance of $2\pi x$, it generates a ring having volume $dV = 2\pi x\, dA$. The volume, generated by revolving A about the y axis, is therefore $V = 2\pi \int_A x\, dA$. Since $\int_A x\, dA = \bar{x} A$, where $\bar{x}$ locates the centroid C of the generating area A, the volume becomes $V = 2\pi \bar{x} A$. In general, though,

$$V = \theta \bar{x} A \qquad\qquad (9\text{–}12)$$

where
V is the volume of revolution
θ is the angle of revolution, measured in radians, $\theta \leqslant 2\pi$
$\bar{x}$ is the distance between the centroid of the generating area and the axis
A is the generating area

The following examples illustrate application of the above two theorems.

Example 9–9

Show that the surface area of a sphere is $4\pi R^2$.

Solution
A sphere is generated by rotating the semicircular *arc* shown in Fig. 9–20 about the x axis. Using Appendix B, it is seen that the centroid of this arc is located at a distance of $\bar{y} = 2R/\pi$ from the x axis of rotation. Since the centroid moves through an angle of $\theta = 2\pi$ rad in generating the sphere, applying Eq. 9–11, the surface area of the sphere becomes

$$A = \theta \bar{y} L; \qquad A = 2\pi\left(\frac{2R}{\pi}\right)\pi R = 4\pi R^2 \qquad \textit{Ans.}$$

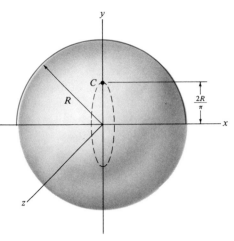

Fig. 9–20

Example 9–10

Determine the mass of concrete needed to construct the arched beam shown in Fig. 9–21. The density of concrete is $\rho_c = 2.1 \text{ Mg/m}^3$.

Solution
The mass of the arch may be determined by using the density ρ_c provided the volume of concrete is known. The cross-sectional area of the beam is composed of two rectangles having centroids at points C_1 and C_2, as shown in Fig. 9–21b. The volume generated by each of these areas is equal to the product of the cross-sectional area and the distance traveled by its centroid. In generating the arch, the centroids move through an angle of $\theta = \pi(90°/180°) = 1.57$ radians. Point C_1 acts at a distance of $d_1 = 4.5 - 0.1 = 4.4$ m from the center of rotation, point O; whereas C_2 acts at $d_2 = 4 + 0.15 = 4.15$ m from O. Applying Eq. 9–12, the total volume is thus

$$\begin{aligned} V &= \Sigma\, \theta \bar{x} A \\ &= (1.57 \text{ rad})(4.4 \text{ m})(0.5 \text{ m})(0.2 \text{ m}) + (1.57 \text{ rad})(4.15)(0.15 \text{ m})(0.3 \text{ m}) \\ &= 0.984 \text{ m}^3 \end{aligned}$$

The required mass of concrete is then

$$m = \rho_c V = (2.1 \text{ Mg/m}^3)(0.984 \text{ m}^3) = 2.07 \text{ Mg} \qquad \textit{Ans.}$$

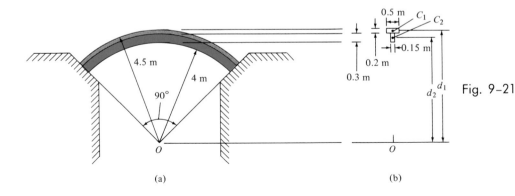

Fig. 9–21

375

Problems

9–49. Using integration, compute both the area and the centroidal distance $\bar{x}$ for the shaded region. Then, using the second theorem of Pappus–Guldinus, compute the volume of the solid generated by revolving the shaded area about the *aa* axis.

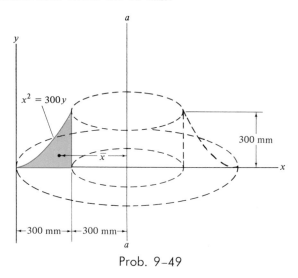

Prob. 9–49

9–50. Determine the surface area and volume of the torus if $d_1 = 400$ mm and $d_2 = 800$ mm.

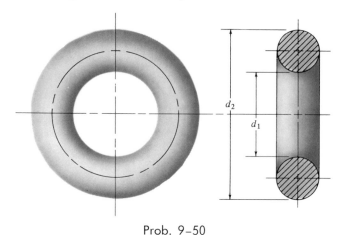

Prob. 9–50

9–50a. Solve Prob. 9–50 if $d_1 = 5$ in. and $d_2 = 8$ in.

9–51. Using integration, determine the area and the centroidal distance $\bar{y}$ of the shaded area. Then, using the second theorem of Pappus–Guldinus, determine the volume of a paraboloid formed by revolving the area about the *x* axis.

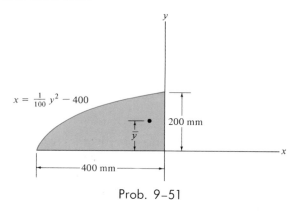

Prob. 9–51

*9–52. Using the first theorem of Pappus–Guldinus, determine the surface area of a cone, using an inclined line segment to generate the cone. See Fig. 9–16.

9–53. A circular V-belt has an inner radius of 600 mm and a cross-sectional area as shown. Determine the volume of material required to make the belt.

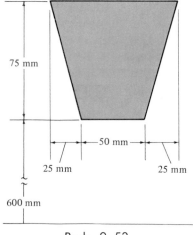

Prob. 9–53

9-54. Determine the height h to which liquid should be poured into the conical cup so that it contacts half the surface area on the inside of the cup.

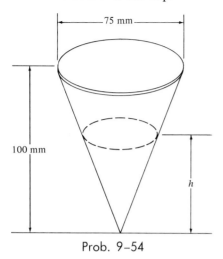

Prob. 9-54

9-55. A circular sea wall is made of concrete. Determine the total mass of the wall if the concrete has a density of $\rho = 2.5 \text{ Mg/m}^3$. Set $a = 3$ m, $b = 10$ m, $c = 1.5$ m, and $r = 20$ m.

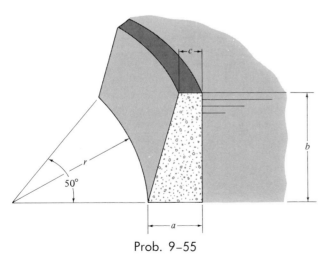

Prob. 9-55

9-55a. Solve Prob. 9-55 if $\rho = 150 \text{ lb/ft}^3$, $a = 15$ ft, $b = 30$ ft, $c = 8$ ft, and $r = 60$ ft.

* **9-56.** Determine the volume of concrete needed to construct the curb.

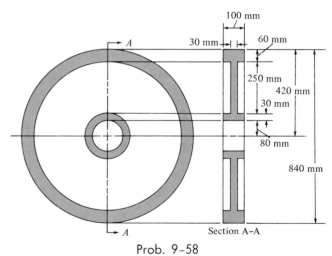

Prob. 9-56

9-57. Determine the surface area of the curb in Prob. 9-56. For the calculation, neglect the area at the ends of the curb.

9-58. A steel wheel has a diameter of 840 mm and a cross section as shown in the figure. Determine the total mass of the wheel if $\rho = 3 \text{ Mg/m}^3$.

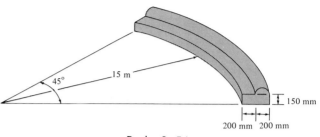

Prob. 9-58

377

9-59. The surface of the water tank is to be painted with noncorrosive paint. If a liter of paint covers 3 m³, determine by approximate means the number of liters of paint required to paint the surface of the tank.

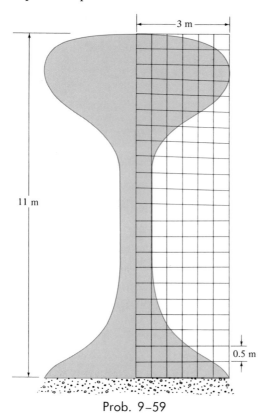

Prob. 9-59

*9-60. Determine both the surface area and the volume of material required to make the casting shown. Set $r_1 = 300$ mm, $r_2 = 100$ mm, and $r_3 = 200$ mm.

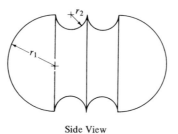

Side View   Front View

Prob. 9-60

*9-60a. Solve Prob. 9-60 if $r_1 = 3$ in., $r_2 = 1$ in., and $r_3 = 2$ in.

*9–5. Fluid Pressure Acting on Submerged Surfaces

The method used to reduce a distributed surface loading to a single resultant force $\mathbf{F}_R$ was outlined in Sec. 4–10. There it was shown that $\mathbf{F}_R$ has a *magnitude* equal to the *volume* under the loading diagram, and a *line of action* which passes through the *centroid of this volume*. In this section, that concept will be applied to finding the resultant force acting on the surface of a body submerged in a fluid.

Pressure Loading. According to Pascal's law, a fluid at rest creates a pressure p at a point that is the *same* in *all* directions. The magnitude of p, measured as a force per unit area, depends upon the mass density ρ of the fluid and the depth z of the point from the fluid surface. The relationship can be expressed mathematically as

$$p = \rho g z \qquad (9\text{–}13)$$

where g is the acceleration of gravity (9.81 m/s²). Equation 9–13 is only valid for fluids that are *incompressible,* as in the case of most liquids. Gases are compressible fluids, and since their density changes significantly with both altitude and temperature, Eq. 9–13 cannot be used.

Consider now what effect the pressure of a liquid has at points A, B, and C, located on the top surface of the submerged plate shown in Fig. 9–22. Since points A and B are both at depth z_2 from the liquid surface, the *pressure* at these points has a magnitude of $p_2 = \rho g z_2$. Likewise, point C is at depth z_1; hence, $p_1 = \rho g z_1$. In all cases, the pressure acts *normal* to the surface area dA located at the point, Fig. 9–22.

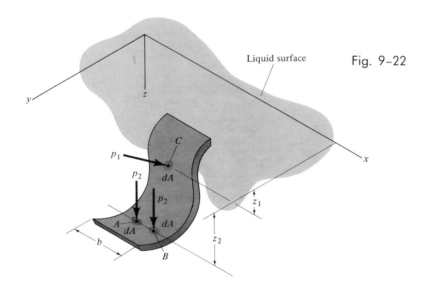

Liquid surface   Fig. 9–22

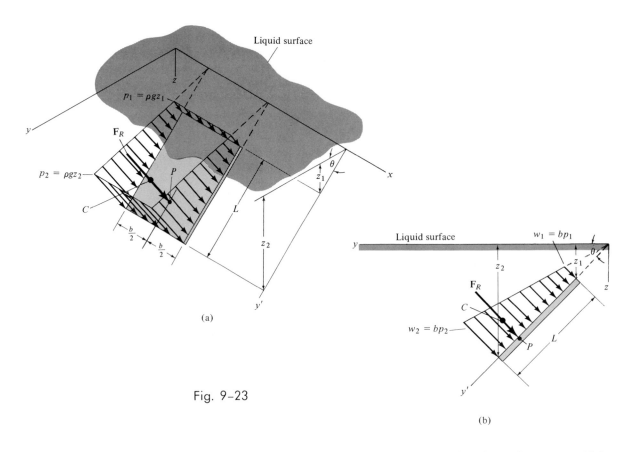

Fig. 9–23

(a)

(b)

Flat Plate of Constant Width. A flat rectangular plate of constant width, which is submerged in a liquid having a mass density ρ, is shown in Fig. 9–23a. The plane of the plate makes an angle θ with the horizontal, such that its top edge is located at depth z_1 from the liquid surface, and its bottom edge is located at depth z_2. Since pressure varies linearly with depth (Eq. 9–13), the distribution of pressure over the plate's surface is represented by a trapezoidal volume of loading. The magnitude of the *resultant force* $\mathbf{F}_R$ is equal to the *volume* of this loading diagram, and has a *line of action* that passes through the volume's centroid, C. Note that $\mathbf{F}_R$ does *not* act at the center of the plate; rather, it acts at point P, called the *center of pressure*.

Since the plate has a *uniform width,* the loading distribution may also be viewed in two dimensions, Fig. 9–23b. Here the loading intensity is measured as force/length and varies linearly from $w_1 = bp_1 = b\rho g z_1$ to $w_2 = bp_2 = b\rho g z_2$. From the theory of Sec. 4–10, the magnitude of $\mathbf{F}_R$ in this case equals the trapezoidal *area,* and $\mathbf{F}_R$ has a *line of action* that passes through the area's *centroid* C. For numerical applications, the area and location of the centroid for a trapezoid are tabulated in Appendix B.

Curved Plate of Constant Width. When the submerged plate is curved, the pressure, acting normal to the plate, continually changes direction, and therefore calculation of the magnitude of $\mathbf{F}_R$ and its location P is more difficult than for a flat plate. Three- and two dimensional views of the loading distribution are shown in Figs. 9–24a and 9–24b, respectively. Here integration can be used to determine both F_R (the volume or area under the loading diagrams) and the location of the centroid C or center of pressure P.

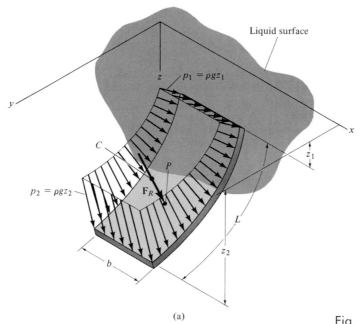

(a)

Fig. 9–24

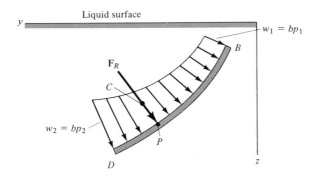

(b)

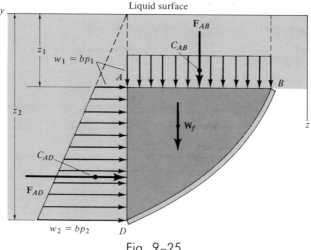

Fig. 9–25

A simpler method exists, however, for calculating the magnitude of $\mathbf{F}_R$ and its location along a curved (or flat) plate having a *constant width*. This method requires separate calculations for the horizontal and vertical *components* of $\mathbf{F}_R$. For example, the distributed loading acting on the top surface of the curved plate *DB* in Fig. 9–24*b* can be represented by the *equivalent loading* shown in Fig. 9–25. Here it is seen that *DB* supports a horizontal force along its *vertical projection AD*. This force, $\mathbf{F}_{AD}$, has a magnitude that equals the area under the trapezoid and acts through the centroid C_{AD} of this area. The distributed loading along the *horizontal projection AB* is constant, since all points lying in this plane are at the same depth from the surface of the liquid. The magnitude of $\mathbf{F}_{AB}$ is simply the area of the rectangle. This force acts through the centroid C_{AB} (or midpoint) of *AB*. In addition to $\mathbf{F}_{AB}$, the curved surface *DB* must also support the downward *weight of fluid* $\mathbf{W}_f$ contained within the block *BDA*. This force has a magnitude of $W_f = (\rho g b)(Area_{BDA})$ and acts through the centroid of *BDA*. Summing the three coplanar forces shown in Fig. 9–25 yields $\mathbf{F}_R = \Sigma \mathbf{F} = \mathbf{F}_{AD} + \mathbf{F}_{AB} + \mathbf{W}_f$. The center of pressure *P* is determined by applying the principle of moments to the force system and its resultant about a convenient reference point. The final results for $\mathbf{F}_R$ and its location *P* will be equivalent to those shown in Fig. 9–24*a*.

Flat Plate of Variable Width. The pressure distribution acting across the top face of a submerged plate having a variable width is shown in Fig. 9–26. The resultant force of this loading equals the volume described by the plate area as its base and linear varying pressure distribution as its altitude. The shaded element shown in Fig. 9–26 may be used if integration is chosen to determine this volume. The element consists of a strip of area $dA = x\,dy'$, located at depth z below the liquid surface. Since a uniform pressure $p = \rho gz$ (force/area) acts on dA, the magnitude of the differential force $d\mathbf{F}$ is equal to $dF = dV = p\,dA = \rho gz(x\,dy')$. Integrating over the entire volume yields

$$F_R = \int_V dV = V \qquad (9\text{–}14)$$

The centroid of V defines the point through which $\mathbf{F}_R$ acts. The center of pressure, which lies on the surface of the plate just below C, has coordinates $P(\bar{x}, \bar{y}')$, defined by the equations

$$\bar{x} = \frac{\int_V \tilde{x}\,dV}{\int_V dV} \qquad \bar{y}' = \frac{\int_V \tilde{y}'\,dV}{\int_V dV}$$

This point should *not* be mistaken for the centroid of the plate *area*.

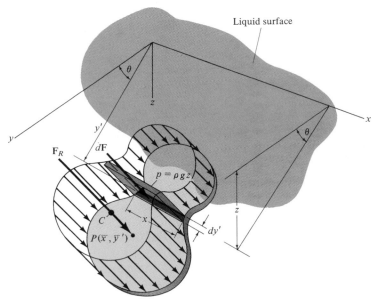

Fig. 9–26

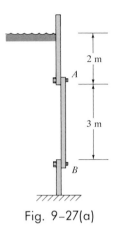

Fig. 9-27(a)

Example 9-11

Determine the magnitude and location of the resultant hydrostatic force acting on the submerged plate *AB* shown in Fig. 9-27a. The plate has a width of 1.5 m; $\rho_w = 1000 \text{ kg/m}^3$.

Solution

Since the plate has a constant width, the distributed loading can be viewed in two dimensions as shown in Fig. 9-27b. The intensity of the load at *A* and *B* is computed as

$$w_A = b\rho_w g z_A = (1.5 \text{ m})(1000 \text{ kg/m}^3)(9.81 \text{ m/s}^2)(2 \text{ m}) = 29.4 \text{ kN/m}$$
$$w_B = b\rho_w g z_B = (1.5 \text{ m})(1000 \text{ kg/m}^3)(9.81 \text{ m/s}^2)(5 \text{ m}) = 73.6 \text{ kN/m}$$

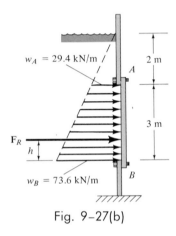

Fig. 9-27(b)

Using Appendix B, the magnitude of the resultant force $\mathbf{F}_R$ created by the distributed load is

$$F_R = (\text{area of trapezoid})$$
$$= \tfrac{1}{2}(3)(29.4 + 73.6) = 154.5 \text{ kN} \qquad Ans.$$

This force acts through the centroid of the area,

$$h = \frac{1}{3}\left(\frac{2(29.4) + 73.6}{29.4 + 73.6}\right)(3) = 1.29 \text{ m} \qquad Ans.$$

measured upward from *B*, Fig. 9-27b.

The same results can be obtained by considering two components of $\mathbf{F}_R$ defined by the triangle and rectangle shown in Fig. 9-27c. Each force acts through its associated centroid and has a magnitude of

$$F_{Re} = (29.4 \text{ kN/m})(3 \text{ m}) = 88.2 \text{ kN}$$
$$F_t = \tfrac{1}{2}(44.2 \text{ kN/m})(3 \text{ m}) = 66.3 \text{ kN}$$

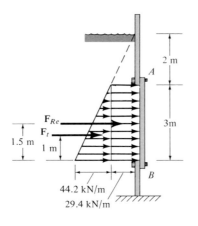

Fig. 9–27(c)

Hence,

$$F_R = F_{Re} + F_t = 88.2 + 66.3 = 154.5 \text{ kN}$$ *Ans.*

The location of $\mathbf{F}_R$ is determined by the principle of moments applied at B, Figs. 9–27b and 9–27c, i.e.,

$$\zeta + (M_R)_B = \Sigma M_B; \quad (154.5)h = 88.2(1.5) + 66.3(1)$$
$$h = 1.29 \text{ m} \qquad Ans.$$

Example 9–12

Determine the magnitude of the resultant hydrostatic force acting on the surface of a sea wall shaped in the form of a parabola as shown in Fig. 9–28a. The wall is 5 m long; $\rho_w = 1020 \text{ kg/m}^3$.

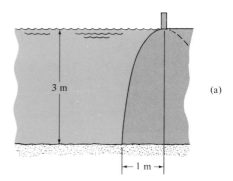

(a)

Solution

The horizontal and vertical components of the resultant force will be calculated, Fig. 9–28b. Since

$$w_B = b\rho_w g z_B = 5 \text{ m}(1020 \text{ kg/m}^3)(9.81 \text{ m/s}^2)(3 \text{ m}) = 150.0 \text{ kN/m}$$

Then

$$F_x = \tfrac{1}{2}(3 \text{ m})(150.0 \text{ kN/m}) = 225.0 \text{ kN}$$

The area of the parabolic sector ABC can be determined using Appendix B. Hence, the weight of water within this region is

$$F_y = (\rho_w g b)(Area_{ABC})$$
$$= (1020 \text{ kg/m}^3)(9.81 \text{ m/s}^2)5 \text{ m}[\tfrac{1}{3}(1 \text{ m})(3 \text{ m})] = 50.0 \text{ kN}$$

The resultant force is, therefore,

$$F_R = \sqrt{F_x^2 + F_y^2} = \sqrt{(225.0)^2 + (50.0)^2}$$
$$= 230.5 \text{ kN} \qquad Ans.$$

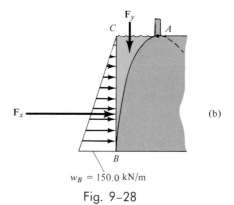

Fig. 9–28

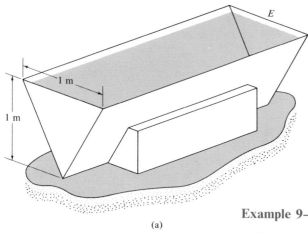

(a)

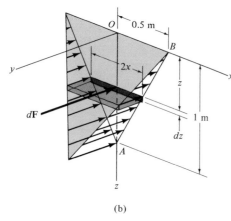

(b)

Fig. 9–29

Example 9–13

Determine the magnitude and location of the resultant force acting on each of the triangular end plates of the water trough shown in Fig. 9–29a, $\rho_w = 1000 \text{ kg/m}^3$.

Solution

The pressure loading acting on the end plate E is shown in Fig. 9–29b. The magnitude of the resultant force $\mathbf{F}$ is equal to the volume of this loading diagram. Choosing the differential volume element shown in the figure yields

$$dF = dV = p\, dA = \rho_w gz(2x\, dz) = 19\,620\, zx\, dz$$

The equation of line AB is

$$x = 0.5(1 - z)$$

Hence, substituting and integrating with respect to z from $z = 0$ to $z = 1$ m yields

$$F = V = \int_V dV = \int_0^1 (19\,620)z[0.5(1 - z)]\, dz$$

$$= 9810 \int_0^1 (z - z^2)\, dz = 1635 \text{ N}$$

or

$$F = 1.64 \text{ kN} \qquad\qquad Ans.$$

This resultant passes through the *centroid of the volume*. Because of symmetry,

$$\bar{x} = 0 \qquad\qquad Ans.$$

Since $\tilde{z} = z$ for the volume element, then

386

$$\bar{z} = \frac{\int_V \bar{z}\, dV}{\int_V dV} = \frac{\int_0^1 z(19\,620)z[0.5(1-z)]\, dz}{1635} = \frac{9810 \int_0^1 (z^2 - z^3)\, dz}{1635}$$

$$\bar{z} = 0.5 \text{ m} \qquad\qquad Ans.$$

Problems

9–61. Determine the magnitude of the resultant hydrostatic force acting on the dam and its location measured from the top surface of the water. The width of the dam is 10 m; $\rho_w = 1.0 \text{ Mg/m}^3$.

Prob. 9–61

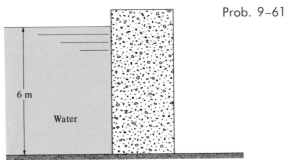

6 m

Water

9–63. The concrete "gravity" dam is held in place by its own weight. If the density of concrete is $\rho_c = 2.5 \text{ Mg/m}^3$, and water has a density of $\rho_w = 1.0 \text{ Mg/m}^3$, determine the smallest width d that will prevent the dam from overturning about its end A.

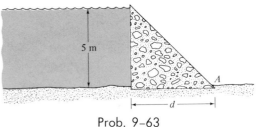

5 m

A

d

Prob. 9–63

9–62. The tank is 1.25 m on each side and 3 m high. If it is filled to a depth of 1 m with water and 2 m with oil, determine the resultant force created by both of these fluids along the side of the tank and its location measured from the top of the tank. $\rho_o = 0.90 \text{ Mg/m}^3$ and $\rho_w = 1.0 \text{ Mg/m}^3$.

Prob. 9–62

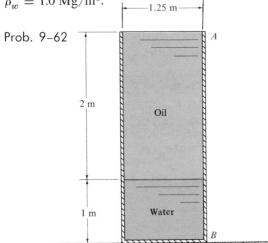

1.25 m

A

2 m

Oil

1 m

Water

B

*9–64. The wind, acting on a square plate, generates a pressure distribution that is parabolic. Determine the magnitude and location of the resultant force.

Prob. 9–64

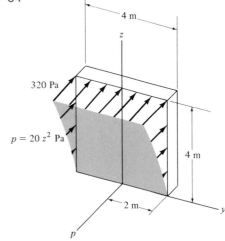

4 m

z

320 Pa

$p = 20 z^2$ Pa

4 m

2 m

y

p

9-65. The rectangular bin is filled with coal, which creates a pressure distribution along wall A that varies as shown. Compute the resultant force created by the coal and specify its location measured from the top surface of the coal. Set $a = 1.5$ m and $p = (500z^3)$ Pa, where z is measured in metres.

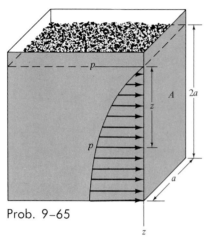

Prob. 9-65

9-65a. Solve Prob. 9-65 if $a = 5$ ft and $p = (4z^3)$ lb/ft^2, where z is measured in feet.

9-66. The form is used to cast concrete columns. Determine the resultant force that wet concrete exerts along the plate A, 0.5 m $\leqslant z \leqslant$ 3 m, if the pressure due to the concrete varies as shown. Specify the location of the resultant force measured from the top of the column.

Prob. 9-66

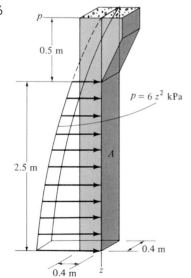

9-67. Determine the magnitude and location of the resultant force acting on each of the cover plates A and B. The density of water is $\rho_w = 1.0$ Mg/m^3.

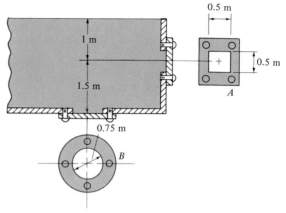

Prob. 9-67

*9-68. Determine the magnitude of the resultant hydrostatic force acting per metre of length on the sea wall; $\rho_w = 1.0$ Mg/m^3.

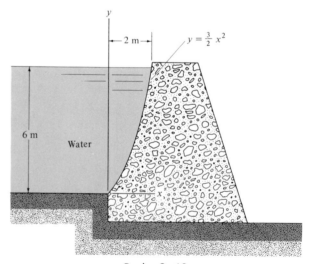

Prob. 9-68

9-69. The storage tank contains oil having a density of $\rho_o = 0.90$ Mg/m^3. If the tank is 1.25 m wide, calculate the resultant force acting on the inclined side AB of the tank caused by the oil and specify its location, measured from A.

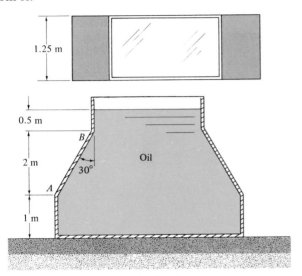

1.25 m

0.5 m

B

2 m

30°

Oil

A

1 m

Prob. 9–69

9–70. Determine the magnitude of the resultant force acting on the gate ABC due to hydrostatic pressure. The gate has a width of $a = 1.5$ m; $\rho_w = 1.0$ Mg/m³. Set $b = 1.5$ m, $c = 1.25$ m, and $d = 2$ m.

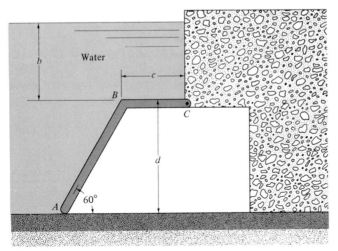

b

Water

c

B

C

d

A

60°

Prob. 9–70

9–70a. Solve Prob. 9–70 if $\rho_w = 62.4$ lb/ft³, $a = 4$ ft, $b = 4$ ft, $c = 3$ ft, and $d = 5$ ft.

9–71. A wind loading creates a positive pressure on one side of a chimney and a negative (suction) pressure on the other side, as shown. If this pressure loading acts uniformly along the chimney's length, determine the magnitude of the resultant force created by the wind.

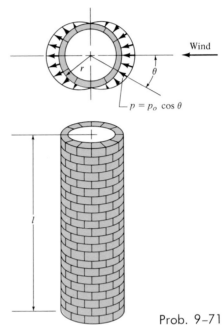

Wind

r

θ

$p = p_o \cos \theta$

l

Prob. 9–71

9–72. The rectangular plate is subjected to a distributed load over its *entire surface*. If the load is defined by the expression $p = p_o \sin (\pi x/a) \sin (\pi y/b)$, where p_o represents the pressure acting at the center of the plate, determine the magnitude and location of the resultant force acting on the plate.

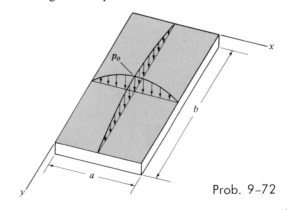

p_o

x

b

a

y

Prob. 9–72

9-73. The semicircular tunnel passes under a river which is 8 m deep. Determine the resultant hydrostatic force acting per metre of length along the length of the tunnel. The tunnel is 6 m wide; $\rho_w = 1.0 \text{ Mg/m}^3$.

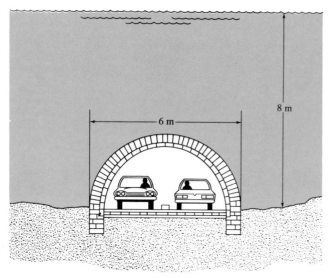

Prob. 9-73

9-74. The tank is filled to the top ($y = 0.5$ m) with water having a density of $\rho_w = 1.0 \text{ Mg/m}^3$. Determine the resultant force of the water pressure acting on the flat end plate C on the tank, and specify its location, measured from the top of the tank.

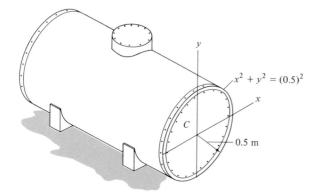

Prob. 9-74

9-75. The semicircular ring supports a uniform distributed load of $w = 800$ N/m. Determine the location and magnitude of the resultant force if $r = 1.5$ m.

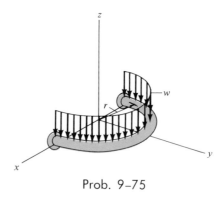

Prob. 9-75

9-75a. Solve Prob. 9-75 if $w = 150$ lb/ft and $r = 6$ ft.

*9-76. The curved beam is subjected to a distributed loading that varies as shown. Determine the resultant force and specify its location.

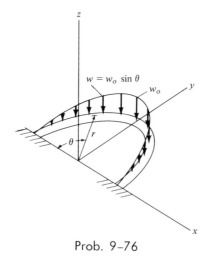

Prob. 9-76

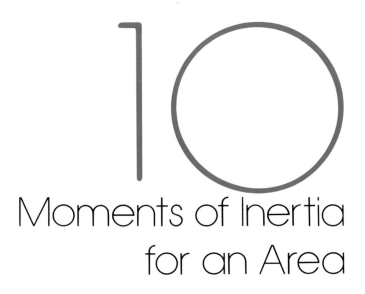

10
Moments of Inertia
for an Area

10-1. Definition of Moments of Inertia for Areas

In the previous chapter the *centroid* of an area has been defined by considering the *first moment* of the area about an axis; i.e., for the computation it is necessary to evaluate integrals of the form $\int_A x\, dA$. There are many important topics in mechanics which require evaluation of an integral of the *second moment* of an area about an axis. These integrals are of the form $\int_A x^2\, dA$, where x is the "moment arm" measured from the element to an axis that is either perpendicular to, or lying in the plane of, the area. Often in engineering practice such integrals are referred to as the *moments of inertia for the area*. The terminology "moment of inertia" as used here is actually a misnomer; however, it has been adopted because of the similarity with integrals of the same form related to mass.*

The moment of inertia of an area originates whenever one computes the moment of a distributed load that varies linearly from the moment axis. A common example of this type of loading occurs in the study of fluid mechanics. It was pointed out in Sec. 9–5 that the pressure p, or force per unit area, exerted at a point located a distance z below the surface of a liquid is $p = \rho g z$, Eq. 9–13, where ρ is the mass density of the liquid. Thus, the magnitude of force exerted by a liquid on the area dA of the submerged plate shown in Fig. 10–1a is $dF = p\, dA = \rho g z\, dA$. The moment of this force about the x axis of the plate is $dM = dF z = \rho g z^2\, dA$ and, therefore, the moment created by the entire distributed (pressure) loading is $M = \rho g \int_A z^2\, dA$. Here the integral represents the moment of inertia of the area of the plate about the x axis.

*The moment of inertia for mass is discussed in *SI Engineering Mechanics: Dynamics*.

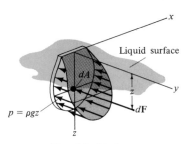

Fig. 10–1(a)

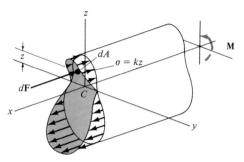

Fig. 10–1(b)

The area moment of inertia also appears when one relates the *normal stress,* or force per unit area, acting on a transverse cross section of an elastic beam to the applied external moment **M,** which causes bending of the beam. From the theory of strength of materials, it can be shown that the normal stress σ (sigma) within the beam varies linearly with its distance from an axis passing through the centroid C of the beam's cross-sectional area; i.e., $\sigma = kz$, Fig. 10–1b. The magnitude of force $d\mathbf{F}$ acting on the area element dA, shown in the figure, is therefore $dF = \sigma \, dA = kz \, dA$. Since this force is located a distance z from the y axis, the moment of $d\mathbf{F}$ about the y axis is $dM = dF z = kz^2 \, dA$. The resulting moment of the entire stress distribution is caused by the applied moment **M,** so for equilibrium it is necessary that $M = k \int_A z^2 \, dA$.

As a third example to illustrate the formulation of area moments of inertia, consider the elastic twisting of a circular shaft about its longitudinal axis, as shown in Fig. 10–1c. The applied external torsional moment **T** causes a *shearing stress* τ (tau) to be distributed over the shaft's cross section which varies linearly along any radial line segment; i.e., $\tau = kr$. The magnitude of the applied torque **T** may be related to the internal

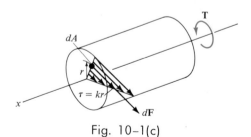

Fig. 10–1(c)

stress distribution by summing moments about the x axis of the shaft, which results in $T = \int_A r \, dF = \int_A r\tau \, dA = k \int_A r^2 \, dA$. Note that the moment of inertia (or integral) in this case differs from that of the other two examples because the "moment" axis is *perpendicular* to the plane of the area, and not lying within it.

Since the moments of inertia for areas frequently appear in design formulas used in fluid mechanics, strength of materials, and structural mechanics, it is important that the engineer become familiar with the methods used to compute these quantities.

Moment of Inertia. Consider the area A shown in Fig. 10-2, which lies in the x-y plane. By definition, the moments of inertia of the differential planar area dA about the x and y axes are $dI_x = y^2 \, dA$ and $dI_y = x^2 \, dA$, respectively. For the entire area the *moment of inertia* is determined by integration, i.e.,

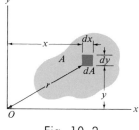

Fig. 10-2

$$I_x = \int_A y^2 \, dA$$

$$I_y = \int_A x^2 \, dA$$

(10-1)

Integrals of this type must be evaluated when solving the fluid-pressure and elastic-bending problems mentioned in reference to Figs. 10-1a and 10-1b.

Also, the second moment of the differential area dA can be formulated about the pole O, or z axis, Fig. 10-2. This is referred to as the polar moment of inertia, $dJ_O = r^2 \, dA$. Here r is the perpendicular distance from the pole (z axis) to the element dA. For the entire area the *polar moment of inertia* is

$$J_O = \int_A r^2 \, dA = I_x + I_y \qquad (10\text{-}2)$$

The relationship between J_O and I_x, I_y is possible since $r^2 = x^2 + y^2$, Fig. 10-2. In the previous discussion, computation for J_O would have been necessary for solving the elastic-torsion problem of Fig. 10-1c.

From the above formulations it is seen that I_x, I_y, and J_O will always be positive, since they involve the product of distance squared and area. Furthermore, the units for moment of inertia involve length raised to the fourth power, e.g., m^4 or mm^4.

10–2. Parallel-Axis Theorem for an Area

If the moment of inertia for an area is known about an axis passing through its centroid, it is possible to determine the moment of inertia of the area about a corresponding parallel axis using the *parallel-axis theorem*. To derive this theorem, consider finding the moment of inertia of the shaded area shown in Fig. 10–3 about the x axis. In this case, a differential element dA of the area is located at an arbitrary distance y from the centroidal $\bar{x}$ axis, whereas the *fixed distance* between the parallel x and $\bar{x}$ axes is defined as d_y. Since the moment of inertia of dA about the x axis is $dI_x = (y + d_y)^2\, dA$, then for the entire area A,

$$I_x = \int_A (y + d_y)^2\, dA$$

$$= \int_A y^2\, dA + 2d_y \int_A y\, dA + d_y^2 \int_A dA$$

The first term on the right represents the moment of inertia of the area about the $\bar{x}$ axis, $\bar{I}_{\bar{x}}$. The second term is zero since the $\bar{x}$ axis passes through the area's centroid C, i.e., $\int_A y\, dA = \bar{y}A = 0$. The final result is therefore

$$I_x = \bar{I}_{\bar{x}} + Ad_y^2 \qquad\qquad (10\text{–}3)$$

A similar expression can be written for I_y, i.e.,

$$I_y = \bar{I}_{\bar{y}} + Ad_x^2 \qquad\qquad (10\text{–}4)$$

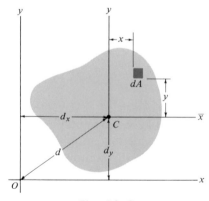

Fig. 10–3

The form of both of these two equations states that the moment of inertia of an area about an axis, I, is equal to the area's moment of inertia about a parallel axis passing through the centroid, $\bar{I}$, plus the product of the area A and the square of the perpendicular distance d between the axes.

In a similar manner, it can be shown that the parallel-axis theorem for determining the polar moment of inertia J_O about an axis perpendicular to the $\bar{x}\text{-}\bar{y}$ plane and passing through the pole O is

$$J_O = \bar{J}_C + Ad^2 \qquad (10\text{-}5)$$

Here $\bar{J}_C$ represents the polar moment of inertia of the area A about the $\bar{z}$ axis which passes through the centroid C and is perpendicular to the $\bar{x}\text{-}\bar{y}$ plane, and d is the distance between points O and C, since it represents the perpendicular distance between the parallel axes, Fig. 10-3.

10-3. Radius of Gyration of an Area

The *radius of gyration* of a planar area is often referred to in structural mechanics and is used in formulas for finding the strength of columns. Provided the area A and moments of inertia I_x, I_y, and J_O are *known*, the radii of gyration k_x, k_y, and k_O can be determined from the formulas

$$k_x = \sqrt{\frac{I_x}{A}}$$

$$k_y = \sqrt{\frac{I_y}{A}} \qquad (10\text{-}6)$$

$$k_O = \sqrt{\frac{J_O}{A}}$$

Note that the general form of these equations is similar to that for finding the moment of inertia of a differential area about an axis. For example, $I_x = k_x^2 A$; whereas for a differential area, $dI_x = x^2 \, dA$.

10–4. Moments of Inertia for an Area by Integration

When the boundaries for a planar area can be expressed by mathematical functions, Eqs. 10–1 may be integrated to determine the moments of inertia for the area.

PROCEDURE FOR ANALYSIS

If the element of area chosen for integration has a differential size in two directions as shown in Fig. 10–2, a double integration must be performed to evaluate the moment of inertia. Most often, however, it is easier to choose an element having a differential size or thickness in only one direction, because then the evaluation requires only a single integration.

Since the moment of inertia specifies an integration of the "second moment" of the element about an axis, it is important that the correct "moment area" be specified. In particular, when the element has a differential size in only *one direction,* this "moment" condition is satisfied only when the element is *properly oriented* with respect to the axis. For example, the rectangular element $dA = y\, dx$ shown in Fig. 10–4 satisfies this requirement when $I_y = \int_A x^2\, dA$ is computed, since all parts of the element lie at the *same* "moment-arm" distance "x" from the y axis. (The element is infinitesimally "thin" in the x direction.)* When computing $I_x = \int_A y^2\, dA$ using this element, an adjustment must be made in the formulation. The element is not infinitesimally thin in the y direction, and thus parts of its entirety lie at *different* distances from the x axis. Using another technique however, it is possible to use this same element to calculate I_x for the area. To do this, it is first necessary to formulate the moment of inertia for the (rectangular) element about its centroidal axis, and then to determine the moment of inertia of the element about the x axis using the parallel-axis theorem. Integration of this result will yield I_x.

The following examples illustrate the above procedures.

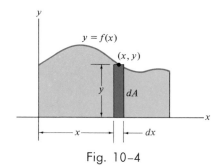

Fig. 10–4

Example 10–1

Determine the radius of gyration for the rectangular area shown in Fig. 10–5a with respect to (a) the centroidal $\bar{x}$ axis, (b) an axis perpendicular to the $\bar{x}$-$\bar{y}$ plane and passing through the centroid O, and (c) an axis parallel to the $\bar{x}$ axis and passing through the base of the rectangle.

---

*In the case of the element $dA = dx\, dy$, Fig. 10–2, the moment arms y and x are appropriate for the formulation of I_x and I_y (Eq. 10–1) since the *entire* element, because of its "smallness," lies at the specified y and x perpendicular distances from the x and y axes.

Solution

Part (a). The radius of gyration $k_{\bar{x}}$ may be found after first obtaining the moment of inertia $\bar{I}_{\bar{x}}$. The differential element shown in Fig. 10–5a is chosen for integration. Because of its location and orientation, the *entire element* is at a distance y from the $\bar{x}$ axis. The area is covered if the elements are summed from $y = -h/2$ to $y = h/2$. Since $dA = b\,dy$, then

$$\bar{I}_{\bar{x}} = \int_A y^2\,dA = \int_{-h/2}^{h/2} y^2(b\,dy) = b\int_{-h/2}^{h/2} y^2\,dy$$
$$= \tfrac{1}{12}bh^3 \tag{1}$$

Applying the first of Eqs. 10–6 yields

$$k_{\bar{x}} = \sqrt{\frac{\bar{I}_{\bar{x}}}{A}} = \sqrt{\frac{\tfrac{1}{12}bh^3}{bh}} = \frac{h}{\sqrt{12}} \qquad\qquad Ans.$$

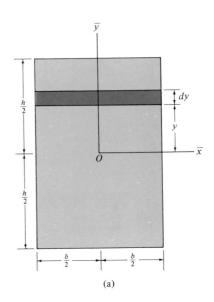

(a)

Part (b). The rectangular moment of inertia $\bar{I}_{\bar{y}}$ may be found by interchanging the dimensions b and h in Eq. (1), in which case

$$\bar{I}_{\bar{y}} = \tfrac{1}{12}hb^3$$

Using Eq. 10–2, the polar moment of inertia about O is

$$\bar{J}_O = \bar{I}_{\bar{x}} + \bar{I}_{\bar{y}} = \tfrac{1}{12}bh(h^2 + b^2)$$

From the third of Eqs. 10–6,

$$k_O = \sqrt{\frac{\bar{J}_O}{A}} = \sqrt{\frac{\tfrac{1}{12}bh(h^2 + b^2)}{bh}} = \sqrt{\frac{h^2 + b^2}{12}} \qquad\qquad Ans.$$

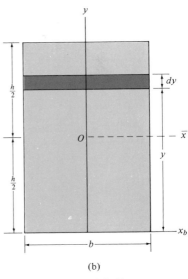

Part (c). The moment of inertia about an axis passing through the base of the rectangle will be computed using the differential element shown in Fig. 10–5b. Here it is necessary to integrate from $y = 0$ to $y = h$. Hence,

$$I_{x_b} = \int_A y^2\,dA = b\int_0^h y^2\,dy = \frac{1}{3}bh^3$$

This same result can be obtained by using Eq. (1) in part (a) and applying the parallel-axis theorem, Eq. 10–3.

$$I_{x_b} = \bar{I}_{\bar{x}} + Ad_y^2$$
$$= \frac{1}{12}bh^3 + bh\left(\frac{h}{2}\right)^2 = \frac{1}{3}bh^3 \tag{2}$$

The radius of gyration is therefore

$$k_{x_b} = \sqrt{\frac{I_{x_b}}{A}} = \sqrt{\frac{\tfrac{1}{3}bh^3}{bh}} = \frac{h}{\sqrt{3}} \qquad\qquad Ans.$$

(b)

Fig. 10–5

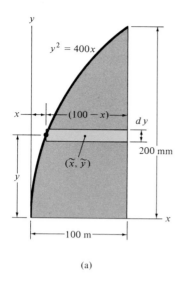

(a)

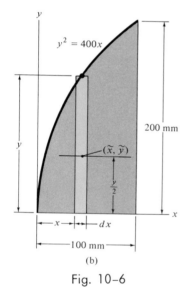

(b)

Fig. 10-6

Example 10-2

Compute the moment of inertia of the shaded area shown in Fig. 10-6a about the x axis.

Solution I

The differential element of area that is parallel to the x axis, as shown in Fig. 10-6a, is chosen for integration. Since the element has a thickness dy and intersects the curve at the arbitrary point (x, y), the area is $dA = (100 - x) \, dy$. Furthermore, all parts of the element lie at a distance y from the x axis. Hence, integrating with respect to y, from $y = 0$ to $y = 200$ mm, yields

$$I_x = \int_A y^2 \, dA = \int_A y^2 (100 - x) \, dy$$

$$= \int_0^{200} y^2 \left(100 - \frac{y^2}{400}\right) dy = 100 \int_0^{200} y^2 \, dy - \frac{1}{400} \int_0^{200} y^4 \, dy$$

$$= 106.7(10^6) \text{ mm}^4 \qquad\qquad Ans.$$

Solution II

It is also possible to choose a differential element that is parallel to the y axis, such as shown in Fig. 10-6b. In this case, however, the entire element does *not* lie at a fixed distance from the x axis, and therefore the parallel-axis theorem must be used to determine the *moment of inertia of the element* with respect to the axis. For a rectangle having a base b and height h, the moment of inertia about its centroidal $\bar{x}$ axis has been computed in Example 10-1. There it was found that $\bar{I}_{\bar{x}} = \frac{1}{12}bh^3$. For the differential element shown in Fig. 10-6b, $b = dx$ and $h = y$, and thus $d\bar{I}_{\bar{x}} = \frac{1}{12} dx \, y^3$. Since the centroid of the element is at $d_y = y/2$ from the x axis, the moment of inertia of the element about the x axis is

$$dI_x = d\bar{I}_{\bar{x}} + dA \, d_y^2$$

$$= \frac{1}{12} dx \, y^3 + y \, dx \left(\frac{y}{2}\right)^2$$

$$= \frac{1}{3} y^3 \, dx$$

(This result can also be concluded from Eq. (2) of Example 10-1). Integrating with respect to x, from $x = 0$ to $x = 100$ mm, yields

$$I_x = \int_A dI_x = \int_A \frac{1}{3} y^3 \, dx$$

$$= \int_0^{100} \frac{1}{3}(400x)^{3/2} \, dx = 2666.7 \int_0^{100} x^{3/2} \, dx$$

$$= 106.7(10^6) \text{ mm}^4 \qquad\qquad Ans.$$

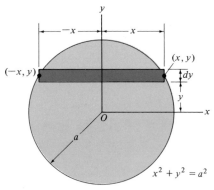

Fig. 10-7(a)

Example 10-3

Determine the moment of inertia with respect to the x axis of the circular area shown in Fig. 10-7a.

Solution I

Using the differential element shown in Fig. 10-7a, since $dA = 2x\,dy$, we have

$$\;\;\;\;\;\;\;\;\;\;\;= |{-}x| + |x|$$

$$I_x = \int_A y^2\,dA = \int_A y^2(2x)\,dy$$

$$= \int_{-a}^{a} y^2(2\sqrt{a^2 - y^2})\,dy = \frac{\pi a^4}{4} \qquad Ans.$$

Solution II

When the differential element is chosen as shown in Fig. 10-7b, the parallel-axis theorem must be used. Why? The centroid for the element lies on the x axis. Hence, $d_y = 0$. Applying Eq. 10-3, noting that $dA = 2y\,dx$, we have

$$dI_x = d\bar{I}_{\bar{x}} + dA\,d_y^2$$

$$= \frac{1}{12}\,dx\,(2y)^3 + 2y\,dx\,(0)$$

$$= \frac{2}{3}y^3\,dx$$

Therefore, integrating with respect to x,

$$I_x = \int_{-a}^{a} \frac{2}{3}(a^2 - x^2)^{3/2}\,dx = \frac{\pi a^4}{4} \qquad Ans.$$

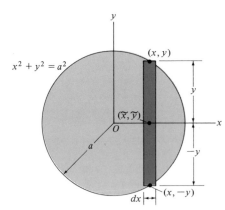

Fig. 10-7(b)

Problems

10-1. The irregular area has a moment of inertia about the AA axis of $20(10^6)$ mm^4. If the total area is $1.2(10^4)$ mm^2, determine the moment of inertia of the area about the BB axis. The DD axis passes through the centroid C of the area.

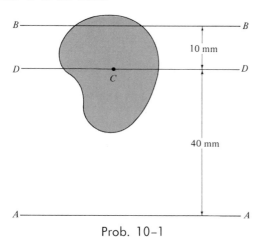

Prob. 10-1

10-2. The polar moment of inertia of the area is $J_C = 15(10^6)$ mm^4, computed about the centroid C. If the moment of inertia about the y axis is $5(10^6)$ mm^4, and the moment of inertia about the x' axis is $12(10^6)$ mm^4, determine the area A.

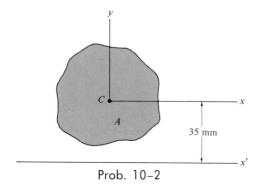

Prob. 10-2

10-3. Determine the moment of inertia of the beam's cross-sectional area about the x axis.

Prob. 10-3

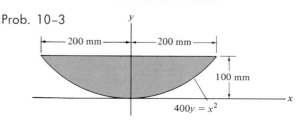

* **10-4.** Determine the moment of inertia of the beam's cross-sectional area about the x axis. Solve the problem in two ways, using rectangular differential elements: (a) having a thickness of dx; (b) having a thickness of dy.

Prob. 10-4

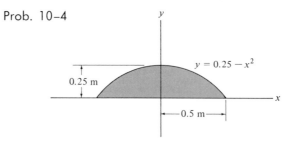

10-5. Determine the moments of inertia I_x and I_y of the cross-sectional area of the plate having a boundary defined by the cosine curve, where $a = 200$ mm.

Prob. 10-5

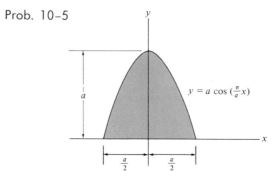

10-5a. Solve Prob. 10-5 if $a = 3$ in.

400

10-6. Compute the moment of inertia of the triangular area about: (a) the x axis; (b) the centroidal $\bar{x}$ axis.

Prob. 10-6

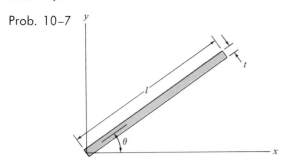

10-7. Determine the moment of inertia of the cross-sectional area of the bar with respect to the x axis. The bar is oriented at an angle θ from the x axis. Assume that $t \ll l$.

Prob. 10-7

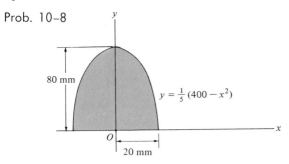

*10-8. The cross-sectional area of a beam is parabolic as shown in the figure. Determine the radius of gyration k_y about the y axis.

Prob. 10-8

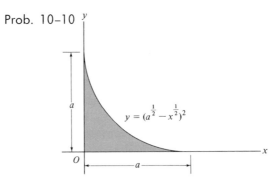

$$y = \tfrac{1}{5}(400 - x^2)$$

10-9. Determine the moment of inertia of the cross-sectional area about the x axis. Then, using the parallel-axis theorem, compute the moment of inertia about the $\bar{x}$ axis passing through the centroid C of the area. $\bar{y} = 60$ mm.

Prob. 10-9

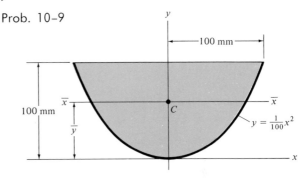

$$y = \tfrac{1}{100}x^2$$

10-10. Determine the radius of gyration k_O of the shaded area about an axis perpendicular to the x-y plane and passing through the origin O. Set $a = 100$ mm.

Prob. 10-10

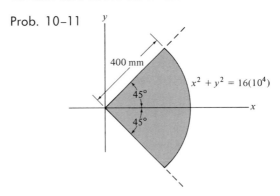

$$y = (a^{\frac{1}{2}} - x^{\frac{1}{2}})^2$$

10-10a. Solve Prob. 10-10 if $a = 1$ ft.

10-11. Determine the moment of inertia of the quarter circular area about the x axis.

Prob. 10-11

$$x^2 + y^2 = 16(10^4)$$

401

*10-12. Determine the moment of inertia about the x axis of an area bounded by the lines $x = 0$, $y = 1$ m, and the parabola $y = 3x^2$.

10-13. Determine the moments of inertia I_x and I_y of the shaded elliptical area. What is the polar moment of inertia about the origin O?

10-14. Determine the moment of inertia for the semi-circular area about the x axis and about the $\bar{x}$ axis that passes through the centroid C.

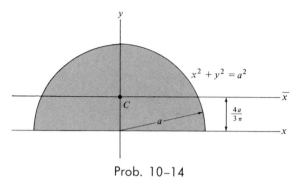

Prob. 10-14

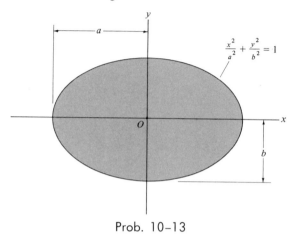

Prob. 10-13

10-5. Moments of Inertia for Composite Areas

A composite area consists of a series of connected "simpler" area shapes, such as semicircles, rectangles, and triangles. Provided the moment of inertia of each of these shapes is known or can be computed about a common axis, then the moment of inertia of the composite area equals the *algebraic sum* of the moments of inertia of all its composite parts.

PROCEDURE FOR ANALYSIS

The following three-step procedure should be used to determine the moment of inertia of a composite area about a reference axis.

Step 1: Using a sketch, divide the area into its composite parts and indicate the perpendicular distance from the *centroid* of each part to the reference axis. (See Appendix B.)

Step 2: The moment of inertia of each part should be computed about its centroidal axis, which is parallel to the reference axis. (See Appendix B.) Then, if the centroidal axis does not coincide with the reference axis, the parallel-axis theorem, $I = \bar{I} + Ad^2$, should be used to determine the moment of inertia of the part about the reference axis.

Step 3: The moment of inertia of the entire area about the reference axis is determined by summing the results of each of its composite parts. In particular, if a composite area has a "hole," the moment of inertia for the composite is found by "subtracting" the moment of inertia for the hole from the moment of inertia of the entire area including the hole.

Example 10-4

Compute the moments of inertia of the beam's cross-sectional area shown in Fig. 10–8a about the *x* and *y* centroidal axes.

Solution

Step 1: The cross section can be considered as three composite rectangular areas *A*, *B*, and *D* shown in Fig. 10–8b. For the calculation, the centroid of each of these rectangles is located in the figure.

Step 2: From Appendix B, or Example 10–1, the moment of inertia of a rectangle about its centroidal axis is $\bar{I} = \frac{1}{12}bh^3$. Hence, using the parallel-axis theorem for rectangles *A* and *D*, the computations are as follows:

Rectangle A:

$$I_x = \bar{I}_{\bar{x}} + Ad_y^2 = \tfrac{1}{12}(100)(300)^3 + (100)(300)(200)^2 = 14.25(10^8) \text{ mm}^4$$
$$I_y = \bar{I}_{\bar{y}} + Ad_x^2 = \tfrac{1}{12}(300)(100)^3 + (100)(300)(250)^2 = 19(10^8) \text{ mm}^4$$

Rectangle B:

$$I_x = \tfrac{1}{12}(600)(100)^3 = 0.50(10^8) \text{ mm}^4$$
$$I_y = \tfrac{1}{12}(100)(600)^3 = 18(10^8) \text{ mm}^4$$

Rectangle D:

$$I_x = \bar{I}_{\bar{x}} + Ad_y^2 = \tfrac{1}{12}(100)(300)^3 + (100)(300)(200)^2 = 14.25(10^8) \text{ mm}^4$$
$$I_y = \bar{I}_{\bar{y}} + Ad_x^2 = \tfrac{1}{12}(300)(100)^3 + (100)(300)(250)^2 = 19(10^8) \text{ mm}^4$$

Step 3: The moments of inertia for the entire cross section are thus

$$I_x = 14.25(10^8) + 0.50(10^8) + 14.25(10^8) = 29(10^8) \text{ mm}^4 \quad Ans.$$
$$I_y = 19(10^8) + 18(10^8) + 19(10^8) = 56(10^8) \text{ mm}^4 \quad Ans.$$

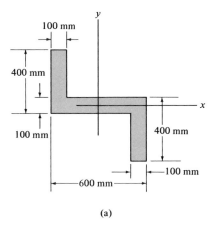

(a)

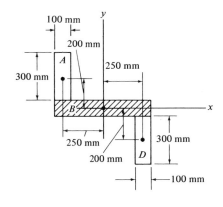

(b)

Fig. 10–8

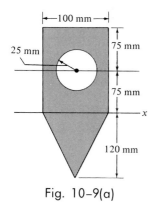

Fig. 10-9(a)

Example 10-5

Compute the moment of inertia of the composite area shown in Fig. 10-9a about the x axis.

Solution

Step 1: For convenience, the composite area may be obtained by *subtracting* the circle from the rectangle and triangle shown in Fig. 10-9b. The centroid of each area is located in the figure.

Step 2: The moments of inertia about the x axis are computed using the parallel-axis theorem and the data contained in Appendix B.

Circle:

$$I_x = \bar{I}_{\bar{x}} + Ad_y^2$$
$$= \tfrac{1}{4}\pi(25)^4 + \pi(25)^2(75)^2 = 11.4(10^6) \text{ mm}^4$$

Rectangle:

$$I_x = \bar{I}_{\bar{x}} + Ad_y^2$$
$$= \tfrac{1}{12}(100)(150)^3 + (100)(150)(75)^2 = 112.5(10^6) \text{ mm}^4$$

Triangle:

$$I_x = \bar{I}_{\bar{x}} + Ad_y^2$$
$$= \tfrac{1}{36}(100)(120)^3 + \tfrac{1}{2}(100)(120)(40)^2 = 14.4(10^6) \text{ mm}^4$$

Step 3: The moment of inertia for the entire composite area is thus

$$I_x = -11.4(10^6) + 112.5(10^6) + 14.4(10^6)$$
$$= 115.5(10^6) \text{ mm}^4 \qquad\qquad Ans.$$

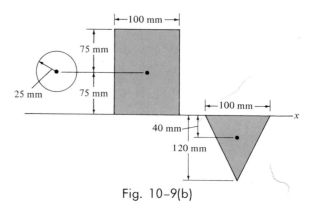

Fig. 10-9(b)

Problems

10-15. Determine the moments of inertia I_x and I_y of the shaded area. The origin of coordinates is at the centroid C. Set $a = 50$ mm and $b = 20$ mm.

Prob. 10–15

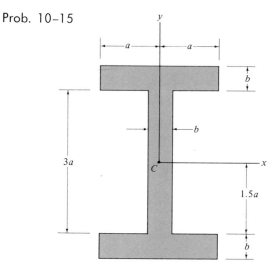

10-15a. Solve Prob. 10–15 if $a = 3$ in. and $b = 1$ in.

* 10-16. Determine the moment of inertia of the beam's cross-sectional area with respect to the $\bar{x}\bar{x}$ axis passing through the centroid, C. $\bar{y} = 87.5$ mm.

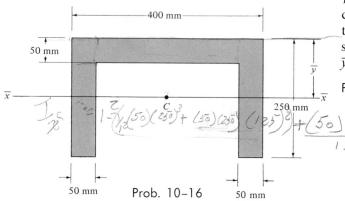

Prob. 10–16

10-17. A column has a built-up cross section of four equal-sized angles, as shown. Neglecting the inertia of the lacing and rivets, determine the radius of gyration of the column about the $\bar{x}$ centroidal axis. Handbook values for the area, moment of inertia, and the location of the centroid for one of the angles are listed in the figure.

Prob. 10–17

$\bar{y}_a = 31$ mm
$A_a = 1.2(10^3)$ mm^2
$I_{a_{\bar{x}_a \bar{x}_a}} = 5.3(10^6)$ mm^4

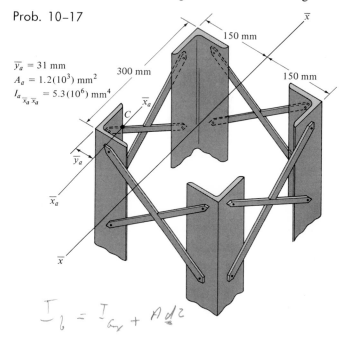

10-18. Determine the moment of inertia of the beam's cross-sectional area with respect to the $\bar{x}\bar{x}$ axis passing through the centroid C of the cross section. Neglect the size of the corner welds at A and B for the calculation. $\bar{y} = 90.5$ mm.

Prob. 10–18

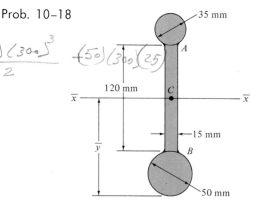

10–19. Determine the moment of inertia of the beam's cross-sectional area with respect to the $\bar{x}\bar{x}$ centroidal axis. Handbook values for the moment of inertia I_C, area A_C, and centroidal location for the channels are indicated in the figure.

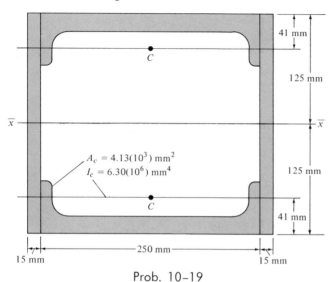

$A_c = 4.13(10^3)$ mm^2
$I_c = 6.30(10^6)$ mm^4

41 mm
125 mm
125 mm
41 mm
250 mm
15 mm
15 mm

Prob. 10–19

* 10–20. Determine the moments of inertia I_x and I_y of the shaded area. Set $r_1 = 100$ mm and $r_2 = 300$ mm.

Prob. 10–20

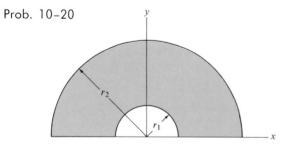

* 10–20a. Solve Prob. 10–20 if $r_1 = 1.5$ in. and $r_2 = 3$ in.

10–21. Determine the moment of inertia of the beam's cross-sectional area with respect to the $\bar{x}\bar{x}$ centroidal axis. Neglect the size of all the rivet heads, R, for the calculation. Handbook values for the area, moment of inertia, and location of the centroid C of one of the angles are listed in the figure.

Prob. 10–21

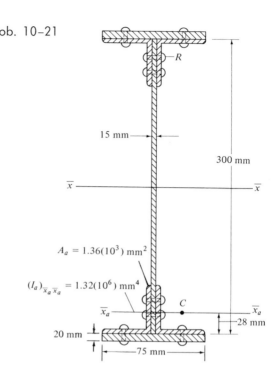

R

15 mm
300 mm

$A_a = 1.36(10^3)$ mm^2

$(I_a)_{\bar{x}_a \bar{x}_a} = 1.32(10^6)$ mm^4

$\bar{x}_a$
20 mm
75 mm
C
28 mm
$\bar{x}_a$

10–22. Determine the moment of inertia of the beam's cross-sectional area with respect to the $\bar{x}$ centroidal axis. Neglect the size of all the rivet heads, R, for the calculation. Handbook values for the area, moment of inertia, and centroids of the channel and one angle located at the top and one at the bottom of the beam are listed in the figure. $\bar{y} = \text{104.4 mm}$.

106.2

Prob. 10–22

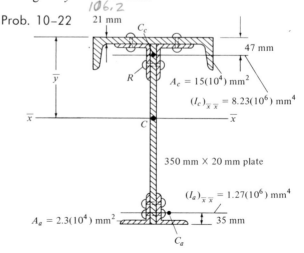

21 mm
C_c
47 mm
$\bar{y}$
R
$A_c = 15(10^4)$ mm^2
$(I_c)_{\bar{x}\bar{x}} = 8.23(10^6)$ mm^4
$\bar{x}$
C
350 mm $\times$ 20 mm plate
$(I_a)_{\bar{x}\bar{x}} = 1.27(10^6)$ mm^4
$A_a = 2.3(10^4)$ mm^2
35 mm
C_a

10–23. Determine the polar moment of inertia of the shaded area about the origin of coordinates, C, located at the centroid.

Prob. 10–23

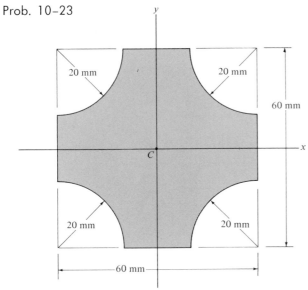

* 10–24. The cross section of the beam consists of a wide-flange beam and a cover plate welded together as shown. Determine the moment of inertia of this section with respect to the $\bar{x}\bar{x}$ centroidal axis. $\bar{y} = 80.1$ mm.

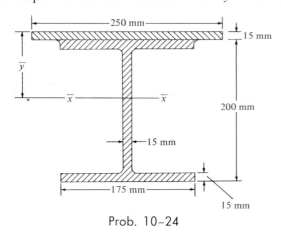

Prob. 10–24

10–25. The composite cross section for the column consists of two cover plates riveted to two channels. Determine the radius of gyration $k_{\bar{x}}$ with respect to the cen-

troidal $\bar{x}\bar{x}$ axis. Each channel has a cross-sectional area of $A_c = 6.53(10^3)$ mm^2 and a moment of inertia $(I_{\bar{x}})_c = 5.75(10^6)$ mm^4. The plates have dimensions of $a = 200$ mm, $t = 15$ mm, and $b = 250$ mm.

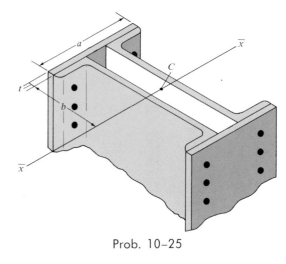

Prob. 10–25

10–25a. Solve Prob. 10–25 if $A_c = 11.8$ in.^{2}, $(I_{\bar{x}})_c = 349$ in.^{4}, $a = 10$ in., $t = 0.5$ in., and $b = 7.5$ in.

10–26. Determine the location $\bar{y}$ of the centroid C of the beam's cross-sectional area. Then compute the moment of inertia of the area about the $\bar{x}\bar{x}$ axis.

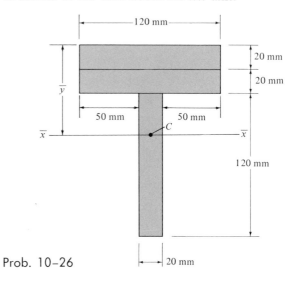

Prob. 10–26

407

10-27. Determine the moment of inertia of the beam's cross-sectional area with respect to the $\bar{x}\bar{x}$ axis passing through the centroid C. Neglect the size of the corner welds at A and B for the calculation. $\bar{y} = 154.4$ mm.

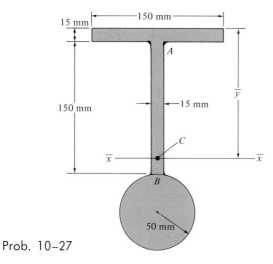

Prob. 10–27

10-29. Determine the moment of inertia of the beam's cross-sectional area with respect to the $\bar{x}\bar{x}$ axis passing through the centroid C of the cross section. *Suggestion:* Use the result of Prob. 10–11.

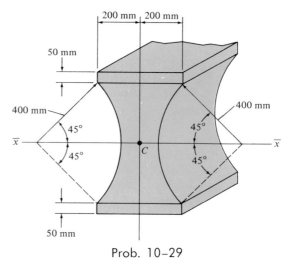

Prob. 10–29

*10-28. Determine the moments of inertia I_x and I_y of the Z-section. The origin of coordinates is at the centroid C.

Prob. 10–28

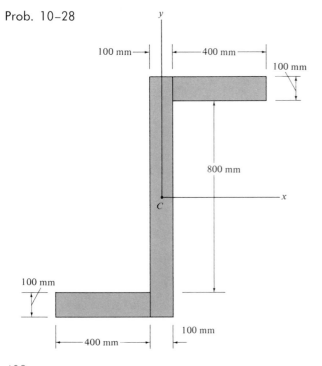

*10-6. Product of Inertia for an Area

In general, the value of the moment of inertia for an area about an axis will depend upon how the axis is inclined with respect to the area. In some applications of structural design it is necessary to know the inclination of those axes which give, respectively, the maximum and minimum moment of inertia for the area. The method for determining this is discussed in the next section. To use this method, however, one must first compute the product of inertia for the area as well as its moments of inertia for a given orientation of the x, y axes.

The product of inertia for an element of area dA, located at a point (x, y) as shown in Fig. 10–10, is defined as $dI_{xy} = xy\,dA$. Thus, for the entire area A, the *product of inertia* is

$$I_{xy} = \int_A xy\,dA \qquad (10\text{–}7)$$

If the element of area chosen has a differential size in two directions, as shown in Fig. 10–10, a double integration must be performed to evaluate I_{xy}. Most often, however, it is easier to choose an element having a differential size or thickness in only one direction in which case the evaluation requires only a single integration. In this case, the coordinates x and y in Eq. 10–7 locate the *centroid of the element,* i.e., $\tilde{x} = x, \tilde{y} = y$. See Example 10–6.

Like the moment of inertia, the product of inertia has units of length raised to the fourth power, e.g., m⁴ or mm⁴. However, since x or y may be a negative quantity, while the element of area is always positive, the product of inertia may be positive, negative, or zero, depending upon the location and orientation of the coordinate axes.

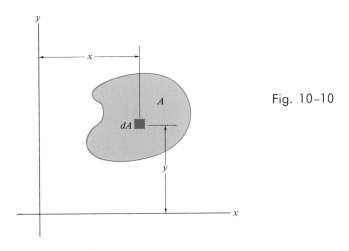

Fig. 10–10

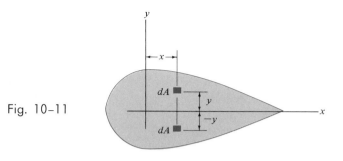

Fig. 10-11

The product of inertia I_{xy} for the area will be *zero* if either the x or y axis is an axis of *symmetry* for the area. To show this, consider the shaded area in Fig. 10–11, where for every element dA located at point (x, y) there is a corresponding element dA located at $(x, -y)$. Since products of inertia for these elements are respectively $xy\, dA$ and $-xy\, dA$, the algebraic sum or integration of the products of inertia for all the elements of area that are chosen in this way will cancel each other. Consequently, the product of inertia for the total area becomes zero.

Parallel-Axis Theorem.   Consider the shaded area shown in Fig. 10–12, where $\bar{x}$ and $\bar{y}$ represent a set of axes passing through the *centroid C* of the area, and x and y represent a corresponding set of parallel axes. Since the product of inertia of dA with respect to the x and y axes is $dI_{xy} = (x + d_x)(y + d_y)\, dA$, then for the entire area,

$$I_{xy} = \int_A (x + d_x)(y + d_y)\, dA$$

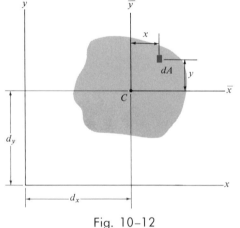

Fig. 10-12

$$= \int_A xy\, dA + d_x \int_A y\, dA + d_y \int_A x\, dA + d_x d_y \int_A dA$$

The first term on the right represents the product of inertia of the area with respect to the centroidal axis, $\bar{I}_{\bar{x}\,\bar{y}}$. The integrals in the second and third terms are zero since the moments of the area are taken about the centroidal axis. Realizing that the fourth integral represents the area A, the final result is thus

$$I_{xy} = \bar{I}_{\bar{x}\,\bar{y}} + Ad_x d_y \qquad (10\text{--}8)$$

The similarity between this equation and the parallel-axis theorem for moments of inertia should be noted. In particular, it is important that the *algebraic signs* for d_x and d_y be maintained when applying Eq. 10–8. As illustrated in Example 10–7, the parallel-axis theorem finds important application in determining the product of inertia of a *composite area* with respect to a set of x, y axes.

Example 10-6

Determine the product of inertia of the triangle shown in Fig. 10–13a (a) with respect to the x, y axes; (b) with respect to the $\bar{x}, \bar{y}$ centroidal axes.

Solution I

Part (a). Consider the differential element that has a thickness dx and area $dA = y\,dx$, Fig. 10–13b. The *centroid* is located at $\tilde{x} = x$, $\tilde{y} = y/2$, so that the product of inertia of the element becomes

$$dI_{xy} = \tilde{x}\tilde{y}\,dA = x\frac{y}{2}(y\,dx) = x\frac{h}{2b}x\left(\frac{h}{b}x\,dx\right)$$

$$= \frac{h^2}{2b^2}x^3\,dx$$

Integrating with respect to x from $x = 0$ to $x = b$ yields

$$I_{xy} = \frac{h^2}{2b^2}\int_0^b x^3\,dx = \frac{b^2h^2}{8} \qquad\qquad Ans.$$

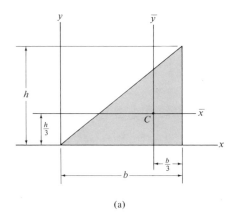

(a)

Solution II

Part (a). Consider a differential element that has a thickness dy and area $dA = (b - x)\,dy$, as shown in Fig. 10–13c. The *centroid* is located at point $\tilde{x} = x + (b - x)/2 = (b + x)/2$, $\tilde{y} = y$, so that the product of inertia of the element becomes

$$dI_{xy} = \tilde{x}\tilde{y}\,dA = \left(\frac{b + x}{2}\right)y(b - x)\,dy = \left(\frac{b + \frac{b}{h}y}{2}\right)y\left(b - \frac{b}{h}y\right)dy$$

$$= \frac{1}{2}y\left(b^2 - \frac{b^2}{h^2}y^2\right)dy$$

Integrating with respect to y from $y = 0$ to $y = h$ yields

$$I_{xy} = \frac{1}{2}\int_0^h y\left(b^2 - \frac{b^2}{h^2}y^2\right)dy = \frac{b^2h^2}{8} \qquad\qquad Ans.$$

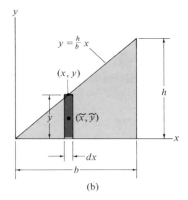

(b)

Part (b). The product of inertia for the triangle with respect to its centroidal axis, Fig. 10–13a, may be computed using the above result and the parallel-axis theorem, Eq. 10–8;

$$I_{xy} = \bar{I}_{\bar{x}\bar{y}} + Ad_xd_y$$

$$\frac{b^2h^2}{8} = \bar{I}_{\bar{x}\bar{y}} + \frac{1}{2}bh\left(\frac{1}{3}h\right)\left(\frac{2}{3}b\right)$$

Hence,

$$\bar{I}_{\bar{x}\bar{y}} = \frac{1}{72}b^2h^2 \qquad\qquad Ans.$$

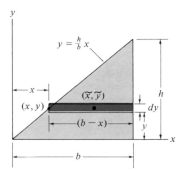

(c)

Fig. 10–13

411

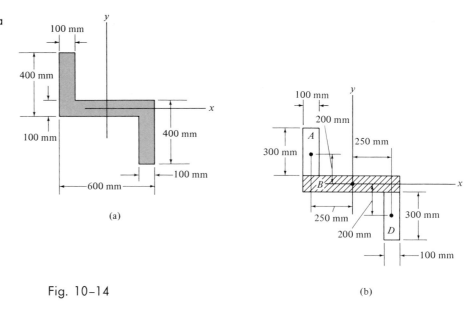

Fig. 10–14

(a)

(b)

Example 10–7

Compute the product of inertia of the beam's cross-sectional area, shown in Fig. 10–14a, about the x and y centroidal axes.

Solution

As in Example 10–4, the cross section can be considered as three composite rectangular areas A, B, and D, Fig. 10–14b. The coordinates for the centroid of each of these rectangles are shown in the figure. Due to symmetry the product of inertia of *each rectangle* is *zero* about a set of $\bar{x}, \bar{y}$ axes that pass through the rectangle's centroid. Hence, application of the parallel-axis theorem to each of the rectangles yields

Rectangle A:

$$I_{xy} = \bar{I}_{\bar{x}\bar{y}} + Ad_x d_y = 0 + (300)(100)(-250)(200) = -15(10^8) \text{ mm}^4$$

Rectangle B:

$$I_{xy} = \bar{I}_{\bar{x}\bar{y}} + Ad_x d_y = 0 + 0 = 0$$

Rectangle D:

$$I_{xy} = \bar{I}_{\bar{x}\bar{y}} + Ad_x d_y = 0 + (300)(100)(250)(-200) = -15(10^8) \text{ mm}^4$$

The product of inertia for the entire cross section is thus

$$I_{xy} = [-15(10^8)] + 0 + [-15(10^8)] = -30(10^8) \text{ mm}^4 \qquad Ans.$$

*10–7. Moments of Inertia for an Area About Inclined Axes

In structural mechanics, it is sometimes necessary to calculate the moments and product of inertia I_u, I_v, and I_{uv} for an area with respect to a set of inclined u and v axes, when the values for θ, I_x, I_y, and I_{xy} are *known*. From Fig. 10–15 the perpendicular distance from the area element dA may be related to the axes of the two coordinate systems by using the *transformation equations*

$$u = x \cos \theta + y \sin \theta$$
$$v = y \cos \theta - x \sin \theta$$

Using these equations, the moments and product of inertia of dA about the u and v axes become

$$dI_u = v^2 \, dA = (y \cos \theta - x \sin \theta)^2 \, dA$$
$$dI_v = u^2 \, dA = (x \cos \theta + y \sin \theta)^2 \, dA$$
$$dI_{uv} = uv \, dA = (x \cos \theta + y \sin \theta)(y \cos \theta - x \sin \theta) \, dA$$

Expanding each expression and integrating, realizing that $I_x = \int_A y^2 \, dA$, $I_y = \int_A x^2 \, dA$, and $I_{xy} = \int_A xy \, dA$, we obtain

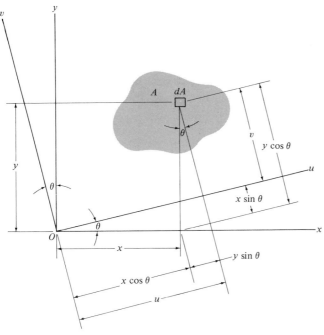

Fig. 10–15

$$I_u = I_x \cos^2 \theta + I_y \sin^2 \theta - 2I_{xy} \sin \theta \cos \theta$$
$$I_v = I_x \sin^2 \theta + I_y \cos^2 \theta + 2I_{xy} \sin \theta \cos \theta$$
$$I_{uv} = I_x \sin \theta \cos \theta - I_y \sin \theta \cos \theta + I_{xy}(\cos^2 \theta - \sin^2 \theta)$$

These equations may be simplified by using the trigonometric identities $\sin 2\theta = 2 \sin \theta \cos \theta$ and $\cos 2\theta = \cos^2 \theta - \sin^2 \theta$, in which case

$$I_u = \frac{I_x + I_y}{2} + \frac{I_x - I_y}{2} \cos 2\theta - I_{xy} \sin 2\theta$$

$$I_v = \frac{I_x + I_y}{2} - \frac{I_x - I_y}{2} \cos 2\theta + I_{xy} \sin 2\theta \qquad (10\text{--}9)$$

$$I_{uv} = \frac{I_x - I_y}{2} \sin 2\theta + I_{xy} \cos 2\theta$$

Note that if the first and second equations are added together, it is seen that the polar moment of inertia about the z axis passing through point O is *independent* of the orientation of the u and v axes, i.e.,

$$J_O = I_u + I_v = I_x + I_y$$

Principal Moments of Inertia. From Eqs. 10–9, it may be seen that I_u, I_v, and I_{uv} depend upon the angle of inclination, θ, of the u, v axes. In structural mechanics, it is sometimes important to determine the orientation of the u, v axes about which the moments of inertia for the area, I_u and I_v, are maximum and minimum. This particular set of axes is called the *principal axes* of the area, and the corresponding moments of inertia with respect to these axes are called the *principal moments of inertia*. In general, there is a set of principal axes for every chosen origin O, although in structural mechanics, the area's centroid is an important location for O.

The angle $\theta = \theta_p$, which defines the orientation of the principal axes for the area, may be found by differentiating the first of Eqs. 10–9 with respect to θ, and setting the result equal to zero. Thus,

$$\frac{dI_u}{d\theta} = -2 \left(\frac{I_x - I_y}{2} \right) \sin 2\theta - 2I_{xy} \cos 2\theta = 0$$

Therefore, at $\theta = \theta_p$,

$$\tan 2\theta_p = \frac{-I_{xy}}{\dfrac{I_x - I_y}{2}} \qquad (10\text{--}10)$$

This equation has two roots, θ_{p_1} and θ_{p_2}, which specify the inclination of the principal axes. Because of the nature of the tangent, the values of $2\theta_{p_1}$

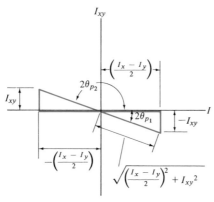

Fig. 10–16

and $2\theta_{p_2}$ are 180° apart, so that θ_{p_1} and θ_{p_2} are 90° apart. Assuming that I_{xy} and $(I_x - I_y)$ are both positive quantities, the sine and cosine of $2\theta_{p_1}$ and $2\theta_{p_2}$ can be obtained from the triangles shown in Fig. 10–16, which are based upon Eq. 10–10.

$$\text{For } \theta_{p_1} \begin{cases} \sin 2\theta_{p_1} = -I_{xy} \bigg/ \sqrt{\left(\dfrac{I_x - I_y}{2}\right)^2 + I_{xy}^2} \\[3mm] \cos 2\theta_{p_1} = \left(\dfrac{I_x - I_y}{2}\right) \bigg/ \sqrt{\left(\dfrac{I_x - I_y}{2}\right)^2 + I_{xy}^2} \end{cases}$$

$$\text{For } \theta_{p_2} \begin{cases} \sin 2\theta_{p_2} = I_{xy} \bigg/ \sqrt{\left(\dfrac{I_x - I_y}{2}\right)^2 + I_{xy}^2} \\[3mm] \cos 2\theta_{p_2} = -\left(\dfrac{I_x - I_y}{2}\right) \bigg/ \sqrt{\left(\dfrac{I_x - I_y}{2}\right)^2 + I_{xy}^2} \end{cases}$$

If these two sets of trigonometric relations are substituted into the first or second of Eqs. 10–9 and simplified, the result is

$$I_{\substack{\text{max} \\ \text{min}}} = \frac{I_x + I_y}{2} \pm \sqrt{\left(\frac{I_x - I_y}{2}\right)^2 + I_{xy}^2} \qquad (10\text{–}11)$$

Depending upon the sign chosen, this result gives the maximum or minimum moment of inertia for the area. Furthermore, if the above trigonometric relations for θ_{p_1} and θ_{p_2} are substituted into the third of Eqs. 10–9, it may be seen that $I_{uv} = 0$; that is, the *product of inertia with respect to the principal axes is zero*. Since it was indicated in Sec. 10–6 that the product of inertia is zero with respect to any symmetrical axis, it therefore follows that *any symmetrical axis represents a principal axis of inertia for the area*.

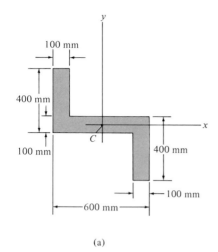

(a)

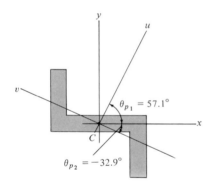

(b)

Fig. 10-17

Example 10-8

Determine the principal moments of inertia for the beam's cross-sectional area shown in Fig. 10–17a with respect to an axis passing through the centroid.

Solution

The moments and product of inertia of the cross section with respect to the x, y axes have been computed in Examples 10–4 and 10–7. The results are

$$I_x = 29(10^8) \text{ mm}^4 \qquad I_y = 56(10^8) \text{ mm}^4 \qquad I_{xy} = -30(10^8) \text{ mm}^4$$

Using Eq. 10–10, the angles of inclination of the principal axes u and v are

$$\tan 2\theta_p = \frac{-I_{xy}}{\dfrac{I_x - I_y}{2}} = \frac{30(10^8)}{\dfrac{29(10^8) - 56(10^8)}{2}} = -2.22$$

$$2\theta_{p_1} = -65.8° \qquad \text{and} \qquad 2\theta_{p_2} = 114.2°$$

Thus, as shown in Fig. 10–17b,

$$\theta_{p_1} = -32.9° \qquad \text{and} \qquad \theta_{p_2} = 57.1°$$

The principal moments of inertia with respect to the u and v axes are determined by using Eq. 10–11. Hence,

$$I_{\substack{max \\ min}} = \frac{I_x + I_y}{2} \pm \sqrt{\left(\frac{I_x - I_y}{2}\right)^2 + I_{xy}^2}$$

$$= \frac{29(10^8) + 56(10^8)}{2} \pm \sqrt{\left[\frac{29(10^8) - 56(10^8)}{2}\right]^2 + [-30(10^8)]^2}$$

$$= 42.5(10^8) \text{ mm}^4 \pm 32.9(10^8) \text{ mm}^4$$

or

$$(I_u)_{max} = 75.4(10^8) \text{ mm}^4 \qquad (I_v)_{min} = 9.6(10^8) \text{ mm}^4 \qquad \textit{Ans.}$$

Specifically, the maximum moment of inertia, $(I_u)_{max} = 75.4(10^8) \text{ mm}^4$, occurs with respect to the u axis since, *by inspection,* most of the cross-sectional area is farthest away from this axis. (To show this mathematically, substitute the data with $\theta = 57.1°$ into the first of Eqs. 10–9.)

*10–8. Mohr's Circle for Moments of Inertia

Equations 10–9 through 10–11 have a graphical solution that is convenient to use and generally easy to remember. Squaring the first and third of Eqs. 10–9 and adding, it is found that

$$\left(I_u - \frac{I_x + I_y}{2}\right)^2 + I_{uv}^2 = \left(\frac{I_x - I_y}{2}\right)^2 + I_{xy}^2 \qquad (10\text{–}12)$$

In a given problem, I_u and I_{uv} are *variables,* and I_x, I_y, and I_{xy} are *known constants.* Thus, Eq. 10–12 may be written in compact form as

$$(I_u - a)^2 + I_{uv}^2 = R^2$$

When this equation is plotted, the resulting graph represents a *circle* of radius

$$R = \sqrt{\left(\frac{I_x - I_y}{2}\right)^2 + I_{xy}^2}$$

having its center located at point (a, O), where $a = (I_x + I_y)/2$. The circle so constructed is called *Mohr's circle,* named after the German engineer Otto Mohr (1835–1918.)

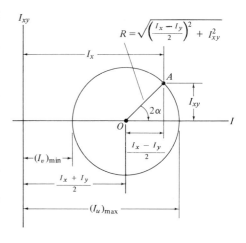

(a)

PROCEDURE FOR ANALYSIS

There are several methods for plotting Mohr's circle as defined by Eq. 10–12. The main purpose in using the circle here is to have a convenient means for transforming I_x, I_y, and I_{xy} into the principal moments of inertia. The following seven-step procedure provides a method for doing this.

Step 1: Establish the x, y axes for the area, with the origin located at the point P of interest, and determine I_x, I_y, and I_{xy} (Fig. 10–18a).

Step 2: Construct a rectangular coordinate system of axes such that the abscissa represents the moment of inertia I, and the ordinate represents the product of inertia I_{xy}, Fig. 10–18b.

Step 3: Determine the center of the circle, O, which is located at a distance $(I_x + I_y)/2$ from the origin, Fig. 10–18b.

Step 4: Plot the "controlling point" A having coordinates (I_x, I_{xy}). By definition, I_x is always positive, whereas I_{xy} will be either positive or negative, Fig. 10–18b.

Step 5: Connect the controlling point A found in *Step 4* with the center of the circle, and determine the distance OA by trigonometry. This distance represents the radius of the circle, Fig. 10–18b.

Step 6: Draw the circle having the radius found in *Step 5*. The points where the circle intersects the abscissa give the values of the principal

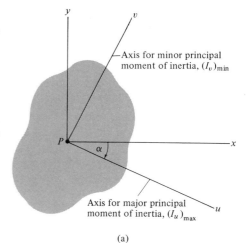

(b)

Fig. 10–18

moments of inertia $(I_v)_{min}$ and $(I_u)_{max}$. Notice that the *product of inertia will be zero at these points,* Fig. 10–18b.

Step 7:   To find the direction of the major principal axis, determine by trigonometry the angle 2α, *measured from the radius OA to the direction line of the positive abscissa,* Fig. 10–18b. This angle represents twice the angle from the x axis of the area in question to the axis of maximum moment of inertia, $(I_u)_{max}$, Fig. 10–18a. Both the angle in the circle, 2α, and the angle on the area, α, *must be measured in the same sense,* as shown in Fig. 10–18. The axis for minimum moment of inertia $(I_v)_{min}$ is perpendicular to the axis for $(I_u)_{max}$.

Using trigonometry, each of the above steps may be verified to be in accordance with the equations developed in Sec. 10–7.

Example 10–9

Using Mohr's circle, determine the principal moments of inertia for the beam's cross-sectional area, shown in Fig. 10–19a, with respect to an axis passing through the centroid.

Solution

The problem will be solved using the seven-step procedure outlined above.

Step 1:   The moments of inertia and the product of inertia have been determined in Examples 10–4 and 10–7 with respect to the x, y axes shown in Fig. 10–19a. The results are $I_x = 29(10^8)$ mm^4, $I_y = 56(10^8)$ mm^4, and $I_{xy} = -30(10^8)$ mm^4.

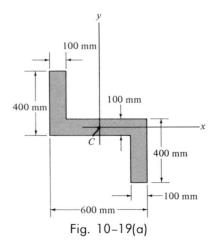

Fig. 10–19(a)

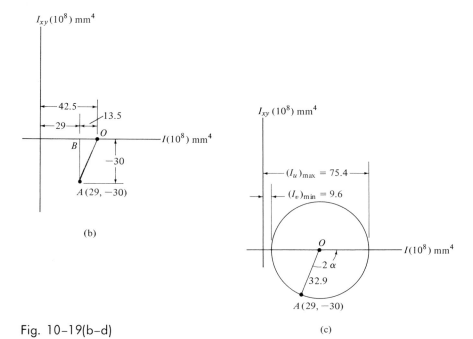

(b)

Fig. 10-19(b-d)

(c)

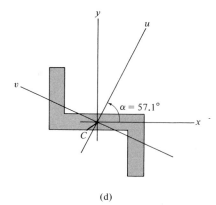

(d)

Steps 2, 3, 4, and 5: The I and I_{xy} axes are shown in Fig. 10-19*b*. The center of the circle, O, lies at a distance $(I_x + I_y)/2 = (29 + 56)/2 = 42.5$ from the origin. When the controlling point $A(29, -30)$ is connected to point O, the radius OA is determined from the triangle OBA using the Pythagorean theorem.

$$OA = \sqrt{(13.5)^2 + (-30)^2} = 32.9$$

Step 6: The circle is constructed in Fig. 10-19*c* and intersects the I axis at points $(75.4, 0)$ and $(9.6, 0)$. Hence,

$$(I_u)_{max} = 75.4(10^8) \text{ mm}^4 \qquad Ans.$$
$$(I_v)_{min} = 9.6(10^8) \text{ mm}^4 \qquad Ans.$$

Step 7: As shown in Fig. 10-19*c*, the angle 2α is determined from the circle by measuring counterclockwise from OA to the direction of the *positive I* axis. Hence,

$$2\alpha = 180° - \sin^{-1}\left(\frac{|BA|}{|OA|}\right) = 180° - \sin^{-1}\left(\frac{30}{32.9}\right) = 114.2°$$

The principal axis for $(I_u)_{max} = 75.4(10^8) \text{ mm}^4$ is therefore oriented at an angle $\alpha = 57.1°$, measured *counterclockwise*, from the *positive x* axis to the *positive u* axis. The v axis is perpendicular to this axis. The results are shown in Fig. 10-19*d*.

Problems

10-30. Determine the product of inertia of the quarter circle with respect to the x and y axes if $a = 150$ mm.

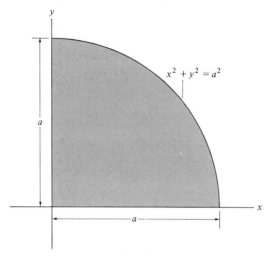

$x^2 + y^2 = a^2$

Prob. 10-30

10-30a. Solve Prob. 10-30 if $a = 2$ in.

10-31. Determine the product of inertia of the cross-sectional area of the bar with respect to the x and y axes. The bar is oriented at an angle θ from the x axis. Assume that $t \ll l$.

Prob. 10-31

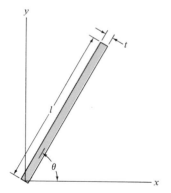

* 10-32. Determine the product of inertia of the shaded portion of the parabola with respect to the x and y axes.

Prob. 10-32

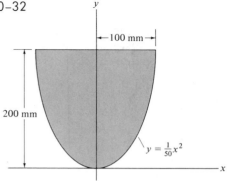

$y = \frac{1}{50}x^2$

10-33. Determine the product of inertia I_{xy} of half the shaded portion of the parabola in Prob. 10-32, bounded by the lines $y = 200$ mm and $x = 0$.

10-34. Determine the product of inertia of the shaded area with respect to the x and y axes.

Prob. 10-34

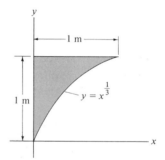

$y = x^{\frac{1}{3}}$

10-35. Determine the product of inertia of the shaded section of the ellipse with respect to the x and y axes. Set $a = 10$ mm.

Prob. 10-35

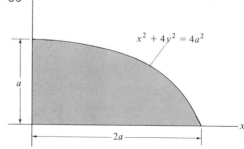

$x^2 + 4y^2 = 4a^2$

10-35a. Solve Prob. 10–35 if $a = 1.5$ in.

* 10-36. Determine the product of inertia for the angle with respect to the x and y axes having their origin located at the centroid C. Assume all corners to be square.

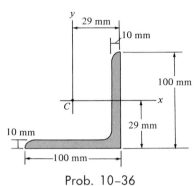

29 mm

10 mm

100 mm

x

C

29 mm

10 mm

100 mm

Prob. 10–36

10-37. Determine the product of inertia of the beam's cross-sectional area with respect to the x and y axes that have their origin located at the centroid C.

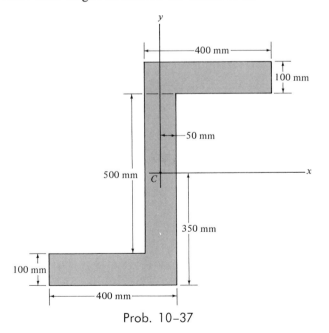

400 mm

100 mm

50 mm

500 mm

C

x

350 mm

100 mm

400 mm

Prob. 10–37

10-38. Determine the product of inertia of the beam's cross-sectional area with respect to the x and y axes that have their origin located at the centroid C.

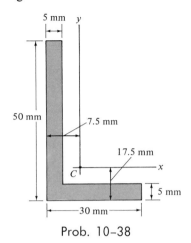

5 mm

50 mm

7.5 mm

17.5 mm

C

x

5 mm

30 mm

Prob. 10–38

10-39. Determine the principal moments of inertia of the composite area with respect to a set of principal axes that have their origin located at the centroid C. Use the equations developed in Sec. 10–7. $I_{xy} = -810(10^3)$ mm⁴.

Prob. 10–39

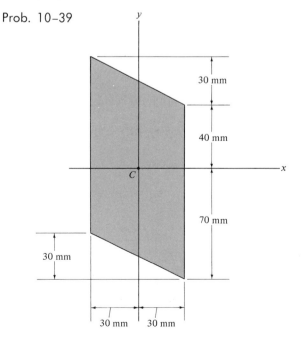

30 mm

40 mm

C

x

70 mm

30 mm

30 mm   30 mm

421

*10-40. Determine the product of inertia of the Z-section with respect to the x and y axes that have their origin located at the centroid C. Set $a = 20$ mm.

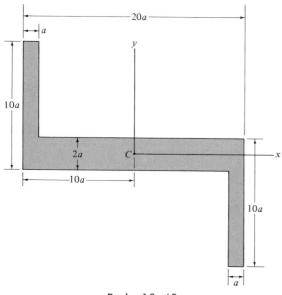

Prob. 10-40

*10-40a. Solve Prob. 10-40 if $a = 0.5$ in.

10-41. Locate the centroid, $\bar{y}$, and compute the moments of inertia I_u and I_v of the channel section. The u and v axes have their origin at the centroid C. For the calculation, assume all corners to be square.

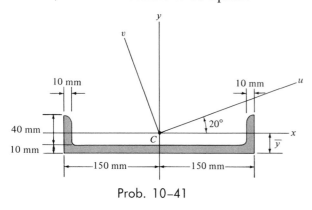

Prob. 10-41

10-42. Determine the product of inertia I_{uv} for the beam's cross-sectional area. For the calculation, assume all corners to be square.

Prob. 10-42

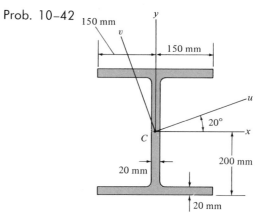

10-43. Determine the principal moments of inertia about the principal axes that have their origin located at the centroid C. Use the equations developed in Sec. 10-7. For the calculation, assume all corners to be square.

Prob. 10-43

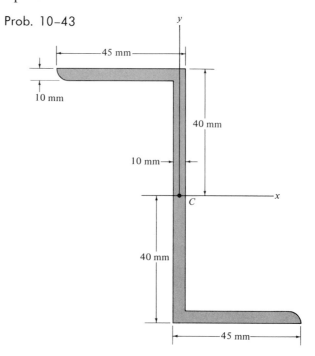

*10-44. Determine the moments of inertia I_u and I_v and the product of inertia I_{uv} for the rectangular area. The u and v axes pass through the centroid C.

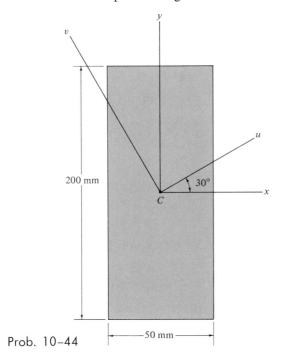

200 mm

50 mm

Prob. 10-44

10-45. Locate the centroid, $\bar{y}$, and determine the orientation of the principal centroidal axes for the composite area. What are the moments of inertia with respect to these axes? Set $a = 60$ mm, $b = 40$ mm, and $c = 90$ mm.

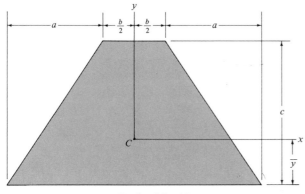

Prob. 10-45

10-45a. Solve Prob. 10-45 if $a = 4$ in., $b = 2$ in., and $c = 6$ in.

10-46. Determine the moments of inertia I_u and I_v and the product of inertia I_{uv} for the semicircular area.

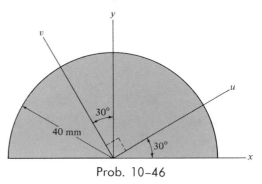

Prob. 10-46

10-47. Compute the moments of inertia I_u and I_v of the shaded area.

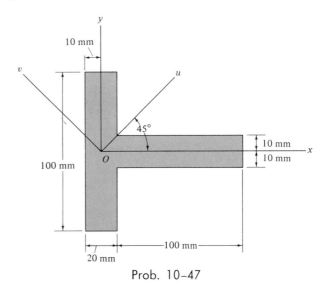

Prob. 10-47

*10-48. Determine the directions of the principal axes with origin located at point O, and the principal moments of inertia for the rectangular area about these axes.

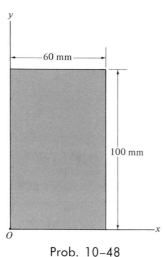

Prob. 10-48

10-49. Determine the principal moments of inertia for the angle with respect to a set of principal axes that have their origin located at the centroid C. Use the equations developed in Sec. 10-7. For the calculation, assume all corners to be square.

Prob. 10-49

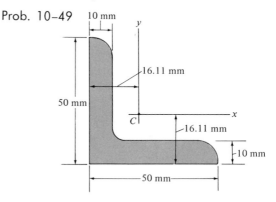

10-50. Construct Mohr's circle for the shaded area in Prob. 10-45.

10-50a. Construct Mohr's circle for the shaded area in Prob. 10-45a.

10-51. Solve Prob. 10-39 using Mohr's circle.

*10-52. Solve Prob. 10-48 using Mohr's circle.

10-53. Solve Prob. 10-43 using Mohr's circle.

10-54. Solve Prob. 10-49 using Mohr's circle.

11

Virtual Work

11-1. Definition of Work and Virtual Work

In the previous chapters, the solution of all equilibrium problems has been presented utilizing the concept of the free-body diagram and application of the equations of equilibrium. It is also possible to solve equilibrium problems using the *principle of virtual work,* which is based on the principles of work and energy. This method of solution is found to be very useful for solving equilibrium problems that involve a series of connected rigid bodies.

Work of a Force.   To understand the principle of virtual work, it is first necessary to define the *work done by a force* **F** when the force undergoes a differential displacement *d***s** along its path *s*, Fig. 11-1*a*. If θ is the angle made between the tails of the force and displacement vectors, the *differential* amount of work done by **F** is a *scalar quantity* defined by the dot product,

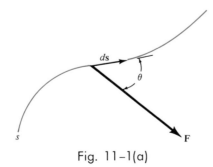

Fig. 11-1(a)

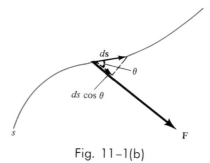

Fig. 11–1(b)

$$dU = \mathbf{F} \cdot d\mathbf{s} = F\,ds\cos\theta \qquad (11-1)$$

From this equation, the work may be interpreted in one of two ways, either as the product of F and the magnitude of the component of the displacement in the direction of $\mathbf{F}$, i.e., $ds\cos\theta$, Fig. 11–1b, or as the product of $d\mathbf{s}$ and the magnitude of the component of force in the direction of $d\mathbf{s}$, i.e., $F\cos\theta$, Fig. 11–1c. If $F\cos\theta$ and $d\mathbf{s}$ are in the *same direction,* the work is *positive;* if they are in *opposite directions,* the work is *negative.* Furthermore, from Eq. 11–1, if $\mathbf{F}$ is *perpendicular* to $d\mathbf{s}$, $\theta = 90°$ and the work done by $\mathbf{F}$ is *zero.*

The basic unit for work, the joule (J), combines the units of force and displacement. Specifically, 1 *joule* is equivalent to the work done by a force of 1 newton which moves 1 metre in the direction of the force $(1\text{ J} = 1\text{ N}\cdot\text{m})$. The moment of a force has the same combination of units $(\text{N}\cdot\text{m})$; however, the concepts of moment and work are in no way related. A moment is a vector quantity, whereas work is a scalar.

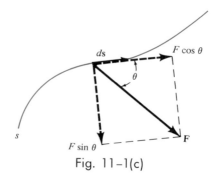

Fig. 11–1(c)

Work of a Couple. The two forces of a couple do work when the couple *rotates* about an axis perpendicular to the plane of the couple. To show this, consider the body in Fig. 11–2a, which is subjected to a couple having a magnitude of $M = Fr$. Any general differential displacement of the body can be considered as a separate translation and rotation. When the body *translates* such that the *component of displacement* along the line of action of each force is $d\mathbf{s}_t$, clearly the "positive" work of one force (Fds_t) *cancels* the "negative" work of the other ($-Fds_t$), Fig. 11–2b. Consider now a differential *rotation* $d\boldsymbol{\theta}$ of the body about an axis perpendicular to the plane of the couple, which intersects the plane at the midpoint O, Fig. 11–2c. (For the derivation, any other point in the plane may also be considered.) As shown, each force undergoes a displacement $ds_\theta = (r/2)\, d\theta$ in the direction of the force, hence from Eq. 11–1, the total work done is

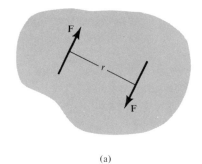

(a)

$$dU = F\left(\frac{r}{2}\, d\theta\right) + F\left(\frac{r}{2}\, d\theta\right) = (Fr)\, d\theta$$

or

$$dU = M\, d\theta \qquad (11\text{–}2)$$

The resultant work is *positive* when the rotational sense of both $\mathbf{M}$ and $d\boldsymbol{\theta}$ is in the *same* direction, and negative when they are in opposite directions. As in the case of the moment vector, the *direction* of $d\boldsymbol{\theta}$ is defined by the right-hand rule, where the fingers of the right hand follow the rotation or "curl," and the thumb indicates the direction of $d\boldsymbol{\theta}$. Hence, in Eq. 11–2, the line of action of $d\boldsymbol{\theta}$ is *parallel* to the line of action of $\mathbf{M}$. This is always the case if movement of the body occurs in the *same plane*. If the body rotates in space, however, the *component* of $d\boldsymbol{\theta}$ in the direction of $\mathbf{M}$ is required; i.e., the work done is defined by the dot product, $dU = \mathbf{M} \cdot d\boldsymbol{\theta}$.

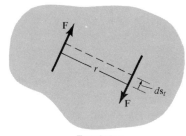

Translation

(b)

Virtual Work. The definitions of work done by both a force and a couple have been presented in terms of *actual movements* expressed by the infinitesimal displacement $d\mathbf{s}$ and rotation $d\boldsymbol{\theta}$. Consider now an *imaginary* or *virtual movement,* which indicates a displacement or rotation that is *assumed* and *does not actually exist*. These movements are first-order differential quantities and will be denoted by the symbols $\delta\mathbf{s}$ and $\delta\boldsymbol{\theta}$ (delta **s** and delta $\boldsymbol{\theta}$), respectively. The *virtual work* done by a force $\mathbf{F}$ undergoing a virtual displacement $\delta\mathbf{s}$ is

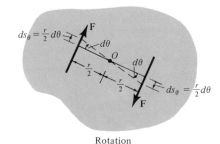

Rotation

(c)

Fig. 11–2

$$\delta U = \mathbf{F} \cdot \delta\mathbf{s} = F \cos\theta\, \delta s \qquad (11\text{–}3)$$

Similarly, when a couple $\mathbf{M}$ undergoes a virtual rotation $\delta\boldsymbol{\theta}$ in the plane of the couple forces, the *virtual work* is

$$\delta U = M\, \delta\theta \qquad (11\text{–}4)$$

11–2. Principle of Virtual Work for a Particle

The principle of virtual work for a particle initially at rest may be stated as follows: *If for a system of forces acting on a particle, the virtual work is zero for every independent virtual displacement of the particle, then the particle is in equilibrium.* Mathematically, this may be expressed as

$$\delta U = 0 \qquad (11\text{–}5)$$

where δU represents the virtual work done by all the forces that act on the particle during any virtual displacement. In three dimensions, an unconstrained particle may have three independent virtual displacements, and therefore three independent virtual-work equations may be written for each of these displacements.

To illustrate application of Eq. 11–5, consider the particle P, shown in Fig. 11–3, which is subjected to the action of three forces, $\mathbf{F}_1$, $\mathbf{F}_2$, and $\mathbf{F}_3$. If the particle is at rest and does not actually move, then if the particle is given an imaginary or virtual displacement $\delta\mathbf{x}$, Fig. 11–3, *only the x components of each force do work.* (The y and z components of force are *perpendicular* to the displacement and therefore do no work.) Hence, the virtual-work equation for this displacement is

$$\delta U = 0; \qquad (F_x)_1\,\delta x + (F_x)_2\,\delta x + (F_x)_3\,\delta x = 0$$
$$[(F_x)_1 + (F_x)_2 + (F_x)_3]\,\delta x = 0$$

Since δx has been *imposed* ($\delta x \neq 0$), the equation is satisfied if the sum of the force components in the x direction equals zero, i.e., $(F_x)_1 + (F_x)_2 + (F_x)_3 = 0$. In other words, requiring $\delta U = 0$ is equivalent to writing the equilibrium equation $\Sigma F_x = 0$ for the force system. In a similar manner, two other virtual-work equations may be written by assuming virtual displacements $\delta\mathbf{y}$ and $\delta\mathbf{z}$ in the y and z directions, respectively. Doing this amounts to satisfying the equilibrium equations $\Sigma F_y = 0$ and $\Sigma F_z = 0$ for the particle.

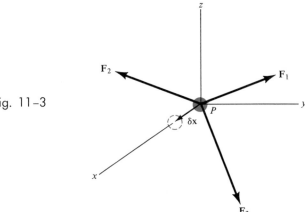

Fig. 11–3

Applying the virtual-work equations is therefore another way of stating both the necessary and sufficient conditions for particle equilibrium. No real advantage is gained, however, in using this method for solution because, as shown above, the virtual displacement in a given direction (δx) appears directly in each term of the virtual-work equation; hence, when factored, it leaves an equation which is the same as that which would be obtained by directly applying a force equilibrium equation to the particle in the same direction ($\Sigma F_x = 0$).

11–3. Principle of Virtual Work for a Rigid Body

Since equilibrium of a rigid body requires that the *resultant* external force and couple acting on the body be equal to zero, the principle of virtual work for a rigid body may be formulated as follows: *If for a system of external forces and couples acting on the rigid body the virtual work is zero for every independent virtual displacement of the body, then the body is in equilibrium.* It is not necessary to include the work done by the *internal forces* acting on the rigid body since the body *does not deform* when subjected to a loading and, furthermore, when the body moves through a virtual displacement, the internal forces occur in equal but opposite collinear pairs, so that the corresponding work done by each pair of forces *cancels.*

When applying the principle of virtual work to problems involving rigid bodies, it is necessary to account for all independent translational and rotational virtual movements of the body. For example, a rigid body subjected to a *three-dimensional force system* can have *six* independent virtual motions, namely, three translations δx, δy, and δz, and three rotations $\delta\alpha$, $\delta\beta$, and $\delta\gamma$, about the x, y, and z axes, respectively. Consequently, an independent virtual-work equation may be written for *each* of these six motions. These equations correspond to the *six* scalar equilibrium equations for the body ($\Sigma F_x = 0$, $\Sigma F_y = 0$, $\Sigma F_z = 0$, $\Sigma M_x = 0$, $\Sigma M_y = 0$, $\Sigma M_z = 0$). If the rigid body is subjected to a *planar system of forces,* it can have only two independent virtual translations on the surface of the plane and one virtual rotation about an axis perpendicular to the plane. These *three* motions will yield *three* independent virtual-work equations which correspond to the *three* equilibrium equations for force and moment ($\Sigma F_x = 0$, $\Sigma F_y = 0$, $\Sigma M_O = 0$).

As in the case of a particle, no added advantage is gained by solving rigid-body equilibrium problems using the principle of virtual work. This is because for each application of the virtual-work equation, the virtual displacement, common to every term, factors out of the equation, leaving an equation that may have been obtained by *direct application* of the equilibrium equations.

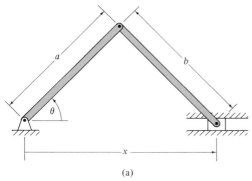

(a)

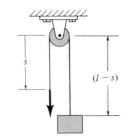

(l is the total vertical length of cord)

(b)

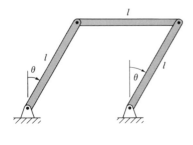

(c)

Fig. 11–4(a–c)

11–4. Principle of Virtual Work for a System of Connected Rigid Bodies

The method of virtual work is most suitable for solving equilibrium problems that involve a system of several *connected* rigid bodies such as shown in Fig. 11–4. For each of these systems, the number of *independent virtual-work equations* that can be written depends upon the number of *independent virtual displacements* that can be made by the system.

Degrees of Freedom. In general, *for a system of connected bodies the number of independent virtual displacements equals the minimum number of independent coordinates needed to specify completely the location of all members of the system.* Each independent coordinate q gives the system a *degree of freedom* that must be consistent with the constraining action of the supports. Thus, an n-degree-of-freedom system requires n independent coordinates q_n to specify the location of all its members with respect to a fixed reference point. For example, if it is assumed that the dimensions of each of the systems in Fig. 11–4 are known, then the link and sliding-block arrangement shown in Fig. 11–4a is an example of a one-degree-of-freedom system. The independent coordinate $q = \theta$ may be used to specify the location of the two connecting links and the block. The coordinate x could also be used as the independent coordinate. However, since the block is constrained to move within the slot, x is not independent of θ; rather it can be related to θ using the cosine law, $b^2 = a^2 + x^2 - 2ax \cos \theta$. Other examples of one-degree-of-freedom systems are given in Figs. 11–4b and 11–4c. The double-link and compound-pulley arrangement, shown in Figs. 11–4d and 11–4e, respectively, are examples of two-degree-of-freedom systems. To specify the location of each link, coordinate angles θ_1 and θ_2 must be known, since a rotation of one link is independent of the rotation of the other. To determine the location of the block attached to the pulley-and-cord arrangement in Fig. 11–4e, it is necessary to know the location of both end points A and B of the cords. Provided the cords are inextensible, and have known lengths, the independent coordinates s_1 and s_2, Fig. 11–4e, may be used to locate these end points.

As discussed in the previous sections, a particle unconstrained to move in space has three degrees of freedom, which may be specified by the three coordinate directions x, y, and z. A rigid body free to move in space has six degrees of freedom; three coordinate directions x, y, and z are needed to specify its translation, and three coordinate angles α, β, and γ specify its rotation. If a rigid body is constrained to move on a plane, it has three degrees of freedom—two translations on the surface of the plane, and one rotation about an axis perpendicular to the plane.

Principle of Virtual Work. The principle of virtual work for a system of rigid bodies whose connections are *frictionless,* may be stated as follows: *A system of connected rigid bodies is in equilibrium provided the virtual work done by all the external forces and couples acting on the system is zero for each independent virtual displacement of the system.* Mathematically, this may be expressed as

$$\delta U = 0 \qquad (11\text{–}6)$$

where δU represents the total virtual work done by all the external forces (and couples) acting on the system during any independent virtual displacement.

It has been pointed out that if a system has n degrees of freedom it takes n independent coordinates q_n to completely specify the location of the system. Hence, for the system it is possible to write n independent virtual-work equations; one for each independent coordinate, while the remaining $n - 1$ coordinates are held fixed.

Application of the principle of virtual work requires more mathematical sophistication, using calculus, than the conventional vector approach, using the equations of equilibrium. However, once the equations of virtual work are written, the solution may be obtained *directly, without* having to dismember the system to obtain relationships between forces occurring at the connections.

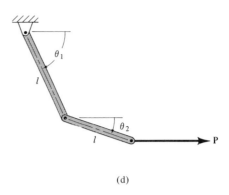

(d)

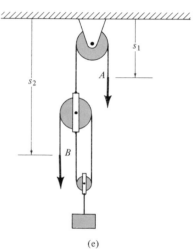

(e)

Fig. 11–4(d, e)

PROCEDURE FOR ANALYSIS

It is recommended that the following three-step procedure be used when applying the method of virtual work to solve problems involving a system of frictionless connected rigid bodies having a single degree of freedom.

Step 1: Draw the free-body diagram of the entire system of connected bodies and define the *independent coordinate q.* Sketch the "deflected position" of the system on the free-body diagram when the system undergoes a *positive* virtual displacement δq. From this, specify the "active" forces and couples that do work.

Step 2: Indicate *position coordinates s_i,* measured from a *fixed point* on the free-body diagram to each of the i number of "active" forces and couples. Each of these coordinate lines should be in the *same direction* as the line of action of the "active" force to which it is directed.

Relate each of the position coordinates s_i to the independent coordinate q; then *differentiate* these expressions in order to express the virtual displacements δs_i in terms of δq.

Step 3: Write the *virtual-work equation* for the system assuming that, whether possible or not, all the position coordinates s_i undergo *positive* virtual displacements δs_i. Using the relations for δs_i obtained in *Step 2*, express the work of *each* "active" force and couple in the equation in terms of the magnitude of the single independent virtual displacement δq. By factoring out this common displacement, one is left with an equation that generally can be solved for an unknown force, couple, or equilibrium position.

If the system contains n degrees of freedom, n independent coordinates q_n must be specified. In this case, follow the above procedure and let *only one* of the independent coordinates undergo a virtual displacement δq_n, while the remaining $n - 1$ coordinates are held fixed. In this way, n virtual-work equations can be written, one for each of the n independent coordinates.

The following examples should help to clarify application of the preceding steps.

Example 11–1

Using the principle of virtual work, determine the angle θ for equilibrium of the two-member linkage shown in Fig. 11–5a. Each member has a mass of 10 kg.

Solution

Step 1: The system has only one degree of freedom, since the location of both links may be specified by the single independent coordinate $(q =) \theta$. As shown on the free-body diagram in Fig. 11–5b, when θ undergoes a *positive* virtual rotation $\delta\theta$, only the active forces, **F** and the two 98.1-N weights, do work. (The reactive forces $\mathbf{D}_x$ and $\mathbf{D}_y$ are fixed, and $\mathbf{B}_y$ does not move along its line of action.)

Step 2: If the origin of coordinates is established at the *fixed* pin support D, the location of **F** and **W** may be specified by the *position coordinates* x_B and y_w, as shown in the figure. Note that in order to compute the work,

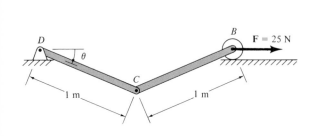

(a)   Fig. 11–5   (b)

these coordinates are in the *same direction* as the lines of action of their associated forces.

Expressing the position coordinates in terms of the independent coordinate θ, we obtain

$$x_B = 2(1 \cos \theta) \text{ m}$$
$$y_w = \tfrac{1}{2}(1 \sin \theta) \text{ m}$$

Differentiating to obtain the relationships between the virtual displacements yields

$$\delta x_B = -2 \sin \theta \; \delta\theta \text{ m} \qquad\qquad (1)$$
$$\delta y_w = 0.5 \cos \theta \; \delta\theta \text{ m} \qquad\qquad (2)$$

It is seen by the *signs* of these equations, and indicated in Fig. 11–5b, that an *increase* in θ (i.e., $\delta\theta$) causes a *decrease* in x_B and an *increase* in y_w.

Step 3: For *positive* virtual displacements δx_B and δy_w, the forces **W** and **F** do positive work since the forces and their corresponding displacements would be in the same direction. Hence, the virtual-work equation for the displacement $\delta\theta$ is

$$\delta U = 0; \qquad W \, \delta y_w + W \, \delta y_w + F \, \delta x_B = 0 \qquad\qquad (3)$$

Substituting Eqs. (1) and (2) into Eq. (3) in order to relate the virtual displacements to the common virtual displacement $\delta\theta$ yields

$$98.1(0.5 \cos \theta \; \delta\theta) + 98.1(0.5 \cos \theta \; \delta\theta) + 25(-2 \sin \theta \; \delta\theta) = 0$$

Notice that the "negative work" done by **F** (force in the opposite direction to displacement) has been *accounted for* in the above equation by the "negative sign" of Eq. (1). Factoring out the *common displacement* $\delta\theta$, we have

$$(98.1 \cos \theta - 50 \sin \theta)\delta\theta = 0$$

Since $\delta\theta \neq 0$,

$$98.1 \cos \theta - 50 \sin \theta = 0$$

or

$$\tan \theta = \frac{98.1}{50}$$

Solving for θ yields

$$\theta = 63.0° \qquad\qquad\qquad\qquad \textit{Ans.}$$

If this problem had been solved using the equations of equilibrium, it would have been necessary to dismember the links and apply three scalar equations to *each* link. Using the principle of virtual work, by means of calculus, this task has been eliminated so that the answer is obtained in a very direct manner.

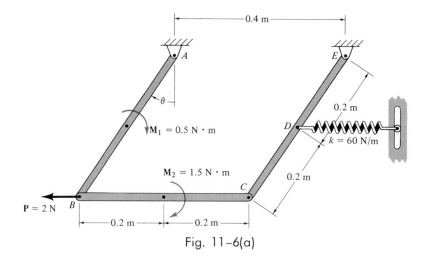

Fig. 11–6(a)

Example 11–2

Using the principle of virtual work, determine the angle θ required to maintain the equilibrium of the mechanism shown in Fig. 11–6a. Neglect the weight of the links. The spring is unstretched when $\theta = 0°$ and it maintains a horizontal position due to the roller.

Solution

Step 1: The mechanism has one degree of freedom, and therefore the location of each member may be specified using the independent coordinate θ. When θ undergoes a *positive* virtual displacement $\delta\theta$, as shown on the free-body diagram in Fig. 11–6b, links AB and EC rotate by the same amount since they have the same length, and link BC only translates. Since a couple does work *only* when it rotates, the work done by $\mathbf{M}_2$ is zero. The reactive forces at A and E do no work since the supports do not translate.

Step 2: The position coordinates x_B and x_D are along the *same* line of action as $\mathbf{P}$ and $\mathbf{F}$ and locate these forces with respect to the *fixed points A* and E. From Fig. 11–6b,

$$x_B = 0.4 \sin \theta \text{ m}$$
$$x_D = 0.2 \sin \theta \text{ m}$$

Thus,

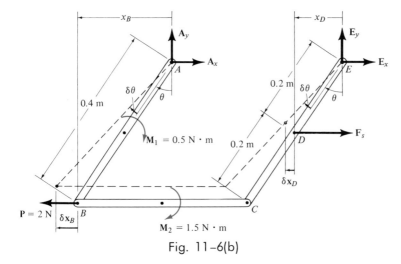

Fig. 11–6(b)

$$\delta x_B = 0.4 \cos \theta \; \delta\theta \text{ m}$$
$$\delta x_D = 0.2 \cos \theta \; \delta\theta \text{ m}$$

Step 3: Applying the equation of virtual work, noting that $\mathbf{F}_s$ is opposite to *positive* δx_D displacement, and hence does negative work, we obtain

$$\delta U = 0; \qquad M_1 \, \delta\theta + P \, \delta x_B - F_s \, \delta x_D = 0$$

Relating each of the virtual displacements to the *common* virtual displacement $\delta\theta$ yields

$$0.5 \, \delta\theta + 2(0.4 \cos \theta \; \delta\theta) - F_s(0.2 \cos \theta \; \delta\theta) = 0$$
$$(0.5 + 0.8 \cos \theta - 0.2 F_s \cos \theta) \, \delta\theta = 0 \qquad (1)$$

For the arbitrary angle θ, the spring is stretched a distance of $x = x_D = (0.2 \sin \theta)$ m; and therefore $F_s = kx = 60 \text{ N/m}(0.2 \sin \theta)$ m $= (12 \sin \theta)$ N. Substituting into Eq. (1) and noting that $\delta\theta \neq 0$, we have

$$0.5 + 0.8 \cos \theta - 0.2(12 \sin \theta) \cos \theta = 0$$

Since $\sin 2\theta = 2 \sin \theta \cos \theta$, then

$$1 = 2.4 \sin 2\theta - 1.6 \cos \theta$$

Solving for θ by trial and error yields

$$\theta = 36.3° \qquad\qquad\qquad \textit{Ans.}$$

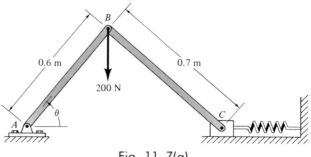

Fig. 11–7(a)

Example 11–3

Using the principle of virtual work, determine the force that the spring must exert in order to hold the mechanism shown in Fig. 11–7a in equilibrium when $\theta = 45°$. Neglect the weight of the members.

Solution

Step 1:   As shown on the free-body diagram, Fig. 11–7b, the system has one degree of freedom, defined by the independent coordinate θ. When θ undergoes a *positive* virtual displacement $\delta\theta$, only $\mathbf{F}_s$ and the 200-N force do work.

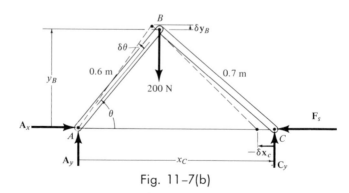

Fig. 11–7(b)

Step 2: The position of $\mathbf{F}_s$ and 200 N is located from the fixed origin at A, by the position coordinates y_B and x_C. From Fig. 11–7b, x_C can be related to θ by using the "law of cosines."

$$(0.7)^2 = (0.6)^2 + x_C^2 - 2(0.6)x_C \cos\theta \qquad (1)$$

The differential of this equation is then

$$0 = 0 + 2x_C \,\delta x_C - 1.2 \,\delta x_C \cos\theta + 1.2x_C \sin\theta \,\delta\theta$$

Thus,

$$\delta x_C = \frac{1.2x_C \sin\theta}{1.2\cos\theta - 2x_C} \,\delta\theta \qquad (2)$$

Also,

$$y_B = 0.6 \sin\theta$$

Therefore,

$$\delta y_B = 0.6 \cos\theta \,\delta\theta \qquad (3)$$

Step 3: When the coordinates y_B and x_C undergo *positive* virtual displacements δy_B and δx_C, forces $\mathbf{F}_s$ and 200 N do *negative work,* since they both act in the opposite direction to $\delta\mathbf{y}_B$ and $\delta\mathbf{x}_C$. Hence, the equation of virtual work becomes

$$\delta U = 0; \qquad -200 \,\delta y_B - F_s \,\delta x_C = 0$$

Substituting Eqs. (2) and (3) into this equation, we obtain

$$-200(0.6 \cos\theta \,\delta\theta) - F_s \frac{1.2x_C \sin\theta}{1.2\cos\theta - 2x_C} \,\delta\theta = 0$$

$$\left[-120 \cos\theta - \frac{F_s(1.2x_C \sin\theta)}{1.2\cos\theta - 2x_C} \right] \delta\theta = 0$$

Hence,

$$F_s = \frac{-120 \cos\theta(1.2\cos\theta - 2x_C)}{1.2x_C \sin\theta}$$

At the required equilibrium position $\theta = 45°$, the corresponding value of x_C can be found by using Eq. (1), in which case

$$x_C^2 - 1.2 \cos 45° \, x_C - 0.13 = 0$$

Solving for the positive root yields

$$x_C = 0.981 \text{ m}$$

Thus,

$$F_s = \frac{-120 \cos 45° \,[1.2 \cos 45° - 2(0.981)]}{1.2(0.981) \sin 45°}$$

$$= 113.5 \text{ N} \qquad\qquad\qquad Ans.$$

Since the answer is positive, it indicates that $\mathbf{F}_s$ was shown in the correct direction in Fig. 11–7b. Hence, the spring must be subjected to a *compressive force* in order to maintain equilibrium of the mechanism.

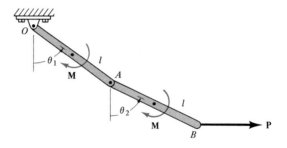

Fig. 11–8(a)

Example 11–4

Using the principle of virtual work, determine the equilibrium position of the two-bar linkage shown in Fig. 11–8a. Neglect the weight of the links.

Solution

The system has two degrees of freedom, since the *independent coordinates* θ_1 and θ_2 must be known to locate the position of both links. The position coordinate x_B, measured from the fixed point O, is used to specify the location of **P**, Figs. 11–8b and 11–8c.

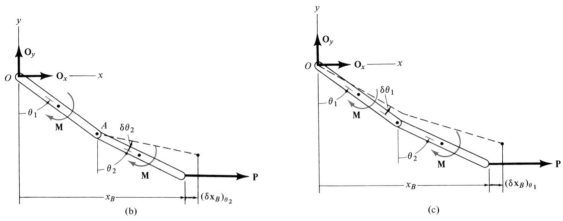

Fig. 11–8(b, c)

If θ_1 is held *fixed* and θ_2 varies by an amount $\delta\theta_2$, as shown in Fig. 11–8b, the virtual-work equation becomes

$[\delta U = 0]_{\theta_2};$ $\qquad\qquad P(\delta x_B)_{\theta_2} - M\,\delta\theta_2 = 0 \qquad\qquad (1)$

where P and M represent the magnitudes of the applied force and couple acting on link AB.

When θ_2 is held *fixed*, and θ_1 varies by an amount $\delta\theta_1$, as shown in Fig. 11–8c, the virtual-work equation becomes

$[\delta U = 0]_{\theta_1};$ $\qquad\quad P(\delta x_B)_{\theta_1} - M\,\delta\theta_1 - M\,\delta\theta_1 = 0 \qquad\qquad (2)$

The *position coordinate* x_B may be related to the independent coordinates θ_1 and θ_2 by the equation

$$x_B = l\sin\theta_1 + l\sin\theta_2 \qquad\qquad (3)$$

To obtain the variation of δx_B in terms of $\delta\theta_2$ it is necessary to take the *partial derivative* of x_B with respect to θ_2 since x_B is a function of both θ_1 and θ_2. Hence,

$$\frac{\partial x_B}{\partial\theta_2} = l\cos\theta_2$$

$$(\delta x_B)_{\theta_2} = l\cos\theta_2\,\delta\theta_2$$

Substituting into Eq. (1), we have

$$(Pl\cos\theta_2 - M)\,\delta\theta_2 = 0$$

Since $\delta\theta_2 \neq 0$, then

$$\theta_2 = \cos^{-1}\left(\frac{M}{Pl}\right) \qquad\qquad \text{Ans.}$$

Using Eq. (3) to obtain the variation of x_B with θ_1 yields

$$\frac{\partial x_B}{\partial\theta_1} = l\cos\theta_1$$

$$(\delta x_B)_{\theta_1} = l\cos\theta_1\,\delta\theta_1$$

Substituting into Eq. (2), we have

$$(Pl\cos\theta_1 - 2M)\,\delta\theta_1 = 0$$

Since $\delta\theta_1 \neq 0$, then

$$\theta_1 = \cos^{-1}\left(\frac{2M}{Pl}\right) \qquad\qquad \text{Ans.}$$

Problems

11-1. Compute the force developed in the spring required to keep the 4-kg rod in equilibrium when $\theta = 30°$.

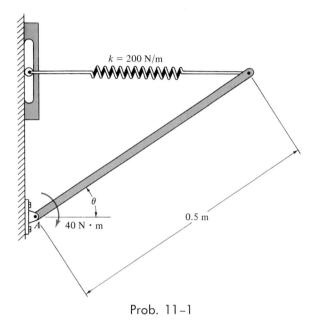

Prob. 11–1

11-2. The pin-connected mechanism is constrained at A by a pin and at B by a smooth sliding block. If $P = 200$ N, determine the angle θ for equilibrium.

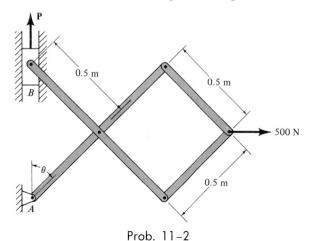

Prob. 11–2

11-3. If the uniform links AB and BC each have a mass of 5 kg, determine the horizontal force **P** required to hold the mechanism in the position shown, $\theta = 40°$. The spring has an unstretched length of 0.6 m.

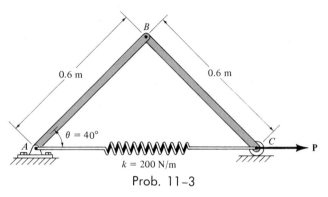

Prob. 11–3

*11-4. Determine the force **P** required to lift the 15-kg block using the differential hoist. The lever arm is fixed to the upper pulley and turns with it.

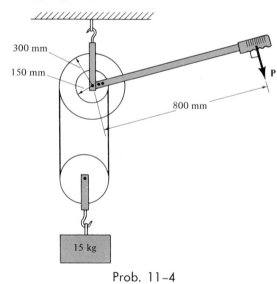

Prob. 11–4

11-5. If each of the three links of the mechanism has a mass of $m = 5$ kg, determine the angle θ for equilibrium. The spring, which always remains horizontal, is unstretched when $\theta = 0°$. Set $k = 800$ N/m and $a = 0.25$ m.

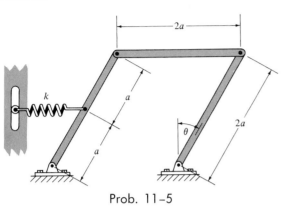

Prob. 11-5

11-5a. Solve Prob. 11-5 if each link weighs $W = 15$ lb, and $k = 45$ lb/ft, $a = 2$ ft.

11-6. The spring has an unstretched length of 0.2 m. Determine the angle θ for equilibrium if the uniform links each have a mass of 5 kg.

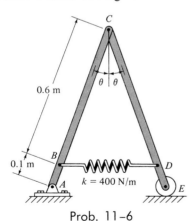

Prob. 11-6

11-7. Determine the angle θ for equilibrium of the frame in Prob. 11-6 if a horizontal force $P = 20$ N acts to the right at C. Use the data in Prob. 11-6.

* 11-8. The members of the mechanism are pin-connected. If a horizontal force of 500 N acts at A, determine the angle θ for equilibrium. The spring is unstretched when $\theta = 90°$.

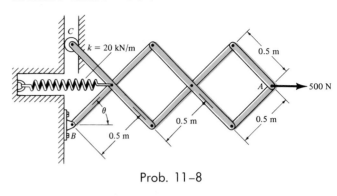

Prob. 11-8

11-9. The *Roberval balance* is in equilibrium when no weights are placed on the pans A and B. If two masses m_A and m_B are placed at *any* location a and b on the pans, show that equilibrium is maintained if $m_A d_A = m_B d_B$.

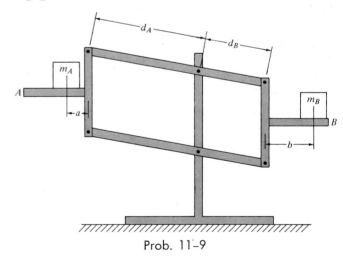

Prob. 11-9

11-10. The uniform rod *OA* has a mass of $m = 5$ kg. When the rod is in the vertical position, $\theta = 0°$, the spring is unstretched. Determine the angle θ for equilibrium if the end of the spring wraps around the periphery of the disk as the disk turns. Set $k = 350$ N/m, $r = 0.2$ m, and $l = 0.6$ m.

Prob. 11–10

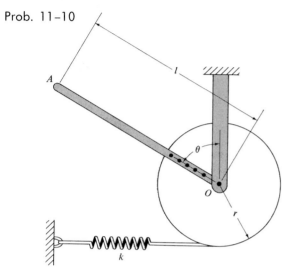

11-10a. Solve Prob. 11–10 if the rod has a weight of $W = 10$ lb, and $k = 35$ lb/ft, $r = 0.5$ ft, $l = 2$ ft.

11-11. Determine the vertical force **P** needed to hold the 6-kg block in equilibrium. Pulley *A* has a mass of 2 kg. Neglect the weight of the cords. *Hint:* Express the total *constant vertical length l* of the cord in terms of the position coordinates s_1 and s_2. The derivative of this equation yields a relation between δs_1 and δs_2.

Prob. 11–11

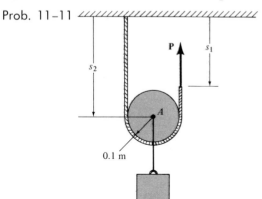

* 11-12. Determine the force **F** acting on the cord which is required to maintain equilibrium of the horizontal 10-kg bar *AB*. *Hint:* Express the total *constant vertical length l* of the cord in terms of the position coordinates s_1 and s_2. The derivative of this equation yields a relation between δs_1 and δs_2.

Prob. 11–12

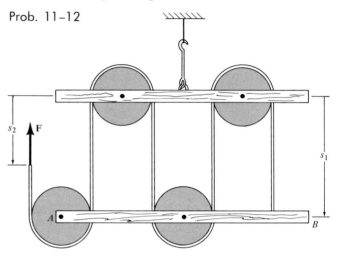

11-13. If a force $P = 125$ N is applied perpendicular to the handle of the toggle press, determine the compressive force developed at C; $\theta = 30°$.

Prob. 11–13

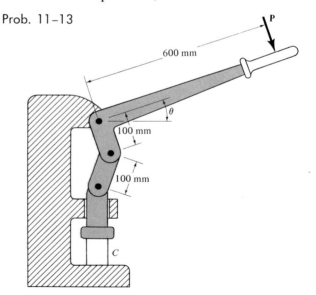

11–14. The piston C moves vertically between the two smooth walls. If the spring has a stiffness of $k = 1.5\,$kN/m and is unstretched when $\theta = 0°$, determine the couple $\mathbf{M}$ that must be applied to AB to hold the mechanism in equilibrium; $\theta = 30°$.

Prob. 11–14

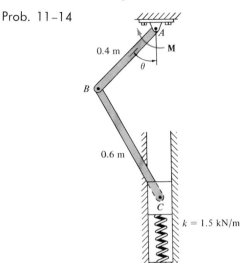

11–15. A disk having a mass of $m_D = 5$ kg is attached to the end of the rod ABC. If the rod is supported by a smooth slider block at C and rod BD, determine the angle θ for equilibrium. Neglect the weights of the rods and the slider. $a = 200\,$mm, $b = 500\,$mm, and $c = 300\,$mm.

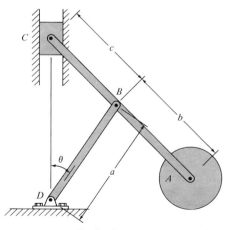

Prob. 11–15

11–15a. Solve Prob. 11–15 if the disk weighs $W = 2$ lb, and $a = 4$ in., $b = 10$ in., $c = 6$ in.

*11–16. The machine shown is used for forming metal plates. It consists of two toggles ABC and DEF, which are operated by a hydraulic cylinder H. The toggles push the movable bar G forward, pressing the plate p into the cavity. If the force which the plate exerts on the head is $P = 6$ kN, determine the force $\mathbf{F}$ in the hydraulic cylinder; $\theta = 30°$, $l = 200\,$mm, and $a = 150\,$mm.

Prob. 11–16

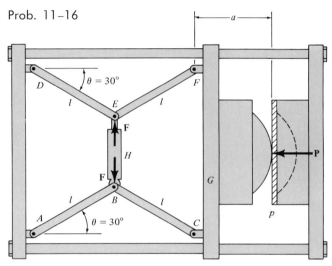

11–17. Rods AB and BC have a center of gravity located at their midpoints. If all contacting surfaces are smooth and BC has a mass of 50 kg, determine the appropriate mass of AB required for equilibrium.

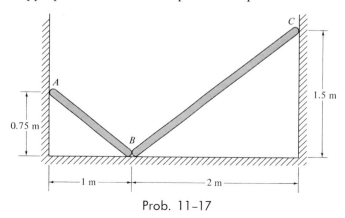

Prob. 11–17

11-18. Determine the force **P** that must be applied to the cord wrapped around the shaft at *C* which is necessary to lift the mass. As the mass is lifted, the pulley rolls on a cord that winds up on shaft *B* and unwinds from shaft *A*.

*11-20. The three-bar mechanism is subjected to a couple $M_A = 60$ N · m. Determine the magnitude of the couple M_D needed to maintain the equilibrium position $\theta = 30°$. Set $a = 400$ mm, $b = 450$ mm, and $c = 300$ mm.

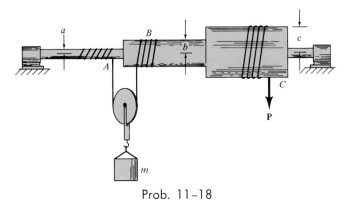

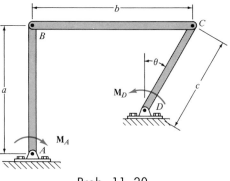

Prob. 11-18

Prob. 11-20

11-19. The truck is weighed on the highway inspection scale. If a known mass *m* is placed a distance *s* from the fulcrum *B* of the scale, determine the mass of the truck m_t if its center of gravity is located at a distance *d* from point *C*. When the scale is empty, the weight of the lever *ABC* balances the scale *CDE*.

*11-20a. Solve Prob. 11-20 if $M_A = 8$ lb · ft, $\theta = 40°$, $a = 1.25$ ft, $b = 1.50$ ft, and $c = 1.0$ ft.

11-21. Determine the horizontal force **F** required to maintain equilibrium of the slider mechanism when $\theta = 60°$.

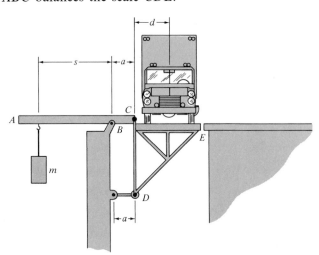

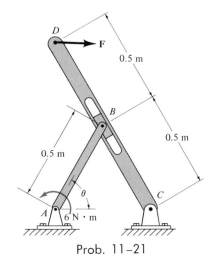

Prob. 11-19

Prob. 11-21

11-22. Determine the angles θ_1 and θ_2 for equilibrium of the two uniform links. Each link has a mass m.

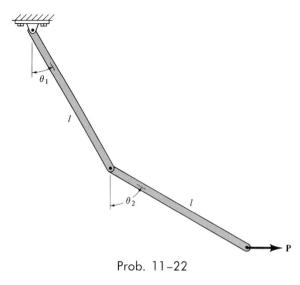

Prob. 11-22

11-23. Determine the horizontal force **P** and vertical force **Q** necessary to maintain the equilibrium position of the two uniform links. Each link has a mass of 6 kg.

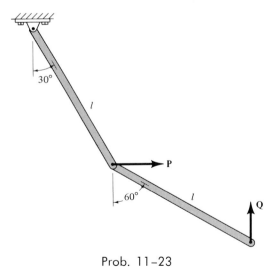

Prob. 11-23

*11-24. The loading arm is used to transfer oil from a ship to land. When the device is in use, the oil flows through the nozzle at C and exits at D as shown by the arrows. If each of the arms AB and BC has a total mass m and they are pin-connected at A, B, and I, determine the mass of blocks J and F that will keep the system in balance for any angle θ or ϕ. J is attached to the extended portion of arm AB and has a center of gravity at G_1; whereas F has a center of gravity at G_2 and is attached to a link that is pin-connected at A and E. Neglect the weight of links AEH, EI, and the connected portions IB of BC and KA of AB. Assume the weight of AB and BC acts through their centers.

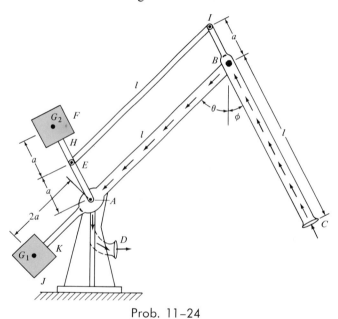

Prob. 11-24

11-25. The chain puller is used to draw two ends of a chain together in order to attach the "master link." The device is operated by turning the screw S, which pushes the bar AB downward, thereby drawing the tips C and D towards one another. If the sliding contacts at A and B are smooth, determine the force $\mathbf{F}$ maintained by the screw at E which is required to develop a drawing tension of $T = 600$ N in the chains. Set $a = 20$ mm and $b = 60$ mm.

11-25a. Solve Prob. 11–25 if $T = 120$ lb, $a = 1.25$ in., and $b = 4$ in.

11-26. Determine the mass of A and B required to hold the 500-g desk lamp in balance for any angle θ or ϕ. Neglect the weight of the mechanism and the size of the lamp.

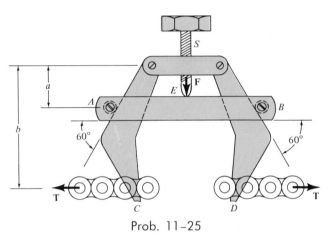

Prob. 11–25

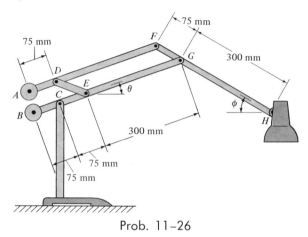

Prob. 11–26

446

The work done by a force **F,** when the force undergoes a *differential displacement d***s,** has been defined by $dU = \mathbf{F} \cdot d\mathbf{s}$ (Eq. 11–1). If the force is displaced over a path that has a *finite length s,* the work is determined by integrating over the path, i.e.,

$$U = \int_s \mathbf{F} \cdot d\mathbf{s} = \int_s F \cos \theta \; ds \qquad (11\text{–}7)$$

To evaluate the integral, it is necessary to obtain a relationship between F and the component of displacement $ds \cos \theta$. In some instances, however, the work done by a force will be *independent* of its path and, instead, will depend only upon the initial and final locations of the force along the path. A force that has this property is called a *conservative force.*

Weight. The weight of a body is a conservative force, since the work done by the weight depends *only* on the body's *vertical displacement.* To show this, consider the body in Fig. 11–9, which is initially at P'. If the body is moved *down* along the arbitrary path A to the dashed position, a vertical distance y, the total work done by **W** is determined from Eq. 11–7. For a given displacement d**s** along the path, the displacement component in the direction of **W** has a magnitude of $dy = ds \cos \theta$, as shown. Since both **W** and d**y** are in the same direction, the work is positive; hence,

$$U = \int_s W \cos \theta \; ds = \int_0^y W \, dy$$

or

$$U = Wy \qquad (11\text{–}8)$$

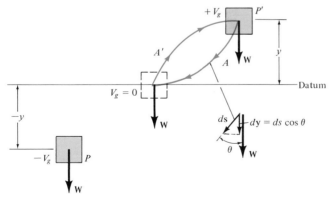

Fig. 11–9

In a similar manner, the work done by the weight when the body moves up a distance y back to P', along the arbitrary path A', is

$$U = -Wy \qquad (11\text{-}9)$$

Why is the work negative?

Elastic Spring. The force developed by an elastic spring ($F_s = kx$) is also a conservative force. If the spring is attached to a body and the body is displaced along *any path,* such that it causes the spring to elongate or compress from a position x_1 to a further position x_2, the work is

$$U = \int_{x_1}^{x_2} \mathbf{F}_s \cdot d\mathbf{x} = \int_{x_1}^{x_2} (-kx)\, dx = -(\tfrac{1}{2}kx_2^2 - \tfrac{1}{2}kx_1^2) \quad (11\text{-}10)$$

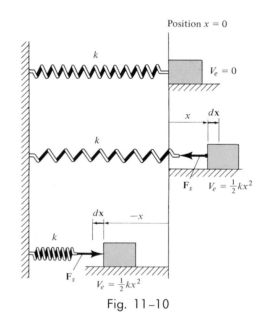

Fig. 11–10

For either extension or compression, the work is negative, since the spring exerts a force $\mathbf{F}_s$ *on the body* that is opposite to the body's displacement $\delta\mathbf{x}$, Fig. 11–10.

Friction. In contrast to a conservative force, consider the force of *friction* exerted on a moving body by a fixed surface. The work done by the frictional force depends upon the path; the longer the path, the greater the work. Consequently, frictional forces are *nonconservative.* The work done by this force is dissipated from the body in the form of heat.

When a conservative force acts on a body, it gives the body the capacity to do work. This capacity, measured as *potential energy, V,* depends upon the location of the body.

Gravitational Potential Energy.   If a body is located at position P', a distance y *above* a fixed horizontal reference or *datum,* Fig. 11–9, the weight **W** of the body has *positive gravitational potential energy* V_g. This energy can be expressed mathematically as

$$V_g = Wy \qquad (11\text{--}11)$$

Here V_g is *positive,* since **W** has the *capacity* for doing positive work when the body is *moved back* to a point on the datum. When the body is *below* the datum at position P, Fig. 11–9, the potential energy is *negative,*

$$V_g = -Wy \qquad (11\text{--}12)$$

since **W** does negative work when the body is *moved back* to the datum.

Elastic Potential Energy.   The *elastic potential energy* V_e that an elastic spring produces on an attached body, when the spring is elongated or compressed from an unstretched position ($x = 0$) to a final position x, is

$$V_e = \tfrac{1}{2}kx^2 \qquad (11\text{--}13)$$

Here V_e is *always positive,* since in the deformed position the spring has the capacity of doing *positive work* in *returning* the body back to the spring's original undeformed position, Fig. 11–10.

Potential-Energy Function.   In the general case, if a body is subjected to *both* gravitational and elastic forces, the *potential energy function* of the body can be expressed as the algebraic sum

$$V = V_g + V_e \qquad (11\text{--}14)$$

where measurement of V depends upon the location of the body with respect to a selected datum in accordance with Eqs. 11–11 through 11–13.

In general, if a system of frictionless connected rigid bodies has a *single degree of freedom* such that its position is defined by the independent coordinate q, then the potential energy for the system may be defined using a potential-energy function $V = V(q)$ as defined by Eq. 11–14. The work done by all the conservative forces acting on the system, in moving the system from point q_1 to q_2, is measured by the *difference in V,* i.e.,

$$U_{1-2} = V(q_1) - V(q_2) \qquad (11\text{--}15)$$

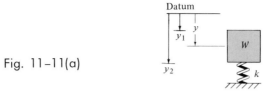

Fig. 11–11(a)

For example, the potential-energy function for a system consisting of a block of weight **W,** supported from a spring, Fig. 11–11a, can be expressed in terms of its independent coordinate y, measured from a fixed datum located at the unstretched length of the spring, i.e.,

$$V = V_g + V_e$$
$$= -Wy + \tfrac{1}{2}ky^2 \qquad (11\text{--}16)$$

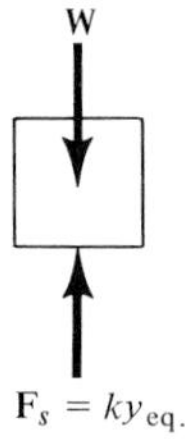

Fig. 11–11(b)

If the block moves from y_1 to a further downward position y_2, then, from Eq. 11–15, the work of **W** and $\mathbf{F}_s$ is

$$U_{1\text{--}2} = V(y_1) - V(y_2) = -W[y_1 - y_2] + \tfrac{1}{2}ky_1^2 - \tfrac{1}{2}ky_2^2$$

*11–7. Potential-Energy Criterion for Equilibrium

System Having One Degree of Freedom. When the displacement of a frictionless connected system is *infinitesimal,* i.e., from q to $q + dq$, Eq. 11–15 becomes

$$dU = V(q) - V(q + dq)$$

or

$$dU = -dV \qquad (11\text{--}17)$$

Furthermore, if the independent coordinate undergoes a *virtual displacement* δq, rather than an actual displacement dq, then $\delta U = -\delta V$. For equilibrium, the principle of virtual work requires $\delta U = 0$ and therefore, provided the potential-energy function for the system is known, this also requires that

$$\delta V = \frac{dV}{dq}\,\delta q = 0 \qquad (11\text{–}18)$$

Hence, *when a system of connected rigid bodies is in equilibrium, the first variation or change in the potential-energy function is zero.* This change is computed by taking the *first derivative* of the potential-energy function and setting it equal to zero. For example, to determine the equilibrium position for the spring and block in Fig. 11–11a, a differential (or virtual) movement δy of the system causes a change in potential energy, δV. Applying the equilibrium criterion to Eq. 11–16 yields

$$\delta V = \frac{dV}{dy}\,\delta y = (W - ky)\,\delta y = 0$$

Since $\delta y \neq 0$, the term in parentheses must be zero. Hence, the equilibrium position $y = y_{eq}$ is

$$y_{eq} = \frac{W}{k}$$

Of course, this *same result* is obtained by applying $\Sigma F_y = 0$ to the forces acting on the free-body diagram of the block, Fig. 11–11b.

System Having n Degrees of Freedom. When the system of connected bodies has n degrees of freedom, the total potential energy stored in the system will be a function of n independent coordinates q_n; i.e., $V = V(q_1, q_2, \ldots, q_n)$. In order to apply the equilibrium criterion $\delta V = 0$, it is necessary to compute the change in potential energy δV by using the "chain rule" of differential calculus; i.e., the total variation or total differential of V must be equal to zero. Thus,

$$\delta V = \frac{\partial V}{\partial q_1}\,\delta q_1 + \frac{\partial V}{\partial q_2}\,\delta q_2 + \cdots + \frac{\partial V}{\partial q_n}\,\delta q_n = 0$$

Since the virtual displacements $\delta q_1, \delta q_2, \ldots, \delta q_n$ are independent of one another, it is necessary that

$$\frac{\partial V}{\partial q_1} = 0, \quad \frac{\partial V}{\partial q_2} = 0, \quad \ldots, \quad \frac{\partial V}{\partial q_n} = 0 \qquad (11\text{–}19)$$

As for the case of virtual work, *it is possible to write n independent potential-energy equations for a system having n degrees of freedom.*

*11-8. Stability of Equilibrium

Once the equilibrium configuration for a body or system of connected bodies is defined, it is sometimes important to investigate the "type" of equilibrium or the stability of the configuration. For example, consider the position of a ball resting at a point on each of the three paths shown in Fig. 11-12. Each situation represents an equilibrium state for the ball. When the ball is at A, it is said to be in *stable equilibrium,* because if it is given a small displacement up the hill, it will always *return* to its original, lowest, position. At A, its total potential energy is a *minimum*. When the ball is at B, it is in *neutral equilibrium*. A small displacement either to the left or right of B will not alter this condition. The ball *remains* in equilibrium in the displaced position, and therefore its potential energy is *constant*. When the ball is at C, it is in *unstable equilibrium*. Here a small displacement will cause the ball's potential energy to be *decreased,* and so it will roll farther *away* from its original, highest position. At C, the potential energy of the ball is a *maximum*.

Fig. 11-12

The above example illustrates that one of three types of equilibrium positions can be specified by a body or system of connected bodies.

1. *Stable equilibrium* occurs when a small displacement of the system causes the system to return to its original position. In this case the original potential energy of the system is a minimum.
2. *Neutral equilibrium* occurs when a small displacement of the system causes the system to remain in its displaced state. In this case the potential energy of the system remains constant.
3. *Unstable equilibrium* occurs when a small displacement of the system causes the system to move farther away from its original position. In this case, the original potential energy of the system is a maximum.

System Having One Degree of Freedom.   For *equilibrium* of a system having a single degree of freedom, defined by the independent coordinate q, it has been shown that the first derivative of the potential-energy function for the system must be equal to zero; i.e., $dV/dq = 0$. If the potential-energy function $V = V(q)$ is plotted, Fig. 11-13, the first derivative (equilibrium position) is represented as the slope, dV/dq, which is zero when the function is maximum, minimum, or an inflection point.

If the *stability* of a body at the equilibrium position is to be investigated, it is necessary to compute the *second derivative* of the potential-energy function and evaluate it at the equilibrium position $q = q_{eq}$. As shown in Fig. 11–13a, if $V = V(q)$ is a *minimum,* then

$$\frac{dV}{dq} = 0, \frac{d^2V}{dq^2} > 0 \qquad \text{stable equilibrium} \qquad (11\text{–}20)$$

If $V = V(q)$ is a *maximum,* Fig. 11–13b, then

$$\frac{dV}{dq} = 0, \frac{d^2V}{dq^2} < 0 \qquad \text{unstable equilibrium} \qquad (11\text{–}21)$$

If the second derivative is zero, it will be necessary to investigate *higher-order* derivatives to determine the stability. In particular, stable equilibrium will occur if the order of the lowest remaining nonzero derivative is *even* and the sign of this nonzero derivative is positive when it is evaluated at $q = q_{eq}$; otherwise, it is unstable.

If the system is in neutral equilibrium, Fig. 11–13c, it is required that

$$\frac{dV}{dq} = \frac{d^2V}{dq^2} = \frac{d^3V}{dq^3} = \cdots = 0 \qquad \text{neutral equilibrium} \qquad (11\text{–}22)$$

since then V must be constant at and around the "neighborhood" of q_{eq}.

System Having Two Degrees of Freedom. A criterion for investigating stability becomes increasingly complex as the number of degrees of freedom for the system increases. For a system having two degrees of freedom, with independent coordinates (q_1, q_2), it may be verified (using the calculus of functions of two variables) that equilibrium and stability occur at a point (q_{1eq}, q_{2eq}) when

$$\frac{\partial V}{\partial q_1} = \frac{\partial V}{\partial q_2} = 0$$

$$\left[\left(\frac{\partial^2 V}{\partial q_1 \, \partial q_2} \right)^2 - \left(\frac{\partial^2 V}{\partial q_1^2} \right) \left(\frac{\partial^2 V}{\partial q_2^2} \right) \right] < 0$$

$$\left(\frac{\partial^2 V}{\partial q_1^2} + \frac{\partial^2 V}{\partial q_2^2} \right) > 0$$

Both equilibrium and instability occur when

$$\frac{\partial V}{\partial q_1} = \frac{\partial V}{\partial q_2} = 0$$

$$\left[\left(\frac{\partial^2 V}{\partial q_1 \, \partial q_2} \right)^2 - \left(\frac{\partial^2 V}{\partial q_1^2} \right) \left(\frac{\partial^2 V}{\partial q_2^2} \right) \right] < 0$$

$$\left(\frac{\partial^2 V}{\partial q_1^2} + \frac{\partial^2 V}{\partial q_2^2} \right) < 0$$

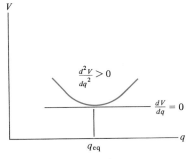

Stable Equilibrium

(a)

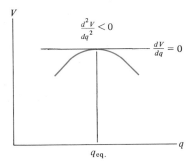

Unstable Equilibrium

(b)

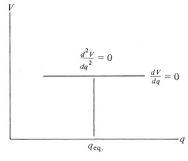

Neutral Equilibrium

(c)

Fig. 11–13

453

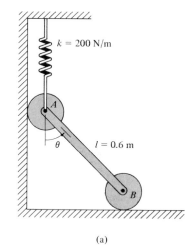

$k = 200$ N/m

A

θ

$l = 0.6$ m

B

(a)

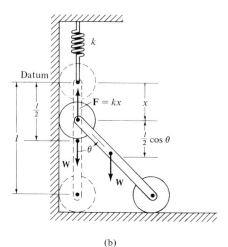

k

Datum

$\frac{1}{2}$

l

W

$F = kx$

x

$\frac{l}{2} \cos \theta$

θ

W

(b)

Fig. 11–14

Using potential-energy methods, both the equilibrium positions and the stability of a body or a system of connected bodies having a single degree of freedom can be obtained by applying the following three-step procedure.

Step 1: Formulate the potential-energy function $V = V_g + V_e$ for the system. To do this, sketch the system so that it is located at some *arbitrary position* specified by the independent coordinate q. A horizontal *datum* is established through a *fixed point,** and the *gravitational potential energy* V_g is expressed in terms of the weight W of each member and its vertical distance y from the datum, $V_g = \pm Wy$ (Eq. 11–11 or 11–12). The elastic potential energy V_e of the system is expressed in terms of the stretch or compression, x, of any connecting spring and the spring's stiffness k, $V_e = \frac{1}{2}kx^2$ (Eq. 11–13). Once V has been established, express the *position coordinates* y and x in terms of the independent coordinate q.

Step 2: The equilibrium position is determined by taking the first derivative of V and setting it equal to zero, $\delta V = 0$ (Eq. 11–18).

Step 3: Stability at the equilibrium position is determined by evaluating the second or higher-order derivatives as indicated by Eqs. 11–20 through 11–22.

The following examples numerically illustrate this procedure.

Example 11–5

The uniform link shown in Fig. 11–14a has a mass of 10 kg and length of 0.6 m. The spring is unstretched when $\theta = 0°$. Using the method of potential energy, determine the angle θ for equilibrium and investigate the stability at the equilibrium position.

Solution

Step 1: The system has one degree of freedom. The datum is established at the top of the link when the *spring is unstretched,* Fig. 11–14b. When the link is located at the arbitrary position, defined by the independent coordinate angle θ, the spring increases its potential energy by stretching, and the weight decreases its potential energy since it moves farther below the datum. The potential-energy function for the system may therefore be written as

$$V = V_e + V_g$$
$$= \frac{1}{2}kx^2 - W\left(x + \frac{l}{2}\cos\theta - \frac{l}{2}\right)$$

Since $l = x + l\cos\theta$ or $x = l(1 - \cos\theta)$, then

*The location of the datum is *arbitrary* since only the *changes* or differentials of V are required for investigation of the equilibrium position and its stability.

$$V = \frac{1}{2}kl^2(1 - \cos\theta)^2 - \frac{Wl}{2}(1 - \cos\theta)$$

Step 2: Having expressed V completely in terms of θ, the first derivative of V determines the equilibrium position s; i.e.,

$$\delta V = 0; \quad \frac{dV}{d\theta}\,\delta\theta = \left[kl^2(1 - \cos\theta)\sin\theta - \frac{Wl}{2}\sin\theta\right]\delta\theta = 0$$

Since $\delta\theta \neq 0$, then

$$l\left[kl(1 - \cos\theta) - \frac{W}{2}\right]\sin\theta = 0$$

This equation is satisfied provided

$$\sin\theta = 0$$
$$\theta = 0° \qquad\qquad Ans.$$

or

$$kl(1 - \cos\theta) - \frac{W}{2} = 0$$

$$\theta = \cos^{-1}\left(1 - \frac{W}{2kl}\right)$$

Substituting the numerical data for k, l, and $W = mg$ yields

$$\theta = \cos^{-1}\left[1 - \frac{10(9.81)}{2(200)(0.6)}\right]$$

$$= 53.8° \qquad\qquad Ans.$$

Step 3: Computing the second derivative of V to determine the type of equilibrium gives

$$\frac{d^2V}{d\theta^2} = kl^2(1 - \cos\theta)\cos\theta + kl^2\sin\theta\sin\theta - \frac{Wl}{2}\cos\theta$$

$$= kl^2(\cos\theta - \cos 2\theta) - \frac{Wl}{2}\cos\theta$$

Substituting the values $\theta = 0°$ and $\theta = 53.8°$ and the numerical values of the constants yields

$$\left.\frac{d^2V}{d\theta^2}\right|_{\theta=0°} = 200(0.6)^2(\cos 0° - \cos 0°) - \frac{10(9.81)(0.6)}{2}\cos 0°$$

$$\frac{d^2V}{d\theta^2} = -29.4 < 0 \quad \text{(unstable equilibrium at } \theta = 0°) \quad Ans.$$

$$\left.\frac{d^2V}{d\theta^2}\right|_{\theta=53.8°} = 200(0.6)^2(\cos 53.8° - \cos 107.6°) - \frac{10(9.81)(0.6)}{2}\cos 53.8°$$

$$\frac{d^2V}{d\theta^2} = 46.9 > 0 \quad \text{(stable equilibrium at } \theta = 53.8°) \quad Ans.$$

Example 11-6

Determine the mass m of the block required for equilibrium of the uniform 10-kg rod shown in Fig. 11–15a when $\theta = 20°$. The spring is unstretched when $\theta = 0°$. Investigate the stability at the equilibrium position.

Solution

Step 1: The potential-energy function for the system may be determined by establishing the datum through point A and measuring the displacement of the rod with the independent coordinate θ, Fig. 11–15b. When $\theta = 0°$, the weight of the block, $W = m(9.81)$, is assumed to be suspended $(y_W)_1$ below the datum. Thus, when the spring is stretched, the weight moves downward a distance $[(y_W)_2 - (y_W)_1]$ *below* the datum. The potential energy function is therefore

$$V = V_e + V_g$$

$$= \tfrac{1}{2}(300)(1.5 \sin \theta)^2 + 98.1\left(\frac{1.5 \sin \theta}{2}\right) - m(9.81)[(y_W)_2 - (y_W)_1] \quad (1)$$

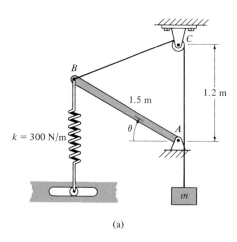

(a)

Fig. 11–15

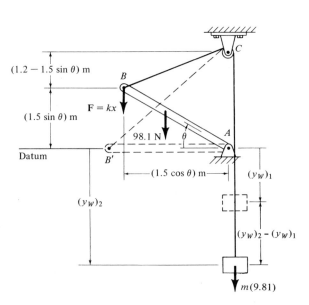

(b)

The distance $[(y_W)_2 - (y_W)_1]$ may be related to θ by measuring the difference in cord lengths $B'C$ and BC. Since

$$B'C = \sqrt{(1.5)^2 + (1.2)^2} = 1.92$$

and

$$BC = \sqrt{(1.5 \cos \theta)^2 + (1.2 - 1.5 \sin \theta)^2}$$
$$= \sqrt{3.69 - 3.60 \sin \theta}$$

then

$$[(y_W)_2 - (y_W)_1] = B'C - BC$$
$$= 1.92 - \sqrt{3.69 - 3.60 \sin \theta}$$

This *same result* can also be obtained by establishing a *separate* datum at the block when it is located at its *initial* (dashed) *position,* Fig. 11–15b. Substituting the above result into Eq. (1) yields

$$V = \tfrac{1}{2}(300)(1.5 \sin \theta)^2 + 98.1 \left(\frac{1.5 \sin \theta}{2}\right)$$
$$- m(9.81)(1.92 - \sqrt{3.69 - 3.60 \sin \theta}) \qquad (2)$$

Step 2: Equilibrium requires

$$\frac{dV}{d\theta} \delta\theta =$$

$$\left\{ 675 \sin \theta \cos \theta + 73.6 \cos \theta - \left[\frac{m(9.81)}{2}\right]\left(\frac{3.60 \cos \theta}{\sqrt{3.69 - 3.60 \sin \theta}}\right) \right\}\delta\theta = 0$$

Substituting $\theta = 20°$ into this expression and simplifying gives

$$\frac{dV}{d\theta}\bigg|_{\theta=20°} \delta\theta = (286.1 - 10.58m)\,\delta\theta = 0$$

Thus,

$$m = \frac{286.1}{10.58} = 27.0 \text{ kg} \qquad\qquad Ans.$$

Step 3: Taking the second derivative of Eq. (2) to determine the type of equilibrium, we obtain

$$\frac{d^2V}{d\theta^2} = 675 \cos 2\theta - 73.6 \sin \theta$$

$$- \left[\frac{m(9.81)}{2}\right]\left(\frac{-1}{2}\right)(3.69 - 3.60 \sin \theta)^{-3/2}(3.60 \cos \theta)^2$$

$$- \frac{m(9.81)}{2}\left(\frac{-3.60 \sin \theta}{\sqrt{3.69 - 3.60 \sin \theta}}\right)$$

For the equilibrium position $\theta = 20°$, $m = 27.0$ kg, hence

$$\frac{d^2V}{d\theta^2} = 792.4 > 0 \qquad \text{(stable equilibrium at } \theta = 20°) \qquad Ans.$$

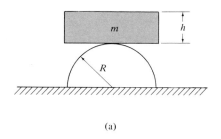

(a)

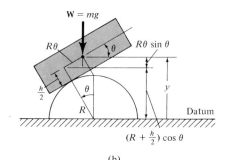

(b)

Fig. 11–16

Example 11–7

A homogeneous block having a mass m rests on the top surface of a cylinder, as shown in Fig. 11–16a. Show that this is a condition of unstable equilibrium if $h > 2R$.

Solution

Step 1: The datum is established at the base of the cylinder as shown in Fig. 11–16b. If the block is displaced by an amount θ from the equilibrium position, the potential-energy function may be written in the form

$$V = V_e + V_g$$
$$= 0 + mgy$$

From Fig. 11–16b,

$$y = \left(R + \frac{h}{2}\right)\cos\theta + R\theta\sin\theta$$

Thus,

$$V = mg\left[\left(R + \frac{h}{2}\right)\cos\theta + R\theta\sin\theta\right]$$

Step 2: For equilibrium,

$$\delta V = \frac{dV}{d\theta}\delta\theta = mg\left[-\left(R + \frac{h}{2}\right)\sin\theta + R\sin\theta + R\theta\cos\theta\right]\delta\theta = 0$$

$$= mg\left(-\frac{h}{2}\sin\theta + R\theta\cos\theta\right)\delta\theta = 0$$

Obviously $\theta = 0°$ is the equilibrium position that satisfies this equation.

Step 3: Taking the second derivative of V yields

$$\frac{d^2V}{d\theta^2} = mg\left(\frac{-h}{2}\cos\theta + R\cos\theta - R\theta\sin\theta\right)\delta\theta$$

At $\theta = 0°$,

$$\left.\frac{d^2V}{d\theta^2}\right|_{\theta=0°} = -mg\left(\frac{h}{2} - R\right)$$

Since all the constants are positive, the block is in unstable equilibrium if $h > 2R$, then $d^2V/d\theta^2 < 0$.

Problems

11-27. If the potential energy for a conservative one-degree-of-freedom system is expressed by the relation $V = (4x^3 - x^2 - 3x + 10)$ J, where x is given in metres, determine the positions for equilibrium and investigate the stability at each of these positions.

* 11-28. If the potential energy for a conservative one-degree-of-freedom system is expressed by the relation $V = (10 \cos 2\theta + 24 \sin \theta)$ J, where $0° < \theta < 180°$, determine the positions for equilibrium and investigate the stability at each of these positions.

11-29. Solve Prob. 11-9 using the principle of potential energy, and investigate the stability at the equilibrium position.

11-30. Solve Prob. 11-5 using the principle of potential energy, and investigate the stability at the equilibrium position.

11-30a. Solve Prob. 11-5a using the principle of potential energy, and investigate the stability at the equilibrium position.

11-31. Solve Prob. 11-10 using the principle of potential energy, and investigate the stability at the equilibrium position.

* 11-32. The uniform rod AB has a mass of 80 kg. If spring DC is unstretched when $\theta = 90°$, determine the angles θ for equilibrium and investigate the stability at each of the equilibrium positions. The spring always acts in the horizontal position because of the roller guide at D.

Prob. 11-32

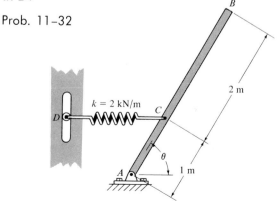

11-33. The uniform beam has a mass of 200 kg. If the contacting surfaces are smooth, determine the angle θ for equilibrium and investigate the stability of the beam when it is in this position. The spring has an unstretched length of 0.5 m.

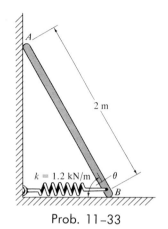

Prob. 11-33

11-34. The uniform rod AB has a mass of 5 kg. If the spring has an unstretched length when $\theta = 60°$, determine the angles θ for equilibrium and investigate the stability at each of the equilibrium positions.

Prob. 11-34

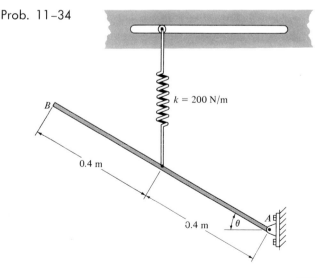

11–35. The spring attached to the mechanism has an unstretched length when $\theta = 90°$. Determine the position θ for equilibrium and investigate the stability of the mechanism at this position. Disk A is pin-connected to the frame at B and has a mass of $m_A = 20$ kg. Set $k = 300$ N/m and $l = 600$ mm.

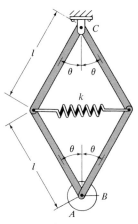

Prob. 11–35

11–35a. Solve Prob. 11–35 if disk A weighs $W_A = 18$ lb, and $k = 16$ lb/ft, $l = 1.25$ ft.

*11–36. Determine the angles θ for equilibrium of the 2-kg disk and investigate the stability at each of the equilibrium positions. Neglect the weight of the rod. The spring is unstretched when $\theta = 0°$.

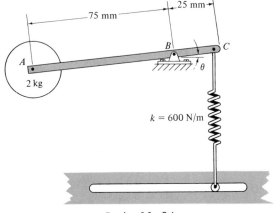

Prob. 11–36

11–37. Two smooth bars of length 500 mm are pin-connected. Determine the angle θ for equilibrium and investigate the stability at this position. The bars each have a mass of 3 kg and the suspended block D has a mass of 4 kg. Cord DC has a total length of 1 m.

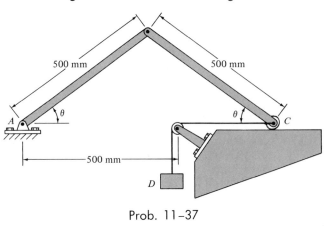

Prob. 11–37

11–38. The uniform link AB has a mass of 3 kg and is pin-connected at both of its end points. The rod BD, having negligible weight, passes through a swivel block at C. If the spring has a stiffness of $k = 100$ N/m and is unstretched when $\theta = 0°$, determine the angle θ for equilibrium of the link. Investigate the stability at the equilibrium position. Neglect the size of the swivel block.

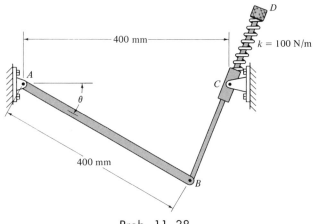

Prob. 11–38

11–39. Determine the spring constant k so that the 3-kg link in Prob. 11–38 is in equilibrium when $\theta = 45°$. The spring is unstretched when $\theta = 0°$.

*11–40. A right circular cone is attached to the hemisphere. If both pieces have the same density $\rho = 7 \text{ Mg/m}^3$, determine the height h of the cone so that the configuration is in neutral equilibrium. $r = 60$ mm.

Prob. 11–40

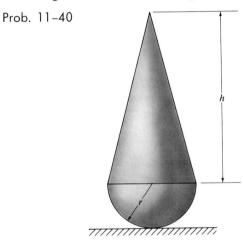

*11–40a. Solve Prob. 11–40 if $\rho = 50 \text{ lb/ft}^3$ and $r = 8$ in.

11–41. If the uniform rod OA has a mass of 12 kg, determine the mass m that will hold the rod in equilibrium when $\theta = 30°$. Point C is coincident with B when OA is horizontal. Neglect the size of the pulley at B.

Prob. 11–41

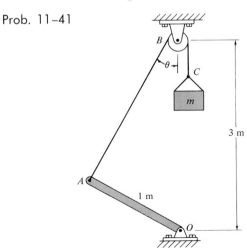

11–42. The cylinder is made of two materials such that it has a mass m and a center of gravity at point G. Show that when G lies directly above the centroid C of the cylinder, the equilibrium position is unstable.

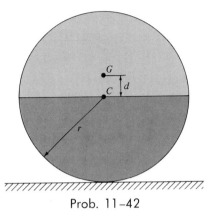

Prob. 11–42

11–43. Determine the position of stable equilibrium for the cylinder in Prob. 11–42.

*11–44. The hemispherical block has a conical cavity cut into it as shown. Determine the depth d of the cavity so that the block balances on the pivot and remains in neutral equilibrium. Set $r = 200$ mm.

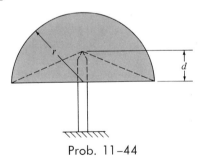

Prob. 11–44

461

11–45. The assembly consists of a hemisphere and cylinder. If the radii of the cylinder and hemisphere are both $a = 50$ mm, determine the cylinder's height h so that the assembly is in neutral equilibrium. The material has a density of $\rho = 2$ Mg/m^3.

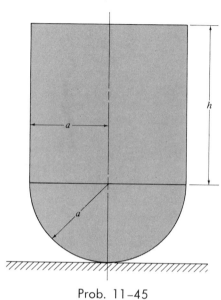

Prob. 11–45

11–45a. Solve Prob. 11–45 if $\rho = 300$ lb/ft^3 and $a = 6$ in.

11–46. A uniform block rests on top of the cylindrical surface. Derive the relationship between the radius r of the cylinder and the dimension b of the block for stable, unstable, and neutral equilibrium. *Hint:* Establish the potential-energy function for *small* angle θ; i.e., approximate $\sin \theta \approx 0$, and $\cos \theta \approx 1 - \theta^2/2$.

Prob. 11–46

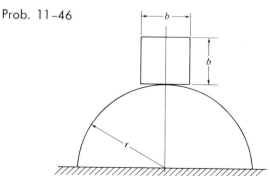

11–47. The homogeneous block has a mass $m = 5$ kg and rests on the smooth corners of two ledges. Determine the angle θ for placement that will cause the block to be stable. $a = 200$ mm and $d = 150$ mm.

Prob. 11–47

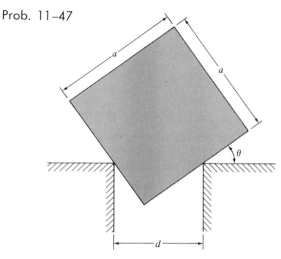

*11–48. If the potential energy for a conservative two-degree-of-freedom system is expressed by the relation $V = (6y^2 + 2x^2)$ J, where x and y are given in metres, determine the position for equilibrium and investigate the stability at this position.

11–49. The two blocks each have a mass of $m = 10$ kg. Determine the stretch of the springs, x_1 and x_2, for equilibrium. Set $k_1 = 300$ N/m and $k_2 = 800$ N/m.

Prob. 11–49

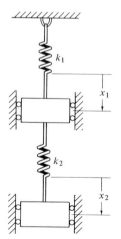

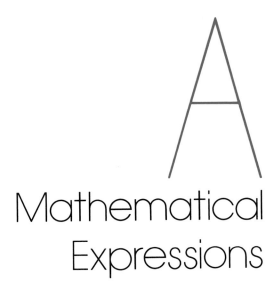

Mathematical Expressions

Quadratic Formula:

If $ax^2 + bx + c = 0$, then $x = \dfrac{-b \pm \sqrt{b^2 - 4ac}}{2a}$

Hyperbolic Functions:

$\sinh x = \dfrac{e^x - e^{-x}}{2}$, $\cosh x = \dfrac{e^x + e^{-x}}{2}$, $\tanh x = \dfrac{\cosh x}{\sinh x}$

Trigonometric Identities:

$\sin^2 \theta + \cos^2 \theta = 1$

$\sin (\theta \pm \phi) = \sin \theta \cos \phi \pm \cos \theta \sin \phi$

$\sin 2\theta = 2 \sin \theta \cos \theta$

$\cos (\theta \pm \phi) = \cos \theta \cos \phi \mp \sin \theta \sin \phi$

$\cos 2\theta = \cos^2 \theta - \sin^2 \theta$

$\cos \theta = \pm \sqrt{\dfrac{1 + \cos 2\theta}{2}}$, $\sin \theta = \pm \sqrt{\dfrac{1 - \cos 2\theta}{2}}$

$\tan \theta = \dfrac{\sin \theta}{\cos \theta}$

$1 + \tan^2 \theta = \sec^2 \theta$

Power-series Expansions:

$$\sin x = x - \frac{x^3}{3!} + \frac{x^5}{5!} - \frac{x^7}{7!} + \cdots$$

$$\cos x = 1 - \frac{x^2}{2!} + \frac{x^4}{4!} - \frac{x^6}{6!} + \cdots$$

$$\sinh x = x + \frac{x^3}{3!} + \frac{x^5}{5!} + \cdots$$

$$\cosh x = 1 + \frac{x^2}{2!} + \frac{x^4}{4!} + \cdots$$

Derivatives:

$$\frac{d}{dx}(u^n) = nu^{n-1}\frac{du}{dx}$$

$$\frac{d}{dx}(uv) = u\frac{dv}{dx} + v\frac{du}{dx}$$

$$\frac{d}{dx}\left(\frac{u}{v}\right) = \frac{v\frac{du}{dx} - u\frac{dv}{dx}}{v^2}$$

$$\frac{d}{dx}(\sin u) = \cos u\frac{du}{dx}$$

$$\frac{d}{dx}(\cos u) = -\sin u\frac{du}{dx}$$

$$\frac{d}{dx}(\tan u) = \sec^2 u\frac{du}{dx}$$

$$\frac{d}{dx}(\cot u) = -\csc^2 u\frac{du}{dx}$$

$$\frac{d}{dx}(\sec u) = \tan u \sec u\frac{du}{dx}$$

$$\frac{d}{dx}(\csc u) = -\csc u \cot u\frac{du}{dx}$$

$$\frac{d}{dx}(\sinh u) = \cosh u\frac{du}{dx}$$

$$\frac{d}{dx}(\cosh u) = \sinh u\frac{du}{dx}$$

Integrals:

$$\int x^n \, dx = \frac{x^{n+1}}{n+1}, \; n \neq -1$$

$$\int \frac{dx}{a+bx} = \frac{1}{b} \ln(a+bx)$$

$$\int \frac{x \, dx}{a+bx^2} = \frac{1}{2b} \ln\left(x^2 + \frac{a}{b}\right)$$

$$\int \frac{dx}{(a+bx)^2} = -\frac{1}{b(a+bx)}$$

$$\int \frac{dx}{a^2 - x^2} = \frac{1}{2a} \ln \frac{a+x}{a-x}, \; a^2 > x^2$$

$$\int \sqrt{a+bx} \, dx = \frac{2}{3b} \sqrt{(a+bx)^3}$$

$$\int x \sqrt{a+bx} \, dx = \frac{-2(2a - 3bx)\sqrt{(a+bx)^3}}{15b^2}$$

$$\int \sqrt{a^2 - x^2} \, dx = \frac{1}{2}\left[x\sqrt{a^2 - x^2} + a^2 \sin^{-1}\frac{x}{a}\right], \; a > 0$$

$$\int \sqrt{x^2 \pm a^2} \, dx = \frac{1}{2}\left[x\sqrt{x^2 \pm a^2} \pm a^2 \ln(x + \sqrt{x^2 \pm a^2})\right]$$

$$\int x\sqrt{a^2 - x^2} \, dx = -\frac{1}{3}\sqrt{(a^2 - x^2)^3}$$

$$\int x^2\sqrt{a^2 - x^2} \, dx = -\frac{x}{4}\sqrt{(a^2 - x^2)^3} + \frac{a^2}{8}\left(x\sqrt{a^2 - x^2} + a^2 \sin^{-1}\frac{x}{a}\right), \; a > 0$$

$$\int x^2 \sqrt{x^2 \pm a^2} = \frac{x}{4}\sqrt{(x^2 \pm a^2)^3} \mp \frac{a^2}{8} x \sqrt{x^2 \pm a^2} - \frac{a^4}{8}\ln(x + \sqrt{x^2 \pm a^2})$$

$$\int \sin x \, dx = -\cos x$$

$$\int \cos x \, dx = \sin x$$

$$\int e^{ax} \, dx = \frac{1}{a}e^{ax}$$

$$\int \sinh x \, dx = \cosh x$$

$$\int \cosh x \, dx = \sinh x$$

B

Geometric Properties of Line, Area, and Volume Elements

Shape and Size	Centroid	Area Moment of Inertia
Circular arc segment $L = 2\theta r$	$\bar{x} = \dfrac{r\sin\theta}{\theta}$	
Quarter and semicircular arcs $L = \frac{\pi}{2}r$, $L = \pi r$	$\bar{y} = \dfrac{2r}{\pi}$	
Rectangular area $A = bh$	$\bar{x} = 0$ $\bar{y} = 0$	$I_x = \frac{1}{12}bh^3$ $I_y = \frac{1}{12}hb^3$
Triangular area $A = \frac{1}{2}bh$	$\bar{y} = \frac{1}{3}h$	$I_x = \frac{1}{36}bh^3$
Trapezoidal area $A = \frac{1}{2}h(a+b)$	$\bar{y} = \dfrac{1}{3}\left(\dfrac{2a+b}{a+b}\right)h$	$I_x = \dfrac{h^3[b^2 + 4ab + a^2]}{36(a+b)}$
Circular sector area $A = \theta r^2$	$\bar{x} = \dfrac{2}{3}\left(\dfrac{r\sin\theta}{\theta}\right)$ $\bar{y} = 0$	$I_x = \frac{1}{4}r^4(\theta - \frac{1}{2}\sin 2\theta)$ $I_y = \frac{1}{4}r^4(\theta + \frac{1}{2}\sin 2\theta)$

y-centroid = $\dfrac{\text{area} \cdot \text{distance} + \text{area} \cdot \text{dist} + ...}{\text{Area} + \text{Area} + \text{Area}}$

Shape and Size	Centroid	Area Moment of Inertia
 Quarter circular area $A = \frac{\pi r^2}{4}$	$\bar{x} = \dfrac{4r}{3\pi}$ $\bar{y} = \dfrac{4r}{3\pi}$	$I_x = \frac{1}{16}\pi r^4$ $I_y = \frac{1}{16}\pi r^4$
 Semicircular area $A = \frac{\pi r^2}{2}$	$\bar{x} = 0$ $\bar{y} = \dfrac{4r}{3\pi}$	$I_x = \frac{1}{8}\pi r^4$ $I_y = \frac{1}{8}\pi r^4$
 Circular area $A = \pi r^2$	$\bar{x} = 0$ $\bar{y} = 0$	$I_x = \frac{1}{4}\pi r^4$ $I_y = \frac{1}{4}\pi r^4$
 Semiparabolic area $A = \frac{2}{3}ab$	$\bar{x} = \frac{2}{5}a$ $\bar{y} = \frac{3}{8}b$	
 Semiparabolic area $A = \frac{ab}{3}$	$\bar{x} = \frac{3}{4}a$ $\bar{y} = \frac{3}{10}b$	
 Parabolic area $A = \frac{4}{3}ab$	$\bar{x} = \frac{2}{5}a$	

Shape and Size	Centroid	Area Moment of Inertia
$A = 2\pi r^2$ Hemispherical shell	$\bar{y} = \frac{1}{2}r$	
$A = \pi r\sqrt{r^2 + h^2}$ Conical shell	$\bar{y} = \frac{1}{3}h$	
$V = \frac{2}{3}\pi r^3$ Hemispherical volume	$\bar{y} = \frac{3}{8}r$	
$V = \frac{1}{3}\pi r^2 h$ Conical volume	$\bar{y} = \frac{1}{4}h$	
$V = \frac{1}{2}\pi r^2 h$ Paraboloid	$\bar{y} = \frac{1}{3}h$	

Area moment of inertia $I_r = I_{Gx} + Ad^2$
$= \frac{1}{12}bh + Area \cdot (distance\ to\ c.)^2$ from top

469

Answers*

1-1. 24 m².

1-2. (a) 6.54 km, (b) 5.20 MN, (c) 62.1 μs.

1-3. (a) 16 MN², (b) 0.90 nm², (c) 8 Ms³.

1-5. yes; m · kg (metre · kilogram), m kg = 1 g.

1-6. 7.41 μN.

1-7. (a) N, (b) nm/s, (c) N/s².

1-9. 0.0209 lb/ft², 101.4 kPa.

1-10. (a) 3.11 slug, (b) 45.4 kg, (c) 444.8 N, (d) 45.4 kg, (e) 73.3 N.

1-11. (a) kg/s, (b) GN/m, (c) km/(kg · s).

2-1. $F_x = 6$ kN, $F_y = 10.39$ kN.

2-2. 101.4°, 980.3 N.

2-3. $F = 117.6$ kN, $F_R = 120.2$ kN.

2-5. 7.48 kN, 19.8°. $\searrow\theta$

2-5a. 9.96 lb, 31.0°. $\searrow\theta$

2-6. 22.6°, 1.03 kN.

2-7. $F_\perp = 459.6$ N, $F_\parallel = 385.7$ N.

2-9. 24.6°, 485.2 N.

2-10. $F_{AC} = 366.0$ N, $F_{AB} = 448.3$ N.

2-10a. $F_{AC} = 256.2$ lb, $F_{AB} = 313.8$ lb.

2-11. (a) $F_{x'} = 34.7$ N, $F_{y'} = 197.0$ N; (b) $F_x = 40.1$ N, $F_{y'} = 217.0$ N.

*Note: Answers to every fourth problem are omitted.

2-13. $\theta = \phi = 60°$.

2-14. (a) 1.59 kN, $\theta = 47.6°$; (b) 1.25 kN, $\theta = 110°$.

2-15. $F_{AB} = 3.58$ kN, $F_{AC} = 2.71$ kN.

2-15a. $F_{AB} = 774.5$ lb, $F_{AC} = 470.8$ lb.

2-17. $F' = 43.3$ N, $F_R = 23.4$ N.

2-18. $\mathbf{F} = \{600\mathbf{i} + 450\mathbf{j}\}$ N.

2-19. (a) 78.1 N, 39.8°, 129.8°, 90°;
(b) 939.4 N, 115.2°, 154.8°, 90°.

2-21. $\mathbf{F}_1 = \{-320\mathbf{i} - 240\mathbf{j}\}$ N, $\mathbf{F}_2 = \{-259.8\mathbf{i} + 150\mathbf{j}\}$ N, $F_R = 586.7$ N, $\alpha = 171.2°$.

2-22. $\mathbf{F}_1 = \{-424.3\mathbf{i} + 424.3\mathbf{j}\}$ N,
$\mathbf{F}_2 = \{-606.2\mathbf{i} - 350\mathbf{j}\}$ N, $\mathbf{F}_3 = \{-500\mathbf{j}\}$ N,
$F_R = 1.12$ kN, $\alpha = 157.6°$, $\beta = 112.4°$.

2-23. $\mathbf{F}_A = \{-34.6\mathbf{i} - 20\mathbf{j}\}$ N, $\mathbf{F}_B = \{34.6\mathbf{i} - 20\mathbf{j}\}$ N,
$\mathbf{F}_C = \{40\mathbf{j}\}$ N, $\mathbf{F}_R = \mathbf{O}$.

2-25. $\mathbf{F}_1 = F_1 \cos\theta\mathbf{i} + F_1 \sin\theta\mathbf{j}$, $\mathbf{F}_2 = \{200\mathbf{i}\}$ N,
$\mathbf{F}_3 = \{-100\mathbf{j}\}$ N. $\theta = 56.3°$, $F_1 = 420.5$ N.

2-25a. $\mathbf{F}_1 = F_1 \cos\theta\mathbf{i} + F_1 \sin\theta\mathbf{j}$, $\mathbf{F}_2 = \{70\mathbf{i}\}$ lb,
$\mathbf{F}_3 = \{-45\mathbf{j}\}$ lb. $\theta = 45.8°$, $F_1 = 272.1$ lb.

2-26. 3.92 kN, 101.0°, 98.8°, 14.2°.

2-27. $F = 87.5$ N, $F_x = 61.85$ N, 45°, 46.7°, 99.9°.

2-29. $\mathbf{F} = \{259.8\mathbf{i} - 150\mathbf{j} + 173.2\mathbf{k}\}$ N, 346.4 N.

2-30. 9.58 kN, 15.5°, 98.3°, 77.0°.

2-30a. 419.3 lb, 55.1°, 35.8°, 83.2°.

2-31. 73.4°, 64.6°, 31.0°.

471

2-33. 31.5 m, 69.6°, 116.4°, 34.4°.

2-34. 7.55 m, 105.4°, 22.0°, 74.6°.

2-35. $\mathbf{F} = \{-150\mathbf{i} - 300\mathbf{j} + 300\mathbf{k}\}$ N, 3 m.

2-35a. $\mathbf{F} = \{-57.0\mathbf{i} - 85.5\mathbf{j} + 71.2\mathbf{k}\}$ lb, 8.77 ft.

2-37. $\mathbf{F} = \{44.7\mathbf{i} + 77.5\mathbf{j} + 178.9\mathbf{k}\}$ N.

2-38. (1.19 m, 1.79 m, 2.09 m).

2-39. $\beta = 115.3°$, 8.66 m, -4.28 m, 2.59 m.

2-41. $\mathbf{F}_R = \{-64.1\mathbf{i} - 166.7\mathbf{j} - 979.5\mathbf{k}\}$ N, 995.6 N, 93.7°, 99.6°, 169.7°.

2-42. $\mathbf{F}_A = \{285.2\mathbf{j} - 93.0\mathbf{k}\}$ N, $\mathbf{F}_C = \{159.3\mathbf{i} + 183.2\mathbf{j} - 59.7\mathbf{k}\}$ N.

3-1. $F_B = 3.46$ kN, $F_D = 4.90$ kN.

3-2. $R = 871.8$ N, $\theta = 36.6°$.

3-3. $T_1 = 42.5$ N, $F = 83.9$ N.

3-5. $F_A = 85.6$ N, $F_B = 59.3$ N.

3-5a. $F_A = 35.7$ lb, $F_B = 30.3$ lb.

3-6. 204.2 mm.

3-7. 709.7 mm.

3-9. CD, 257.1 N.

3-10. $N_A = 41.9$ N, $N_B = 50.0$ N.

3-10a. $N_A = 22.5$ lb, $N_B = 26.8$ lb.

3-11. $F = 28.3$ N, $\theta = 113°$.

3-13. $F_{BA} = 297.5$ N, $\theta = 58.5°$.

3-14. $T_A = 52.9$ mN, $\theta = 19.1°$, $T_B = 34.6$ mN, $m = 4.08$ g.

3-15. 1.32 m.

3-15a. 1.46 ft.

3-17. 15.6 kg.

3-18. 756 mm.

3-19. 158.5 N/m.

3-21. 2.45 m.

3-22. $\theta = 36.9°$, $F_A = 78.5$ N, $F_B = 58.9$ N.

3-23. $N_x = 20$ N, $N_y = 40$ N, $N_z = 138.1$ N.

3-25. $F_B = F_C = 630.6$ N, $F_O = 1401.4$ N.

3-25a. $F_B = F_C = 122.2$ lb, $F_O = 350.5$ lb.

3-26. 779.7 N.

3-27. $P = 1098.1$ N, $F = 503.0$ N, $R = 407.0$ N.

3-29. $F_B = 69.7$ N, $F_C = 57.0$ N, $F_O = 209.0$ N.

3-30. 333 mm.

3-30a. 0.378 ft.

3-31. $F_B = 120.5$ N, $F_C = 127.1$ N, $F_D = 31.8$ N.

3-33. $P = 1.61$ kN, $\alpha = 136.2°$, $\beta = 128.3°$, $\gamma = 72.0°$.

3-34. $F_B = 776.4$ N, $F_D = F_E = 1003.9$ N.

3-35. $F_E = 49.1$ N, $F_F = 277.5$ N.

3-35a. $F_E = 33.3$ lb, $F_F = 141.4$ lb.

3-37. 410.2 mm.

3-38. $T_B = T_C = 149.1$ N, $F_A = 298.3$ N.

3-39. $F_B = 1.92$ kN, $F_C = 1.04$ kN, $F_D = 0.632$ kN.

3-41. $F_A = 1442$ N, $F_B = 1057$ N, $F_C = 1961.5$ N, $F_E = 981.0$ N, $F_O = 2.19$ kN.

3-42. $F_A = 193.0$ N, $F_D = 89.8$ N, $F_C = 198.2$ N.

3-43. (190 mm, 12.3 mm).

3-45. (272 mm, 465 mm).

3-45a. (0.637 ft, 1.73 ft).

4-1. (a) $-12\mathbf{i} + 6\mathbf{k}$, (b) $20\mathbf{i}$, (c) $6\mathbf{i} - 10\mathbf{j} - 30\mathbf{k}$.

4-2. (a) $12\mathbf{i} - 6\mathbf{j} - 16\mathbf{k}$, (b) $34\mathbf{i} - 32\mathbf{j} + 6\mathbf{k}$.

4-3. $-0.6\mathbf{i} - 0.8\mathbf{j}$.

4-5. $\{-60(10^3)\mathbf{i} + 45(10^3)\mathbf{j} - 120(10^3)\mathbf{k}\}$ mm².

4-5a. $\{-0.24\mathbf{i} + 0.12\mathbf{j} - 0.32\mathbf{k}\}$ ft².

4-6. $\{-16.03\mathbf{i} - 32.1\mathbf{k}\}$ N · m.

4-7. (a) 171.8 N · m↰, (b) 190.8 N.→

4-9. 139.9 N · m. ↴

4-10. $\{-254.9\mathbf{i} + 147.2\mathbf{j}\}$ N · m.

4-10a. $\{-207.84\mathbf{i} + 120\mathbf{j}\}$ lb · ft.

4-11. $\{99.3\mathbf{i} + 148.9\mathbf{j}\}$ N · m.

4-13. 16.82 kN · m↴, $\{-16.82\mathbf{k}\}$ kN · m.

4-14. 7.11 N · m↴, $\{-7.11\mathbf{k}\}$ N · m.

4-15. $\mathbf{M}_B = \{30\mathbf{i} + 22.5\mathbf{j} + 75\mathbf{k}\}$ N · m, $\mathbf{M}_C = \{-32.5\mathbf{i} + 22.5\mathbf{j} + 12.5\mathbf{k}\}$ N · m.

4-15a. $\mathbf{M}_B = \{15\mathbf{i} + 21.25\mathbf{j} + 52.5\mathbf{k}\}$ lb · ft,
$\quad \mathbf{M}_C = \{-15\mathbf{i} + 21.25\mathbf{j} + 32.5\mathbf{k})$ lb · ft.

4-17. 28.6°.

4-18. 4.27 N · m, $\alpha = 95.2°$, $\beta = 109.8°$, $\gamma = 20.6°$.

4-19. $\alpha = 90.6°$, $\beta = 62.6°$, $\gamma = 152.6°$.

4-21. $M_A = 1962(8.5 + x)\cos\theta$, $(M_A)_{max} = 26.5$ ɪ N · m.

4-22. $\{-1097.9\mathbf{i} + 1463.8\mathbf{j} - 845.2\mathbf{k}\}$ N · m, 4.03 m.

4-23. 66.1 N.

4-25. 200 N.

4-25a. 53.3 lb.

4-26. $\mathbf{M}_c = \{40\mathbf{i} - 30\mathbf{j}\}$ N · m, 50 N · m.

4-27. 82.7 N.

4-29. $F = 75$ N, $P = 100$ N.

4-30. 288.7 N.

4-30a. 41.6 lb.

4-31. $\{84.8\mathbf{i} - 127.3\mathbf{j} + 234.8\mathbf{k}\}$ N · m.

4-33. $\mathbf{M}_O = \{-22\mathbf{i} + 36\mathbf{j} - 128\mathbf{k}\}$ N · m,
$\quad \mathbf{F} = \{60\mathbf{i} - 70\mathbf{j} - 30\mathbf{k}\}$ N.

4-34. $\mathbf{F}_R = \{86.6\mathbf{i} - 50\mathbf{j}\}$ N, $(\mathbf{M}_R)_o = \{-82.0\mathbf{k}\}$ N · m,
$\quad \mathbf{M}_{R_B} = \{-16.7\mathbf{k}\}$ N · m.

4-35. 100 N ←, 15 N · m ↱; 100 N ←, 40 N · m. ↰

4-35a. 60 lb ←, 24 lb · ft ↱; 60 lb ←, 120 lb · ft. ↰

4-37. 10 kN ↓, 3 kN · m. ↰

4-38. 341.7 N ⦝ 47.0°, 99.5 N · m. ↲

4-39. $F_R = 500$ N ⬃36.9°, $M_{R_A} = 10$ N · m. ↲

4-41. $\mathbf{F}_R = \{600\mathbf{i} - 200\mathbf{k}\}$ N,
$\quad \mathbf{M}_{R_O} = \{-200\mathbf{i} + 500\mathbf{j} - 600\mathbf{k}\}$ N · m.

4-42. $\mathbf{F}_R = \{74.1\mathbf{i} + 25.9\mathbf{j} - 148.3\mathbf{k}\}$ N,
$\quad \mathbf{M}_{R_O} = \{7.5\mathbf{i} + 444.9\mathbf{j} + 81.4\mathbf{k}\}$ N · m.

4-43. O, 43.6 N ⦝60.0°.

4-45. 5.25 kN ⬉30°, 1.73 m.

4-45a. 875 lb ⬉30°, 5.37 ft.

4-46. $F_C = 600$ N, $F_D = 500$ N.

4-47. 23 kN↓, $x = -0.351$ m, $y = 0.477$ m.

4-49. 717.1 N, $x = 2.38$ m, $y = 5.92$ m.

4-50. $\mathbf{F}_R = \{193.2\mathbf{i} + 100\mathbf{j} + 248.2\mathbf{k}\}$ N,
$\quad \mathbf{M}_{R_O} = \{266.9\mathbf{i} - 159.1\mathbf{k}\}$ N · m.

4-50a. $\mathbf{F}_R = \{77.3\mathbf{i} + 25\mathbf{j} + 39.3\mathbf{k}\}$ lb,
$\quad \mathbf{M}_{R_O} = \{42.4\mathbf{i} - 141.6\mathbf{k}\}$ lb · ft.

4-51. 0.35 m, $\mathbf{F}_{R_w} = \{-40\mathbf{i}\}$ N, $\mathbf{M}_{R_w} = \{-40\mathbf{i}\}$ N · m.

4-53. 8 kN↓ 2 m from A.

4-54. 6 kN↓, 1.875 m.

4-55. $F_R = 8$ kN↑, $M_A = 24$ kN · m. ↱

4-55a. $F_R = 52(10^3)$ lb↑, $M_A = 1.014(10^6)$ lb · ft.↱

4-57. $F_R = 3$ kN↓, $M_{R_O} = 2.25$ kN · m. ↱

4-58. $F_R = 2w_OL/\pi$, $\bar{x} = L/2$.

4-59. $F_R = 142.4$ N, $M_R = 293.1$ N · m.

4-61. $\mathbf{F}_R = \{-108\mathbf{i}\}$ N, $\mathbf{M}_{R_O} = \{-194.4\mathbf{j} - 54\mathbf{k}\}$ N · m.

4-62. $b = 1.69$ m, $a = 2.65$ m.

4-63. 19.62 kN, 3.88 m.

4-67. $3x + 5y + 6z - 20 = 0$.

4-69. 85.3 N · m.

4-70. $\{-6.21\mathbf{k}\}$ N · m.

4-70a. $\{-5.18\mathbf{k}\}$ lb · ft.

4-71. 87.42 N · m.

4-73. 23.5°.

4-74. 80.3°.

4-75. 3091.0 N · m.

4-75a. 965.9 lb · ft.

5-1. N_A, N_B.

5-2. A_x, A_y, T_{BC}.

5-3. A_x, A_y, T.

5-5. C_x, C_y, T_{AB}.

5-6. B_x, A_x, A_y.

5-7. A_x, B_y, F_C.

5-9. A_x, A_y, T_{CE}.

5-10. A_x, A_y, F_B.

5-11. A_x, A_y, M_A, F_B.

5-13. $R = 6.71$ N, $P = 18.4$ N, $T = 17.5$ N, $F = 29.7$ N.

5-14. $F_{BD} = 2548.7$ N, $A_x = 2548.7$ N, $A_y = 490.5$ N.

5-15. $N_{EC} = 85.0$ N, $N_{EB} = 85.0$ N, $N_{EA} = 85.0$ N, $N_D = 147.2$ N.

5-15a. $N_{EC} = 17.32$ lb, $N_{EB} = 17.32$ lb, $N_{EA} = 17.32$ lb, $N_D = 30$ lb.

5-17. $B_y = 1600$ N, $A_x = 800$ N, $A_y = 900$ N.

5-18. $T = 686.7$ N, $F_B = 603.7$ N, $\theta = 55.3°$.

5-19. $T = 193.5$ N, $A_x = 86.5$ N, $A_y = 170.3$ N.

5-21. $T = 277.4$ N, $A_x = 240.2$ N, $A_y = 651.8$ N.

5-22. $F_B = 4.71$ kN, $F_A = 3.92$ kN, $N_C = 784.8$ N.

5-23. 2.26 Mg.

5-25. $N_B = 9.90$ kN, $A_y = 4.00$ kN, $A_x = 7.60$ kN.

5-25a. $N_B = 3253.2$ lb, $A_y = 1000$ lb, $A_x = 2380$ lb.

5-26. $m = 16.0$ kg.

5-27. 1640 N pushing, 213.4 N pulling.

5-29. $T = 882.9$ N.

5-30. $30.8°$.

5-30a. $30.8°$.

5-31. $\theta = 0°$, $T = 523.2$ N, $F_B = 915.6$ N.

5-33. $T = 1.57$ kN, $F = 2.78$ kN, $N_A = 0$.

5-34. 31.4 kN.

5-35. $N_B = 342.8$ N, $A_y = 342.8$ N, $A_x = 35.1$ N.

5-35a. $N_B = 205.4$ lb, $A_y = 205.4$ lb, $A_x = 10.2$ lb.

5-37. $F = 108.3$ N, $\bar{x} = 0.341$ m, $m = 9.43$ kg.

5-38. $B_x = 400$ N, $F_A = 721.1$ N.

5-39. 0.464 m, 1.214 m.

5-41. 8.47 Mg.

5-42. $P = 245.2$ N, $A_y = 0$, $B_x = 367.9$ N, $B_z = 245.2$ N, $A_x = 122.7$ N, $A_z = 245.2$ N.

5-43. $A_x = 0$, $A_y = 0$, $A_z = B_z = C_z = 5.33$ kN.

5-45. $N_B = 78.5$ N, $T = 58.9$ N, $A_x = 78.5$ N, $A_y = 58.9$ N, $A_z = 196.2$ N.

5-45a. $N_B = 12.5$ lb, $T = 7.5$ lb, $A_x = 12.5$ lb, $A_y = 7.5$ lb, $A_z = 30$ lb.

5-46. $F_A = 10.34$ kN, $\alpha = 90°$, $\beta = 90°$, $\gamma = 0°$; $M_A = 6.35$ kN $\cdot$ m, $\alpha = 61.9°$, $\beta = 28.1°$, $\gamma = 90°$.

5-47. $C_x = C_y = 0$, $C_z = 760$ N, $A_z = 400$ N, $B_z = 240$ N.

5-49. $A_z = 100.3$ N, $B_z = 156.3$ N, $C_z = 135.8$ N.

5-50. 111.0 N/m.

5-50a. 9.95 lb/ft.

5-51. 196.2 N, 80.9 N $\cdot$ m.

5-53. $A_x = A_y = 0$, $A_z = 700$ N, $F = 1000$ N, $(M_A)_x = 10$ N $\cdot$ m, $(M_A)_z = 0$.

5-54. $F = 242.8$ N, $B_y = 328.0$ N, $B_z = 140.3$ N, $A_x = 226.3$ N, $A_y = 305.3$ N, $A_z = 29.8$ N.

5-55. $T_{BD} = T_{CD} = 116.7$ N, $A_x = 66.7$ N, $A_y = 0$, $A_z = 100.0$ N.

5-55a. $T_{BD} = T_{CD} = 17.7$ lb, $A_x = 8.34$ lb, $A_y = 0$, $A_z = 45.0$ lb.

5-57. $A_y = 230.25$ N, $A_z = 69.8$ N, $B_x = 107.0$ N, $B_y = 230.25$ N, $C_x = 107.0$ N, $C_z = 230.25$ N.

5-58. $T = 607.9$ N, $F = 1080.1$ N, $A_x = 190.8$ N, $A_y = 1884.0$ N, $A_z = 253.3$ N.

5-59. $P = 8.09$ N, $A_x = 5.43$ N, $A_y = 71.0$ N, $B_x = 5.94$ N, $B_y = 73.1$ N, $B_z = 196.2$ N.

6-1. $F_{AB} = 3273.7$ N (C), $F_{AE} = 6636.8$ N (T), $F_{BE} = 3273.7$ N (T), $F_{BC} = 3273.7$ N (C), $F_{DC} = 8273.4$ N (C), $F_{DE} = 4136.7$ N (T), $F_{CE} = 8274.0$ N (T).

6-2. $F_{DC} = 6987.7$ N (C), $F_{DE} = 3125.0$ N (T), $F_{EF} = 3125.0$ N (T), $F_{EC} = 5000$ N (T), $F_{AB} = 9782.8$ N (C), $F_{AF} = 4375.0$ N (T), $F_{BF} = 8750.0$ N (T), $F_{BC} = 4375.0$ N (C), $F_{FC} = 1767.8$ N (T).

6-3. $F_{IB} = 10$ kN (T), $F_{BH} = 14.14$ kN (C), $F_{HD} = 28.29$ kN (C), $F_{JI} = F_{CH} = F_{FG} = F_{FE} = 0$.

6-5. $F_{AD} = 549.6$ N (T), $F_{AB} = 377.6$ N (T), $F_{DB} = 266.3$ N (T), $F_{DC} = 549.7$ N (T), $F_{BC} = 754.2$ N (C).

6-5a. $F_{AB} = 707.1$ lb (T), $F_{AD} = 1030.8$ lb (T), $F_{DC} = 1030.8$ lb (T), $F_{BC} = 1414.2$ lb (C), $F_{DB} = 500$ lb (T).

6-6. $F_{AD} = 3.58$ kN (T), $F_{AB} = 3.2$ kN (C), $F_{CB} = 3.2$ kN (C), $F_{CD} = 3.58$ kN (T), $F_{BD} = 3.2$ kN (C).

6-7. $F_{EC} = 7$ kN (T), $F_{EF} = 7$ kN (C), $F_{CF} = 1.12$ kN (C).

6-9. $F_{BL} = 0$, $F_{LC} = 1118.0$ N (C), $F_{CK} = 500.0$ N (T).

6-10. $F_{BF} = 828.2$ N (C), $F_{DF} = 0$.

6-10a. $F_{BF} = 207.1$ lb (C), $F_{DF} = 0$.

6-11. $F_{BC} = 3.464$ kN (C), $F_{BD} = 11.46$ kN (C),
$F_{CA} = F_{CD} = 6.69$ kN (T), $F_{AB} = 13.46$ kN (C)
(corresponding values on the other side of the truss).

6-13. $F_{QF} = 0$, $F_{EL} = 5.0$ kN (T).

6-14. $F_{EB} = 4.77$ kN (T), $F_{ED} = 4.77$ kN (C).

6-15. $F_{FC} = 10$ kN (C), $F_{GC} = 7.88$ kN (T).

6-15a. $F_{FC} = 1200.0$ lb (C), $F_{GC} = 888.6$ lb (T).

6-17. $F_{LD} = 0$, $F_{BC} = 75$ kN (T).

6-18. $F_{HG} = 12.5$ kN (C), $F_{CF} = 8.84$ kN (T), $F_{GC} = 0$.

6-19. $F_{BE} = 21.2$ kN (T), $F_{EF} = 25$ kN (C).

6-21. $F_{GF} = 7.83$ kN (C), $F_{GD} = 1.80$ kN (C).

6-22. $F_{OE} = 3.36$ kN (T), $F_{LE} = 0$, $F_{LK} = 9.75$ kN (C).

6-23. $F_{CJ} = 1501.1$ N (C), $F_{CB} = 1732.1$ N (T).

6-25. $F_{CD} = 6.67$ kN (T), $F_{CF} = 4.17$ kN (T),
$F_{GF} = 10.31$ kN (C).

6-25a. $F_{CD} = 1600$ lb (T), $F_{CF} = 1000$ lb (T),
$F_{GF} = 2473.9$ lb (C).

6-26. $F_{FE} = 3$ kN (T), $F_{BC} = 8.49$ kN (C), $F_{FB} = 9$ kN (C).

6-27. $F_{BJ} = 0$, $F_{CH} = 3245.0$ N (T).

6-29. $F_{CB} = 0$, $F_{CD} = 0$, $F_{CF} = 2$ kN (C), $F_{BD} = 0$,
$F_{BA} = 2$ kN (C), $F_{EA} = 0$, $F_{EF} = 0$, $F_{FA} = 0$,
$F_{DA} = 0$, $F_{DE} = 3$ kN (C), $F_{DF} = 0$.

6-30. $F_{CD} = 566.7$ N (T), $F_{CA} = F_{CB} = 400.7$ N (C),
$F_{BD} = 283.2$ N (T), $F_{BA} = 0$, $F_{AD} = 283.2$ N (T).

6-30a. $F_{CA} = F_{CB} = 122.5$ lb (C), $F_{CD} = 173.2$ lb (T),
$F_{BD} = 86.6$ lb (T), $F_{BA} = 0$, $F_{AD} = 86.6$ lb (T).

6-31. $F_{BC} = 141.6$ N (C), $F_{AB} = 583.9$ N (T),
$F_{AC} = 1132.8$ N (C).

6-33. $B_x = 500$ N, $B_y = 0$, $A_x = 500$ N, $A_y = 1000$ N,
$C_x = 500$ N, $C_y = 500$ N.

6-34. $A_x = 166.7$ N, $A_y = 388.9$ N, $C_x = 166.7$ N,
$C_y = 111.1$ N.

6-35. $A_x = A_y = C_x = C_y = 1.67$ kN.

6-35a. $A_x = A_y = C_x = C_y = 125$ lb.

6-37. $A_x = 166.7$ N, $A_y = 1166.7$ N, $B_x = 166.7$ N,
$B_y = 833.3$ N, $C_x = 1333.3$ N, $C_y = 833.3$ N.

6-38. $F_C = 9.56$ kN, $F_B = 6.21$ kN.

6-39. $A_x = 2060.1$ N, $A_y = 1569.6$ N, $B_x = 1667.7$ N,
$B_y = 1962.0$ N, $C_x = 2060.1$ N, $C_y = 1962.0$ N.

6-41. $F_B = 500.0$ N, $A_x = 0$, $A_y = 650$ N.

6-42. 666.7 N.

6-43. (a) 24.5 N, (b) 32.7 N, (c) 10.9 N.

6-45. $\theta = 14.6°$.

6-45a. $\theta = 14.6°$.

6-46. $F_C = 23.54$ kN, $B_x = 20.39$ kN, $B_y = 5.89$ kN.

6-47. $A_x = 2982.2$ N, $A_y = 235.4$ N, $B_x = 2982.2$ N,
$B_y = 549.4$ N, $C_x = 2982.2$ N, $C_y = 1334.2$ N.

6-49. $F_{AC} = 1154.1$ N, $F_{BC} = 520.2$ N.

6-50. $F_C = 1500$ N, $D_x = 1500$ N, $D_y = 1200$ N.

6-50a. $F_C = 400$ lb, $D_x = 400$ lb, $D_y = 300$ lb.

6-51. $A_x = 1419.3$ N, $A_y = 419.3$ N, $B_x = 2882.9$ N,
$B_y = 750$ N, $D_x = 2882.9$ N, $D_y = 1750$ N.

6-53. $F_{EF} = 8175.0$ N (T), $F_{AD} = 158.13$ kN (C).

6-54. 956.5 N.

6-55. $N_C = 122.6$ N, $B_x = 122.6$ N, $B_y = 147.1$ N.

6-55a. $N_C = 33.3$ lb, $B_x = 33.3$ lb, $B_y = 40$ lb.

6-57. 3.67 kg.

6-58. 785.3 N.

6-59. $P = 513.1$ N, $F_{CE} = 5007.8$ N.

6-61. 20.5 N.

7-1. $A_D = 0$, $V_D = 2.49$ kN, $M_D = 16.46$ kN $\cdot$ m,
$A_C = 2.48$ kN, $V_C = 2.48$ kN, $M_C = 4.96$ kN $\cdot$ m.

7-2. $A_D = 219.6$ N, $V_D = 219.6$ N, $M_D = 54.9$ N $\cdot$ m,
$A_E = 80.4$ N, $V_E = 0$, $M_E = 112.5$ N $\cdot$ m.

7-3. $A_D = 837.3$ N, $V_D = 16.3$ N, $M_D = 16.3$ N $\cdot$ m.

7-5. $A_A = 900$ N, $V_A = 0$, $M_A = 112.5$ N $\cdot$ m.

7-5a. $A_A = 230$ lb, $V_A = 0$, $M_A = 57.5$ lb $\cdot$ ft.

7-6. $A_E = -252.3$ N, $V_E = 238.2$ N, $M_E = -59.6$ N $\cdot$ m,
$A_D = -337.0$ N, $V_D = 0$, $M_D = 98.7$ N $\cdot$ m.

7-7. $A_E = 0$, $V_E = 250$ N, $M_E = 187.5$ N $\cdot$ m,
$A_D = -750$ N, $V_D = 0$, $M_D = 0$.

7-9. $A_D = 0$, $V_D = 8$ kN, $M_D = 4.8$ kN $\cdot$ m, $A_E = 0$,
$V_E = -2$ kN, $M_E = 4.8$ kN $\cdot$ m, $A_C = -1.73$ kN,
$V_C = -3.67$ kN, $M_C = 2.09$ kN $\cdot$ m.

7-10. $A_D = -750$ N, $V_D = 250$ N, $M_D = 125$ N $\cdot$ m.

7-10a. $A_D = -300$ lb, $V_D = 100$ lb, $M_D = 200$ lb $\cdot$ ft.

7-11. $x = 3.9$ m ($M_E = 17.55$ kN $\cdot$ m).

7-13. $(0 \leqslant x < 5)$ $V = 9.5 - 2x$, $M = 9.5x - x^2$;
$(5 < x < 8)$ $V = -0.5$, $M = 25 - 0.5x$;
$(8 < x \leqslant 10)$ $V = -10.5$, $M = 105 - 10.5x$.

7-14. $(0 \leqslant x < 4)$ $V = 500$, $M = 500x - 6600$;
$(4 < x \leqslant 12)$ $V = 500$, $M = 500x - 6000$.

7-15. $(0 \leqslant x < 2)$ $V = 2.22$, $M = 2.22x$;
$(2 < x \leqslant 4.5)$ $V = -1.78$, $M = 8 - 1.78x$.

7-15a. $(0 \leqslant x < 5)$ $V = 1166.7$, $M = 1166.7x$;
$(5 < x \leqslant 12)$ $V = -833.3$, $M = 10\,000 - 833.3x$.

7-17. $(0 \leqslant x < 3)$ $V = 729.17 - 500x$,
$M = 729.17x - 250x^2$;
$(3 < x \leqslant 3.5)$ $V = 1750 - 500x$,
$M = 1750x - 250x^2 - 3062.5$.

7-18. $V = 0$, $M = -784.8$.

7-19. $(0 \leqslant x < 2)$ $V = 2.25$, $M = 2.25x - 4$;
$(2 < x \leqslant 3)$ $V = 3.25 - 1.5x$,
$M = 3.25x - 0.75x^2 - 3$.

7-21. $(0 \leqslant x < 3)$ $V = 6 - 2x$, $M = 6x - x^2 - 9.6$;
$(3 < x \leqslant 5)$ $V = 0$, $M = -0.6$.

7-22. $(0 \leqslant x < 2)$ $V = 900 - 400x$, $M = 900x - 200x^2$;
$(2 < x \leqslant 4)$ $V = -500$, $M = 2000 - 500x$.

7-23. $(0 \leqslant x < 3)$ $V = 291.67 - 500x$,
$M = 291.67x - 250x^2$;
$(3 < x < 4)$ $V = 2750 - 500x$,
$M = -2166.7x - 250x^2 + 7375$;
$(4 < x \leqslant 4.5)$ $V = 750$, $M = 750x - 3375$.

7-25. $(0 \leqslant x \leqslant 3)$ $V = 250 - 83.33x^2$,
$M = 250x - 27.78x^3$.

7-25a. $(0 \leqslant x \leqslant 18)$ $V = 360 - 3.33x^2$,
$M = 360x - 1.11x^3$.

7-26. $(0 \leqslant x < 3)$ $V = 2.25 - 1.5x + 0.25x^2$,
$M = 2.25x - 0.75x^2 + 0.0833x^3$;
$(3 < x \leqslant 6)$ $V = -2.25 + 1.5x - 0.25x^2$,
$M = -0.0833x^3 + 0.75x^2 - 2.25x + 4.5$.

7-27. $V = 0.5x - 0.333x^2$, $M = 0.25x^2 - 0.111x^3$.

7-29. $A = 249.8\theta \cos \theta$, $V = 249.8\theta \sin \theta$,
$M = 499.6\theta[\cos \theta - \cos (\theta/2)]$.

7-30. $V_{\max} = 600$ N, $M_{\max} = 450$ N $\cdot$ m.

7-30a. $V_{\max} = 1500$ lb, $M_{\max} = 3750$ lb $\cdot$ ft.

7-31. $V_{\max} = 6$ kN, $M_{\max} = -9.6$ kN $\cdot$ m.

7-33. $V_{\max} = 1250$ N, $M_{\max} = -1350$ N $\cdot$ m.

7-34. $V_{\max} = -900$ N, $M_{\max} = 568.02$ N $\cdot$ m.

7-35. $V_{\max} = -500$ N, $M_{\max} = 288.7$ N $\cdot$ m.

7-35a. $V_{\max} = -720$ lb, $M_{\max} = 2494.2$ lb $\cdot$ ft.

7-37. $V_{\max} = -1.5$ kN, $M_{\max} = -0.75$ kN $\cdot$ m.

7-38. $V_{\max} = \pm 20$ kN, $M_{\max} = 4.2$ kN $\cdot$ m.

7-39. $V_{\max} = -9.5$ kN, $M_{\max} = 15.625$ kN $\cdot$ m.

7-41. $V_{\max} = -280$ N, $M_{\max} = -31$ N $\cdot$ m.

7-42. $V_{\max} = 10$ kN, $M_{\max} = -5$ kN $\cdot$ m.

7-43. $V_{\max} = 1508.3$ N, $M_{\max} = 2675.0$ kN $\cdot$ m.

7-45. $V_{\max} = 17.875$ kN, $M_{\max} = -48$ kN $\cdot$ m.

7-45a. $V_{\max} = 17\,500$ lb, $M_{\max} = -4\,800$ lb $\cdot$ ft.

7-46. 1.13 kN/m.

7-47. 1.30 MN.

7-48. 22.75 kN (for one cable).

7-49. $y = \dfrac{h}{L^3}x^3$, $T_{\max} = (w_o L/2)\sqrt{1 + (L/3h)^2}$.

7-50. 5.00 m, 27.45 m.

7-50a. 7.56 ft, 62.5 ft.

7-51. 18.53 m, 492.0 N.

7-53. 2981.4 N, 82.09 m, 3216.9 N.

7-54. 30.20 m, 376.5 N.

7-55. 506.89 N, 59.73 m.

7-55a. 184.9 lb, 50.2 ft.

7-58. 41.14 m.

8-1. 560.4 N.

8-2. 74.3°.

8-3. $\theta = 26.6°$, $P = 356.2$ N.

8-5. 6 boxes.

8-5a. 8 boxes.

8-6. 839.7 kg.

8-7. 2255.0 kg.

8-9. 1.05 m.

8-10. 28.1°.

8-10a. 28.1°.

8-11. 240.2 N.

8-13. slipping occurs.

8-14. 597.9 N.

8-15. 0.75 m.

8-15a. 2.5 ft.

8-17. 1401.4 N.

8-18. 95.7 N.

8-19. 1379.7 N.

8-21. 216.9 N.

8-22. 21.8°.

8-23. 2732.8 N.

8-25. 50 m.

8-25a. 133.3 ft.

8-26. 21 books.

8-27. 22.6°.

8-29. 22.6°.

8-30. 31.6 mm.

8-30a. 0.211 ft.

8-31. 17.1°.

8-33. 159.2 N.

8-34. 220.7 N.

8-35. 160 mm, 39.2 N.

8-35a. 8 in., 20 lb.

8-37. 238.0 N.

8-38. 20.4°.

8-41. 9.41 kN.

8-42. (a) 64.2 N, (b) 6.87 N.

8-43. 58.6 N · m.

8-45. 0.316.

8-45a. 0.627.

8-46. 16.1 N · m.

8-47. 18.0 N · m.

8-49. 91.9 N · m.

8-50. 229.8 N · m.

8-50a. 57.3 lb · ft.

8-51. $\frac{1}{2}\mu PR$.

8-53. 8.48 N · m.

8-54. $T_A = 568.4$ N, $N_S = 949.6$ N, $F = 189.9$ N.

8-55. 19.1 N.

8-55a. 6.67 lb.

8-57. 436.4 N.→

8-58. 266.5 N.

8-59. $T_A = 1.95$ kN, $T_B = 0.759$ kN.

8-61. 0.384 m.

8-62. 318.3 N.

8-63. 75.4 N · m, 0.214 m³.

8-65. 33.9 N · m.

8-65a. 23.1 lb · ft.

8-66. 53.3°.

8-67. 3.36 N · m.

8-69. 53.0°.

8-70. 470.9 N.

8-70a. 64.0 lb.

8-71. 1.35 mm.

8-73. $P = W(a_A + a_B)/2r$.

8-74. 176.6 N.

9-1. 0.792 m.

9-2. 5.0 m.

9-3. $\bar{x} = 1.30$ m, $\bar{y} = 2.30$ m.

9-5. $\bar{x} = 1.44$ m, $\bar{y} = 2.22$ m, $\bar{z} = 2.67$ m.

9-6. $\bar{x} = 124.0$ mm, $\bar{y} = 0$.

9-7. 0.546 m.

9-9. $\bar{x} = 0.20$ m, $\bar{y} = 0.20$ m.

9-10. 60 mm.

9-10a. $\bar{y} = 1.20$ ft, $\bar{x} = 0$.

9-11. 0.40 m.

9-13. $\bar{x} = 0$, $\bar{y} = 8.49$ mm.

9-14. $\bar{x} = 0.40$ m, $\bar{y} = 1.0$ m.

9-15. $\bar{x} = -1.5$ m, $\bar{y} = 0.25$ m, $\bar{z} = -0.30$ m.

9-15a. $\bar{x} = -2.25$ ft, $\bar{y} = 0.75$ ft, $\bar{z} = -0.30$ ft.

9-17. πah, $\frac{2}{3}h$.

9-18. $\bar{x} = \frac{a}{2}$, $\bar{y} = \bar{z} = 0$.

9-19. $\bar{x} = 0.4a$, $\bar{y} = \bar{z} = 0$.

9-21. $\bar{x} = \bar{z} = 0$, $\bar{y} = \frac{3}{4}h$.

9-22. 1.33 m.

9-23. 84.7 mm.

9-25. $\bar{x} = 0$, $\bar{y} = 200$ mm, $\bar{z} = 35.7$ mm.

9-25a. $\bar{x} = 0$, $\bar{y} = 6$ in., $\bar{z} = 1.39$ in.

9-26. $0.675a$.

9-27. $\bar{z} = h/4$.

9-29. $\bar{x} = 147.4$ mm, $\bar{y} = 178.1$ mm, $\bar{z} = 49.1$ mm.

9-30. $\bar{x} = \bar{y} = 56.1$ mm, $\bar{z} = 68.0$ mm.

9-30a. $\bar{x} = \bar{y} = 1.12$ in., $\bar{z} = 1.36$ in.

9-31. 77.3 mm.

9-33. 90.46 mm.

9-34. 154.4 mm.

9-35. $\bar{x} = 0$, $\bar{y} = 40.97$ mm.

9-35a. $\bar{x} = 0$, $\bar{y} = 2.46$ in.

9-37. $\bar{x} = 0$, $\bar{y} = 261.9$ mm.

9-38. 80.1 mm.

9-39. $\bar{x} = 0$, $\bar{y} = 272$ mm.

9-41. $\bar{x} = \bar{y} = 0$, $\bar{z} = 114.7$ mm.

9-42. $\bar{y} = 483.0$ mm, $\bar{z} = 67.7$ mm.

9-43. 62.5 mm.

9-45. $\bar{x} = 71.4$ mm, $\bar{y} = 135.7$ mm, $\bar{z} = 142.9$ mm.

9-45a. $\bar{x} = 0.737$ in., $\bar{y} = 1.34$ in., $\bar{z} = 1.42$ in.

9-46. $\bar{x} = 23.5$ mm, $\bar{y} = 0$, $\bar{z} = 22.6$ mm.

9-47. $\bar{x} = \bar{y} = 0$, $\bar{z} = 53.3$ mm.

9-49. $A = 30(10^3)$ mm², $V = 70.7(10^6)$ mm³.

9-50. $A = 1.184$ m², $V = 0.0592$ m³.

9-50a. $A = 96.23$ in.², $V = 36.09$ in.³

9-51. $A = 53\ 333.3$ mm², $V = 25.133(10^6)$ mm³.

9-53. $22.68(10^6)$ mm³.

9-54. 70.7 mm.

9-55. 1071.7 Mg.

9-55a. $3.12(10^6)$ lb.

9-57. 14.50 m².

9-58. 82.7 kg.

9-59. 45.9 l.

9-61. 1.77 MN, 4 m.

9-62. 50.27 kN, 2.01 m.

9-63. 2.24 m.

9-65. $F_R = 15.19$ kN, $\bar{z} = 2.4$ m.

9-65a. $F_R = 50(10^3)$ lb, $\bar{z} = 8.0$ ft.

9-66. $F_R = 21.5$ kN, $\bar{z} = 2.26$ m.

9-67. $F_{R_1} = 2.45$ kN $\rightarrow$; $F_{R_2} = 10.83$ kN$\downarrow$, center of plate B; $d' = 1.02$ m from liquid surface.

9-69. 38.23 kN, 0.898 m.

9-70. 101.6 kN.

9-70a. 11 173.5 lb.

9-71. $4rlp_o$.

9-73. 332.19 kN/m.

9-74. 3.85 kN, 0.625 m.

9-75. $F = 3769.9$ N, $\bar{x} = 0$, $\bar{y} = 0.955$ m.

9-75a. $F = 2827.4$ lb, $\bar{x} = 0$, $\bar{y} = 3.82$ ft.

10-1. $2.0(10^6)$ mm⁴.

10-2. 1632.7 mm².

10-3. $114.3(10^6)$ mm⁴.

10-5. $I_x = 226.4(10^6)$ mm⁴, $I_y = 48.2(10^6)$ mm⁴.

10-5a. $I_x = 11.46$ in.⁴, $I_y = 2.44$ in.⁴

10-6. (a) $I_{x_b} = \dfrac{bh^3}{12}$, (b) $I_{\bar{x}} = \dfrac{bh^3}{36}$.

10-7. $I_x = \frac{1}{3}tl^3 \sin^2 \theta$.

10-9. (a) $57.14(10^6)$ mm⁴, (b) $9.15(10^6)$ mm⁴.

10-10. 37.8 mm.

10-10a. 0.378 ft.

10-11. $1826.55(10^6)$ mm⁴.

10-13. $I_x = \frac{\pi}{4}ab^3, I_y = \frac{\pi}{4}ba^3, J_O = \frac{\pi ab}{4}(a^2 + b^2)$.

10-14. $I_x = \frac{\pi a^4}{8}, I_{\bar{x}} = 0.1098a^4$.

10-15. $I_x = 34.66(10^6)$ mm⁴, $I_y = 3.43(10^6)$ mm⁴.

10-15a. $I_x = 361.75$ in.⁴, $I_y = 36.75$ in.⁴

10-17. 136.3 mm.

10-18. $20.70(10^6)$ mm⁴.

10-19. $109.95(10^6)$ mm⁴.

10-21. $196.9(10^6)$ mm⁴.

10-22. $3.21(10^9)$ mm⁴.

10-23. $926.6(10^3)$ mm⁴.

10-25. 146.57 mm.

10-25a. 6.22 in.

10-26. $13.76(10^6)$ mm⁴.

10-27. $95.90(10^6)$ mm⁴.

10-29. $8.31(10^9)$ mm⁴.

10-30. $63.3(10^6)$ mm⁴.

10-30a. 2.0 in.⁴.

10-31. $I_{xy} = \frac{1}{6}l^3t \sin 2\theta$.

10-33. $66.67(10^6)$ mm⁴.

10-34. 0.0625 m⁴.

10-35. $5(10^3)$ mm⁴.

10-35a. 2.53 in.⁴

10-37. $3.60(10^9)$ mm⁴.

10-38. $-28.12(10^3)$ mm⁴.

10-39. $\theta = 8.70°, I_{max} = 7.27(10^6)$ mm⁴, $I_{min} = 1.86(10^6)$ mm⁴.

10-41. $\bar{y} = 10.26$ mm, $I_u = 5.06(10^6)$ mm⁴, $I_v = 34.79(10^6)$ mm⁴.

10-42. $135.34(10^6)$ mm⁴.

10-43. $I_{max} = 1.560(10^6)$ mm⁴, $I_{min} = 0.163(10^6)$ mm⁴.

10-45. $\bar{y} = 36$ mm, $I_x = 5.35(10^6)$ mm⁴, $I_y = 10.20(10^6)$ mm⁴.

10-45a. $\bar{y} = 2.33$ in., $I_x = 92.0$ in.⁴, $I_y = 156.0$ in.⁴

10-46. $I_u = I_v = 1.005(10^6)$ mm⁴.

10-47. $I_u = I_v = 5.33(10^6)$ mm⁴.

10-49. $I_{max} = 307.50(10^3)$ mm⁴, $I_{min} = 85.28(10^3)$ mm⁴.

10-50. $I_{min} = 5.35(10^6)$ mm⁴, $I_{max} = 10.20(10^6)$ mm⁴.

10-50a. $I_{min} = 92.0$ in.⁴, $I_{max} = 156.0$ in.⁴

10-51. $I_{max} = 7.27(10^6)$ mm⁴, $I_{min} = 1.86(10^6)$ mm⁴.

10-53. $I_{max} = 1.56(10^6)$ mm⁴, $I_{min} = 0.163(10^6)$ mm⁴.

10-54. $I_{max} = 307.50(10^3)$ mm⁴, $I_{min} = 85.28(10^3)$ mm⁴.

11-1. 193.98 N.

11-2. 75.1°.

11-3. 34.62 N.

11-5. 0°, 11.19°.

11-5a. 0°, 48.2°.

11-6. 10.2°.

11-7. 11.5°.

11-10. 0°, 31.1°.

11-10a. 0°, 50.6°.

11-11. 39.24 N.

11-13. 750 N.

11-14. 42.57 N · m.

11-15. 0°, 33.1°.

11-15a. 0°, 33.1°.

11-17. 50.0 kg.

11-18. $P = [(b - a)/2c] mg$.

11-19. $m_t = (m)(s/a)$.

11-21. 13.9 N.

11-22. $\theta_1 = \tan^{-1}(2P/3mg)$. $\theta_2 = \tan^{-1}(2P/mg)$.

11-23. $P = 33.98$ N, $Q = 29.43$ N.

11-25. 1558.8 N.

11-25a. 332.6 lb.

11-26. $m_A = 2$ kg, $m_B = 2.5$ kg.

11-27. $x = 0590$ stable, $x = -0.424$ unstable.

11-29. neutral equilibrium.

11-30. $\theta = 0°$ stable, $\theta = 11.19°$ unstable.

11-30a. $\theta = 0°$ stable, $\theta = 48.2°$ unstable.

11-31. $\theta = 0°$ unstable, $\theta = 31.1°$ stable.

11-33. 36.4° unstable.

11-34. $\theta = 14.65°$ stable, $\theta = 90°$ unstable.

11-35. 36.6° stable.

11-35a. 39.2° stable.

11-37. 20.6° unstable.

11-38. 20.2° stable.

11-39. 36.81 N/m.

11-41. 5.29 kg.

11-43. G directly below C, $\theta = 180°$ stable.

11-45. 35.4 mm.

11-45a. 4.24 in.

11-46. $r > b/2$ stable, $r < b/2$ unstable, $r = b/2$ neutral; $\theta = 0°$.

11-47. 45°.

11-49. $x_1 = 0.654$ m, $x_2 = 0.123$ m.

Index